Brian H. Kaye

Characterization of Powders and Aerosols

WILEY-VCH

Brian H. Kaye

Characterization of Powders and Aerosols

WILEY-VCH

Weinheim · New York · Chichester · Brisbane · Singapore · Toronto

Professor Emeritus Brian H. Kaye
Laurentian University
Ramsey Lake Road
Sudbury, Ontario P3E 2C6
Canada

Library of Congress Card No. applied for

A catalogue record for this book is available from the British Library

Die Deutsche Bibliothek – Cataloguing-in-Publication Data
Kaye, Brian H.:
Characterization of powder and aerosols / Brian H. Kaye. – 1. Aufl. – Weinheim ; New York ; Chichester ; Brisbane ; Singapore ; Toronto : Wiley-VCH, 1999
ISBN 3-527-28853-8

Printed on acid-free and chlorine-free paper

Composition: Text- und Software-Service Manuela Treindl, D-93059 Regensburg
Printing: strauss offsetdruck, D-69509 Mörlenbach
Bookbinding: Wilhelm Osswald + Co., D-67433 Neustadt
Printed in the Federal Republic of Germany

Preface

I first started working with powders in 1955. In the 43 years since that initial activity there has been a multitude of developments of instruments and sources of information on the performance of these instruments. Back in 1955 the Coulter Counter was becoming well known and the height of sophistication was the Photosedimendometer. I began my studies using an Andreason bottle and moved on to study the possibility of using divers and developed all the way to fractals. In the period covered by my activities in particle size analysis the type of book required for people active in the field has changed. When I began my work there was no journal devoted to the subject but as of now we have three journals, Aerosol Science, Particle and Particle Systems Characterization and Particle Technology. I was involved in setting up of both of these latter journals and they have both grown into many volumes. Also in the early days there was difficulty in finding information on the performance of instruments whereas today many manufactures provide comprehensive notes on operational variables with their machines. The availability of the journal information and literature from manufacturers means that the role of potential textbooks has changed. In this book we have tried to set out the basic methods for characterizing powders and aerosols and have tried to indicate the questions that the investigator should use when trying to choose a method for his particular needs. The inter-method comparison of data generated in particle size is still a complex problem and one of the useful features of this book is the provision of many graphs showing the relative performance of different machines in assessing powder properties.

The question of particle shape is a complex problem and we are still at the stage where we are developing methods to see if we can characterize adequately the range of shapes within a powder and their effect on the powder system and/or the aerosol system. It is becoming apparent that some complex problems will require more than one method of characterization thus if one was inhaling a complex soot particle the aerodynamic diameter which governs the penetration of the lung is one parameter whereas the fractal structure is another needed to assess the potential health hazard of the inhaled aerosol particle.

A problem facing the investigator in powder technology is that many of the earlier publications use methodologies to characterize the powders that are no longer available. To enable the analyst to assess the information presented in earlier publications we have reviewed the physical principles and have set out the problems associated with some of the classical instruments such as the micromerograph which for many years was a standard method in the powder metallurgy industry but is now only of historic interest. Sometimes the problems associated with methods are posed by the cessation of manufacturing of a given procedure. Thus the M.S.A. Centrifuge method was very widely used in occupational health and safety but the manufactures decided to discontinue the manufacturing of equipment so for continuity of interpretation the method has been outlined. Emphasis has been placed on references to enable the reader to recover detailed information for their own investigations. Unfortunately normal systems of training in industry such as pharmaceuticals, chemical engineering, and powder metallurgy do not present a great deal of information on characterization procedures and because methods have developed in different subjects different scientists tend to use different words for the same concept. Therefore we have attempted to clarify some of the vocabulary which has been used in different fields of endeavor which generate information of interest to a wider audience of scientists than those who have immediately carried out the work.

Any author has his own biases when writing a book and since we have been very active at Laurentian University in developing shape methodologies this aspect of powder technology has been fully covered in this text.

Hopefully the advanced reader will find references to work relevant to their own studies and student reader will find this book a useful introduction to methods for characterizing powders and aerosols.

Acknowledgments

Many students have contributed to the development of this book and the typing of the script. I thank then the following people who have been particularly active: Cherie Turbitt Daoust, Lorna Mac Lod, Heather Eberhardt and my two daughters Sharon Kaye and Alison Kaye have also contributed to the text preparation. Cherie undertook the difficult task of copyright clearance and the help of Garry Clark in preparing the diagrams and in general proof reading the scripts have been invaluable. I also wish to thank the manufactures of the various machines who have been most helpful in providing data and material describing their instruments. In particular Morris Wed of Malvern Instruments was most helpful in supplying of literature on diffractometers. I also wish to thank the personal at Wiley-VCH especially Barbara Böck, for encouraging me to finish this project.

Laurentian University
B. H. Kaye

Table of Contents

1 Basic Concepts in Characterization Studies, Representative Samples and Calibration Standards

1.1 Who Needs to Characterize Powders and Spray Systems?

The list of industries using powders, or processes in which there is a substance used as spray or a mist, is long and increasing. My first exposure to the problems of powder technology began in 1955 when I studied the characterization of powders used to fabricate parts of nuclear weapons. One study involved the metal beryllium which was used in powder form. The production of dense beryllium required powders having a specific size and shape distribution. Beryllium powder is however a respirable health hazard and to characterize the powder in a safe atmosphere required the development of new methods of characterizing powders.

After working with beryllium I moved on to study nuclear reactor fabrication. In this study I worked on determining the surface area, size and shape distributions of uranium dioxide and plutonium dioxide powders used to fabricate fuel rods. Looking back I see that my initiation into powder technology was a baptism of fire since all of these powders were extremely toxic and dangerous. The technology that I studied in those years is currently very applicable to the study of modern ceramic materials and powder metallurgical routes to finished products [1, 2].

After my studies of the technology for creating nuclear weapons I soon became involved in studying the fallout from nuclear weapons tests and similar problems of occupational diseases, such as pneumoconiosis and silicosis caused by the inhalation of fineparticles. The study of respirable hazards in industry and from nuclear fallout requires detailed knowledge of the shape and size of fineparticles [3, 4].

The same type of information required to predict the respirable hazard for grains of powder is also vital to the success of therapeutic aerosol technology in which drugs are delivered to the lungs in aerosol form [5]. The same technical information is used by military experts to design the delivery of biological warfare agents, such as clouds of toxic dust. The other side of the military problem is to design filters which will protect military personnel against these toxic clouds of fineparticles; a task requiring detailed size, shape and aerodynamic behaviour information for the aerosol fineparticles. Other industrial activities where detailed knowledge of the size and shape distributions of powder grains are important include industries involved in food processing, cosmetics, paint, pesticide manufacture and delivery, pharmaceutical products, and the manufacture of explosives, abrasive powders, metal powders used in the creation of magnetic tape, and the dry inks used in xerographic copiers.

Size characterization studies have often evolved in parallel in many of these industries and sometimes there is vocabulary confusion because of the different perspectives of scientists from the various industries. We will attempt to develop and use a consistent terminology as we study the multitudes of powders used in various industries.

1.2 The Physical Significance of Size Measurements

If one is concerned with the characterization of dense smooth spheres, the concept of size is elementary and straight forward. If however one must deal with some of the powder grains found in industry, exactly what is meant by size has to be defined very carefully. Consider for example the carbonblack profile shown in Figure 1.1(a) [6].

One measure of the structure of the carbonblack profile is it's circle of equal area as shown in Figure 1.1(b). Another simple descriptor, which has been widely used to describe such objects is the Aspect Ratio. This is the length, defined as the longest dimension of the profile, divided by the width of the profile (right angles to the length measurement.) This is a dimensionless number which is defined as a geometric index of shape. Many different geometric shape factors have been described by different workers [7–11].

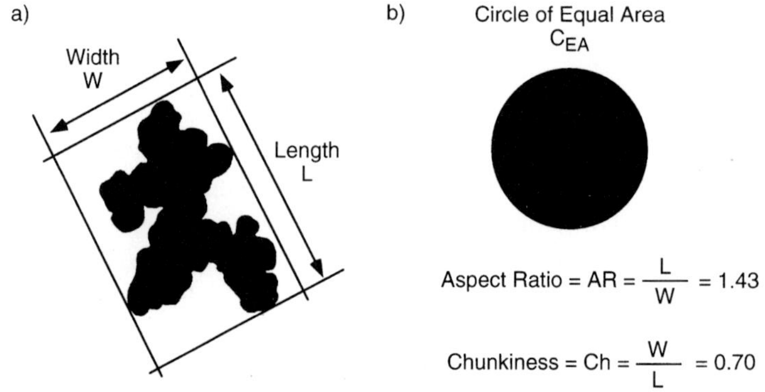

Figure 1.1. To specify the size and shape of a complex fineparticle, many equivalent and operational parameters may be required, as demonstrated by the parameters required to describe a carbonblack profile originally described by Medalia [6]. a) Simple, classical dimensions of a carbonblack profile. b) Typical size and shape descriptors of the profile of (a).

The reciprocal of the Aspect Ratio has recently become quite widely used to describe the shape of fineparticles. The reciprocal quantity is called the Chunkiness of the fineparticle. (The physical significance of this measure will be discussed in Chapter 2.)

Relating the equivalent measure of a fineparticle to its physical properties is not always easy and for this reason what is known as an operational diameter of the fineparticle is sometimes used. Thus, the equivalent area of the carbonblack of Figure 1.1(a) is probably related to the opacity of the fineparticle when it is used as a pigment. However, if it is to be used to be part of a defensive smoke screen in a military operation the opacity of the profile, with respect to scattered light, has to be measured and in this situation some of the diffractometer measurements discussed in Chapter 6, may be a more direct measure of the operational behavior of the profile.

Soot fineparticles produced by a combustion processes are similar in structure to the carbonblack profile of Figure 1.1(a). When one is looking at the dispersal dynamics of a smoke and/or the health hazards of the smoke fineparticles, one must use an operational diameter known as the aerodynamic diameter. The aerodynamic diameter is the size of the smooth dense sphere of unit density which has the same dynamic behavior as the soot particle. Several procedures for measuring the aerodynamic diameter of airborne fineparticles will be discussed in various chapters of this book.

When looking at a complex profile such as that of Figure 1.1(a) one can sometimes clearly identify subunits in the structure of an agglomerate. In some instances workers report the size distribution of the subunits in the agglomerate as the operational size of the fineparticle system but this can be confusing and lead to difficulty interpreting the data. Thus in Figure 1.2(a) a set of fineparticles captured on a whisker filter and studied by Schafer and Pfeifer are shown [12]. The size distribution of the fineparticles on the filter whiskers were studied by two methods. The distributions reported by Schafer and Pfeifer are shown in Figure 1.2(b). It is quite surprising that the image analysis data shows much smaller fineparticles than those that are obviously visible under a microscope in the array of Figure 1.2(a). The reason for this is that Schafer and Pfeifer measured what they called "obvious units" contributing to clusters which they claimed were formed on the filters as capture trees [13]. Deciding whether a cluster of smaller fineparticles has grown on the filter fiber or existed in the aerosol being filtered is a value judgment for which different scientists would reach different conclusions. In the case of the study reported by Schafer and Pfeifer the decision as to the reality of the structure of the cluster is not critical since they were studying alumina fineparticles used to create visible trails in wind tunnel experiments. However, looking at a typical cluster such as that shown enlarged in Figure 1.2(c), if the study had been on the health hazard of the dust, the hazard would be very different if the cluster was a single entity of the size of 3 microns long or if it was in fact 20 or 30 small particles less than half a micron in size.

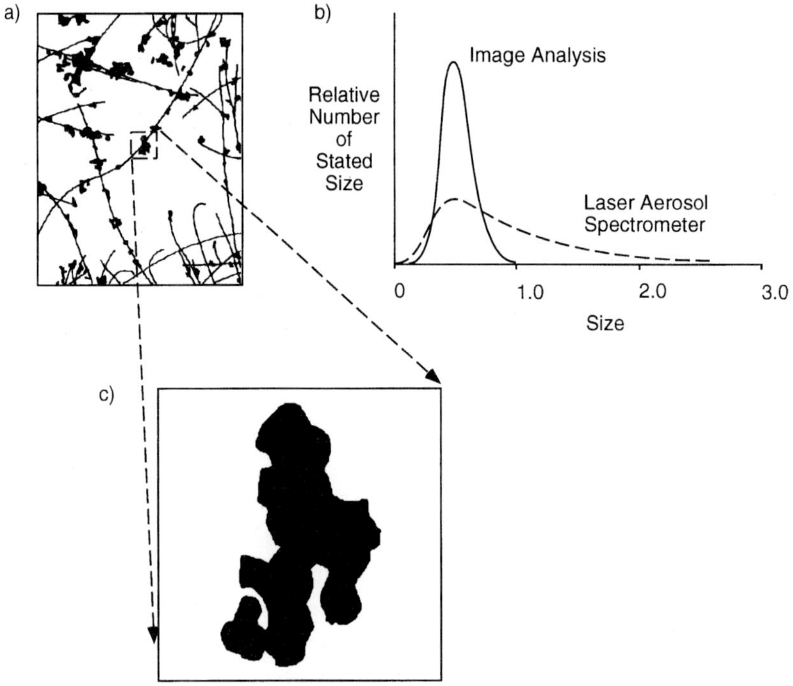

Figure 1.2. The decision as to what constitutes a separate fineparticle can lead to very different descriptions of a fineparticle population, as demonstrated by the data reported by Schafer and Pfeifer [12]. a) Low magnification field of view of fineparticles captured in the fibres of a filter. b) Size distributions by two different methods of the fineparticles of (a). c) A typical agglomerate which Schafer and Pfeifer describe as constituted from "obvious" subunits which they report as the effective unit in their image analysis size distribution.

The difficulties of using image analysis in health hazard studies is demonstrated by the profile of Figure 1.2(c). Predicting the aerodynamic diameter from the perceived physical structure of the profile is very difficult. (See discussion of the aerodynamic profiles of complex fineparticles in Chapter 6.) In the discussion so far of the profiles of the Figures 1.1 and 1.2, the term agglomerate has been used without definition. Unfortunately in powder technology literature the terms agglomerate and aggregate are used somewhat indiscriminately. One author's agglomerate may be another author's aggregate. In this book the term agglomerate is used to describe a structure which is strong enough to persist throughout the handling of the fineparticle in the process of interest. The term aggregate on the other hand is used to describe a temporary cluster which breaks down during the processing of the material. This is a logical use of the two terms since agglomerate means "made into a ball" whereas aggregate means behaving like a flock of sheep. Anyone who has watched the behavior of a flock of sheep knows that the flock assembly disintegrates as soon as the dog and the shepherd walk away. Thus when looking at titanium dioxide powder taken out of a bag, the powder is often clustered into aggregates as large as 20 microns in diameter but when dispersed by high shear forces into a paint these agglomerates will breakdown into individual fineparticles of one micron or less.

When selecting a method of size characterization to study a powder, one should try to use an analytical procedure to disperse the powder resulting in fineparticles that will be the same operative size as those in the process under study. Thus, if we were to have a cluster of fineparticles which persisted throughout a pharmaceutical processing operation, it would be inappropriate to use a sizing procedure which used dispersing forces strong enough to rip the cluster apart. This aspect of size characterization will be discussed throughout the text when discussing the various characterization procedures.

Again, when choosing a method of size characterization, one should always choose a method close to the operational context for which the information is required. Thus if one wants to study the dust movement into and out of the lung one should use a method that actually measures the aerodynamic size of the fineparticle.

Sometimes it is necessary to measure several size description parameters for a more complete description of a fineparticle in the operational context. For example, if one is studying a soot fineparticle having a structure similar to that of the profile of Figure 1.1(a), one needs to know the aerodynamic diameter to predict the movement in the atmosphere and/or into or out of the lung; however to look at the health hazard of the fineparticle one needs to measure the structure and the surface of the fineparticle. Thus, an open textured, fluffy soot fineparticle would have a small aerodynamic diameter the magnitude of which would give very little indication of the probability of lodging on the surface of the lung or to the possibility of capturing the soot fineparticle in a respirator or filter. For such purposes, one would have to measure the physical dimensions of the profile such as the length and chunkiness.

Two other parameters which would be useful when evaluating potential health hazards of fineparticles, such as the soot profile of Figure 1.1(a), are the fractal dimensions of the structure and the texture of the profile. The fractal dimension of a boundary is a concept from the subject of applied fractal geometry [14, 15]. Fractal geom etry, invented by Mandelbrot [16], describes the ruggedness of objects in various dimensions of space. (As will be pointed out in the various discussions in the use of the term fractal in powder science, the word fractal dimension can mean different things, in this case the word fractal dimension describes the rugged structure of the boundary of a profile.) To describe the ruggedness of lines in two dimensional space, the fractal dimension is a fractional addendum to the topological dimension of a line, which is 1, as illustrated for the various lines of Figure 1.3. It can be seen that this fractal addendum increases as the ruggedness, i. e. the ability of the line to fill space, increases.

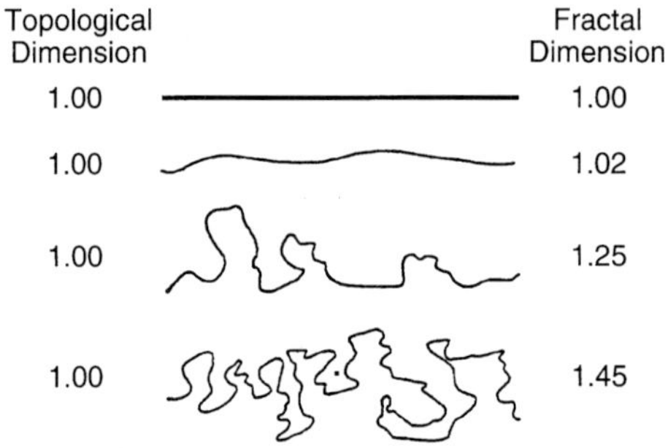

Figure 1.3. The fractal dimension of a profile can be used to describe the ruggedness of a fineparticle profile. The fractal dimension consists of a fractional number, which is related to the ruggedness or space filling ability of a profile, added to the topological dimension of a line or other structure [13].

We will show in Chapter 2 that the carbonblack profile of Figure 1.1(a) has two fractal dimensions, one describing the gross structure of the agglomerate and the other the texture. The magnitude of the structural fractal dimension is about 1.32. The structural fractal dimension of the agglomerate is useful information concerning the way in which the agglomerate formed in the smoke in which it was created. The other fractal dimension used to describe the carbon black agglomerates, called the textural fractal dimension, describes the texture of the agglomerate. This parameter has information on the way in which the subunits are packed together to form the agglomerate [17]. The techniques for measuring the fractal dimensions of profiles such as that of Figure 1.1(a) will be described in detail in Chapter 2.

Because the various methods for characterizing aspects of a complex structure explore different aspects of that structure, the data generated from a given study of the system may not correlate directly with data generated by another technique. From time to time in the body of the text the differences in the data generated by different studies of the same type of population by various methods will be discussed. In the final chapter we will collect together various comparative studies illustrative of the usefulness of the information generated by different size characterization techniques. Predicting the physical properties of a powder system from the size distribution study is not usually a direct procedure. For this reason in Chapter 9 we will look at assessing by direct study, physical properties of powder systems such as the flow of a powder system, the packing of a powder assembly, and permeability/porosity of compressed powder systems.

1.3 Standard Powders for Calibrating Powder Measurement Techniques

Sometimes the interpretation of data generated in a method for studying the size of a fineparticle can be carried out using physical relationships. Thus when studying the sedimentation of a fineparticle in a viscous fluid, the Stokes diameter of the fineparticle can be established using known values of viscosities and densities along with measured falling speeds and a well known formula developed by Stokes (see Chapter 4). However, in other techniques, the physical significance of data generated by a method is interpreted by carrying out calibrations using standard fineparticles. For example, when looking at the size of fineparticles using a stream counter, such as the HIAC system described in Chapter 6, the instrument is calibrated using standard latex spheres. The data generated for a particular powder is then reported in terms of the size of the equivalent spheres which would represent the fineparticles.

Standard latexes, and other reference materials, are available from various organizations [18–24]. One of the calibrations standards available to fineparticle scientists are latex spheres which were made on board the space shuttle in 1985. Because these spheres were formed in the absence of gravity they are perfectly spherical. The National Bureau of Standard makes available standard reference material in the form of ten micron microspheres mounted on glass slides. In the first type of slide a few thousand microspheres are deposited as a regular array on a glass microscope slide. In the other type, the fineparticles are randomly distributed [18]. A series of standard non-spherical fineparticles have been prepared by the Community Bureau of Reference Commission of the European Community for use in comparing the performance of size methods. These reference powders are known as BCR standards and several publications are available describing the use of such reference materials [19].

1.4 Representative Samples

Often in the laboratory one is given a sample of a few grams taken from a large supply of powder. It should be self obvious that if this sample is not representative of the original bulk supply of powder then one is wasting time characterizing the sample in the laboratory. Unfortunately this fundamental step in powder technology is often overlooked sometimes simply because the laboratory is separated in time and space from the original bulk supply of powder. Several times in my career I have been in charge of laboratories providing size analysis data to other groups. When such a position I have always insisted that I would not touch a sample until I knew how the sample had been obtained. This demand often led to discussions with the group requesting information which resulted in new sampling strategies being put into place to ensure that the small sample of powder delivered to the laboratory was representative of the population to be characterized. The sampling of a very large quantity of powder is a difficult task and several companies have developed special equipment for taking samples from powder streams and from powder storage devices. The literature in this subject should be consulted by those faced with the decision of taking samples from very large quantities of powders (i. e. many tons of powders) [25–32]. Note that in the references the term ASTM refers to publication by the American Society for Testing Materials and BSI stands for the British Standards Institute.

When taking samples from a large quantity of powder it should always be recognized that handling of the powder may have caused segregation. The simple act of pouring powder from a storage device into a large canister can create segregation since, as the air moves out of the container as the powder is moving in, finer particles can be flushed upwards to the top of the container.

One widely used technique for sampling different regions of a powder supply is the thief sampler shown in Figure 1.4(a) [33, 34]. In this device a hollow tube with a point is provided with several entry ports along its length. An inner tube that fits smoothly into the outer tube is also provided with entry ports to a series of sectional containers along its length. To use the thief the inner tube is placed in the outer tube with the ports in a position where no powder may enter the inner tube. The tube is then thrust into the powder supply and the inner tube rotated until powder can enter the compartments. The handle of the inner tube is then twisted further to close the ports and the sampler withdrawn from the powder. This equipment is useful for non-abrasive powders such as flour and other food powders but can be quickly rendered inoperative if used with an abrasive power. This is because if any of the abrasive powder is caught between the two tubes, the abrasive grains bind the two tubes together. Literature on many different sampling devices is to be found in references 33 through 41.

In the laboratory a widely used sampling method is the spinning riffler shown in Figure 1.4(b) [32]. This sampler was developed in response to conflicts over the accuracy of size analysis data at a time that I was operating a service lab providing particle size analysis to various groups at the Atomic Weapons Research Establishment, England. In this sampling procedure the powder to be sampled is fed through a chute or funnel into a rotating set of containers. For any one cup to contain a representative sample the time of flow of the powder supply divided by a time of rotation should be at least 100. The efficiency of this sampling device has been established by many experiments. It is available from several instrument companies, see references 33 and 42 for their names.

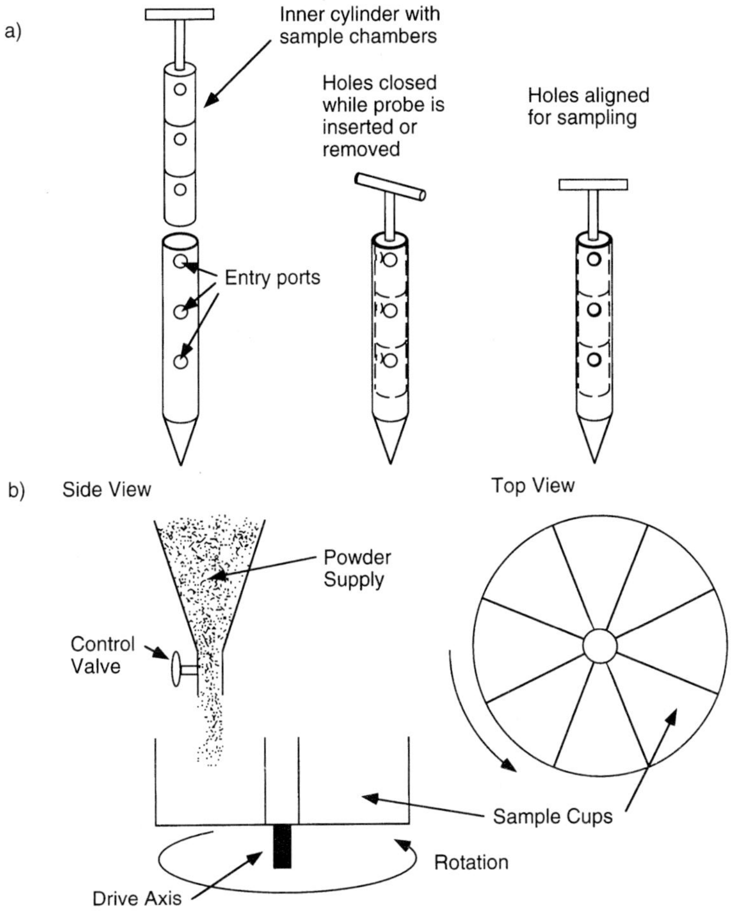

Figure 1.4. Several devices are available for taking a small representative sample from a large supply of powder [31, 32, 34]. a) A thief sampler consists of two concentric cylinders. b) The spinning riffler can efficiently produce small, representative samples.

Although the spinning riffler has proved to be an efficient sampling device it does have some disadvantages. First, if the powder is very fine, the spinning action of the riffler basket may cause some of the fines in the powder to blow away. Secondly, if the powder is cohesive, it may not flow readily through the feed funnel into the riffler basket. For situations when one is concerned with analyzing a sample of a cohesive powder, which will not be returned to the parent population of powder, adding a small amount of silica flow agent will make the powder flow into the riffler and such an addition would not normally cause problems in the analytical method (see discussion of flowagents in reference 43).

Another disadvantage of the riffler system is that a small sample size (as required by some modern size characterization methods) requires many successive rifflings and the cleaning of the apparatus can be tedious and time consuming.

A recently developed technique for obtaining a representative sample of a powder which overcomes some of these problems is shown in Figure 1.5. This new system exploits the idea that if one can mix a quantity of powder so that it is a homogeneous population, any small sample taken at random from the population is a representative sample. Since the sampling chamber used in this new method is sealed any problems due to very fine powders and/or cohesive powders are eliminated [44]. The sampling device is called a free fall tumbling mixer/ sampler device. A commercial version of the equipment, known as the AeroKaye™ mixer/sampler, is available from Amherst Process Instruments Inc. [45]. The powder to be sampled is placed in a container equipped with a lid carrying a sample cup as shown in Figure 1.5(a). When assembled

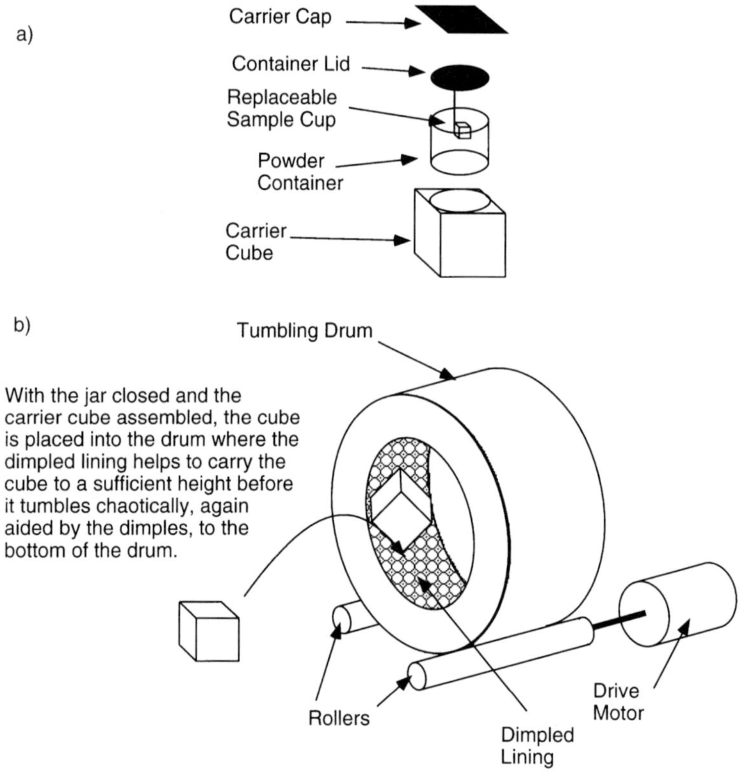

Figure 1.5. In the free-fall tumbling mixer, a powder to be sampled is homogenized in a sample jar which tumbles chaotically in a rotating drum. After the tumbling, any sample taken at random from the powder is a representative sample. It is convenient to take random samples using a sampling cup mounted to the lid of the jar [43, 44].

the sample cup will be within the body of the powder in the container when the container is upright. The sample container, complete with lid and the required sample cups, is placed in a carrying cube as shown in Figure 1.5(a). When the carrier cube is placed in the rotating tumbler drum where it tumbles chaotically. As a consequence the powder in the container tumbles chaotically and is thoroughly mixed and/or homogenized. When the cube is removed from the tumbler and the sample cup is retrieved, it will contain the desired representative sample of the powder in the container.

Several different sizes of carrying cubes are available and various models of the rotating drum facilitate sampling from small amounts, less than one hundred grams, to amounts of several hundreds of grams. One of the advantages of this sampling device is that the size of the sampling cup can be changed to obtain a sample of the required size directly. Thus, for use with a time-of-flight aerosol spectrometer, a stream method which will be discussed in detail in Chapter 6, a sample of less than one gram can be obtained directly using a small sample cup. On the other hand, if a larger amount is required for another method of characterization, a larger sample cup can be used. Several sampling cups of different sizes can be mounted on the lid at the same time.

The performance of this equipment has been demonstrated by making measurements on the samples using the Aerosizer®, described in Chapter 6 [45], as illustrated by the data of the graph of Figure 1.6. In these tests a cohesive fine calcium carbonate powder was placed in the mixing chamber which was tumbled in the rotating drum system. A one cubic centimeter cup was used to take a sample of powder which was then characterized using the Aerosizer®. After a second period of tumbling another sample was taken and it can be seen that the two sets of data are virtually indistinguishable.

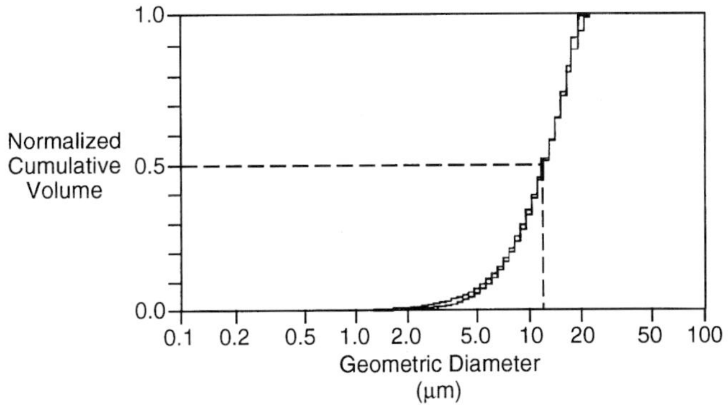

Figure 1.6. Two samples of calcium carbonate from a bulk sample homogenized in the free-fall tumbler of Figure 5, taken several minutes apart, exhibit virtually identical size distributions.

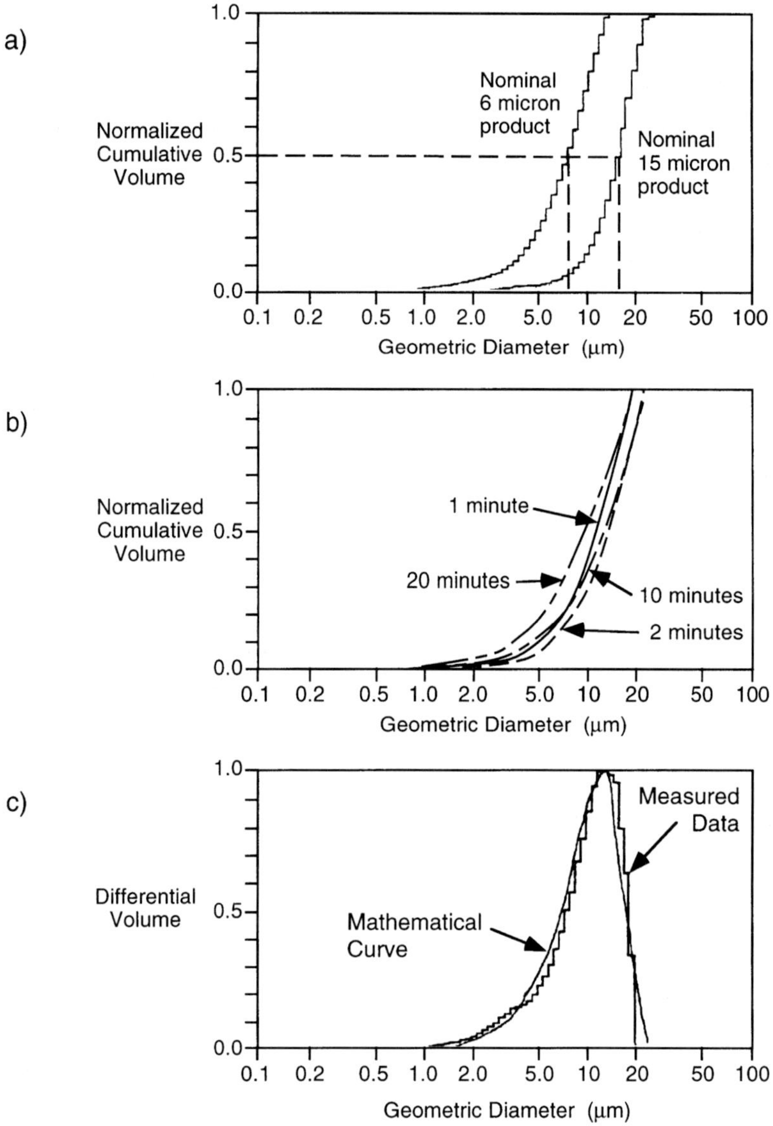

Figure 1.7. The free-fall tumbling mixer can work efficiently with initially completely segregated cohesive powders [44]. a) Size distributions of the two original calcium carbonate powders. b) Progress of mixing for a sample containing 25 % by weight of 6 μm product and 75 % by weight of the 15 μm product. Note that the two powders were initially completely segregated in the mixing chamber. c) After 20 minutes of mixing the measured size distribution of the mixture was virtually identical to the mathematically calculated size distribution based on the size distributions of the two individual powders.

As mentioned earlier the basic concept of the new sampling device is that it mixes powders so well that one can take any small subsection of the material as a representative sample. To demonstrate the mixing efficiency of the device two fine cohesive calcium carbonate powders were placed in the container in the proportions 25 % of a nominally 6 micron calcium carbonate and 75 %, 15 micron calcium carbonate. The size distribution of the two powders are shown in Figure 1.7(a). The portions of the two powders were placed as layers in the container. Samples were taken at various times after mixing by tumbling was initiated and the size distributions of the samples are shown in Figure 1.7(b). The mixture did not immediately achieve homogeneity but within 20 minutes the measured size distribution of the material was virtually indistinguishable from the calculated size distribution of the mixed powders as illustrated by the data of Figure 1.7(c). It should be appreciated that the mixing together of two such fine cohesive powders is a very difficult task and the fact that it was achieved within 20 minutes is in itself noteworthy. Using normal powders to be sampled in the laboratory, much shorter tumbling times would be adequate. It is relatively cheap to provide the technologists with different disposable sampling cups and many different mixing chambers can be placed in carrier cubes to facilitate the use of the standard bottles used in any given laboratory. The only caution is that the container to be used in the carrier cubes should be a relatively squat configuration and should never be filled more than half full to facilitate the random motion of the powder during the chaotic tumbling of the carrier cube. Sometimes if one is working the large grained, free-flowing powder it may be necessary to place a randomizing paddle in the mixing chamber. Useful information on other powder sampling equipment is available from the manufacturers cited in references 33 and 46.

1.5 Representative Samples from Suspensions and Aerosol Clouds

One can often take a sample from a suspension using a pipette but in such instances one must be aware that the rate of suction can bias the results. This factor will be discussed in more detail in Chapter 4.

When taking a sample from a liquid suspension process stream, a useful and efficient sampling device is the Isolock® sampler shown in Figure 1.8 [47]. The Isolock® sampler is fitted with a retractable piston consisting of two parts. In the passive position shown in Figure 1.8(a) the front of the piston sticks into the flowing suspension where it has the useful purpose of creating turbulence in the suspension which facilities efficient sampling of the suspension. The back section of the piston seals the sampling bottle/pipe system from the flowing suspension. When activated, the piston is

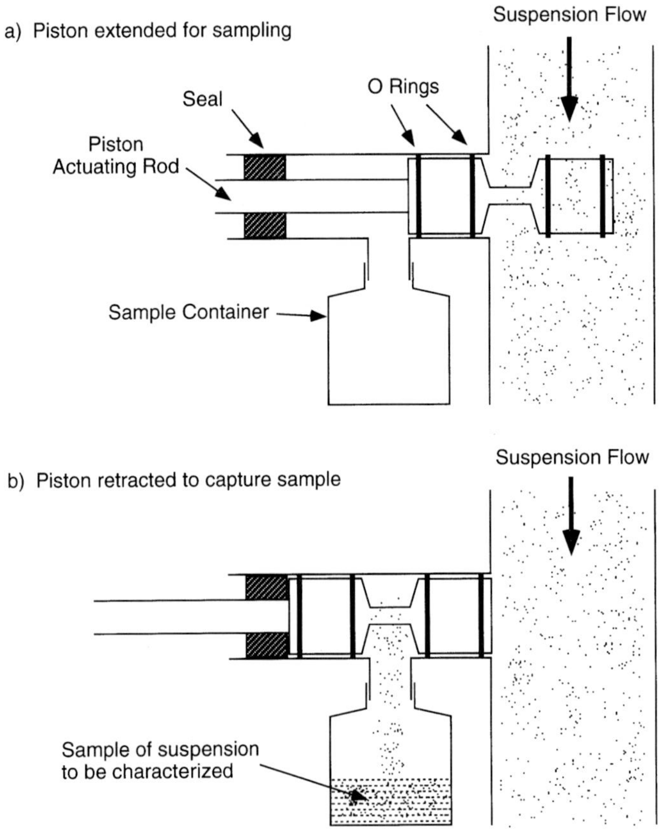

Figure 1.8. The Isolock sampler is used to take samples of suspensions or slurries containing fineparticles to be characterized [47].

withdrawn so that some sample flows into the bottle before the retreating front of the piston seals the entry orifice as shown in Figure 1.8(b). In a typical industrial process the piston would be activated relatively rapidly to take a small portion of suspension into the holding bottle. The piston would be activated several times over an appropriate time period so that the eventual sample filling the bottle consists of several small sub-samples drawn from the process pipe over a period of time. When taking samples from a flowing suspension one should always be careful to create turbulence immediately in front of the sampling device. One should never operate a device such as the Isolock® sampler near a bend in the pipe since all bends create centrifugal forces which tend to segregate the particles flowing in the suspension by size.

To obtain samples of fineparticles constituting air pollution and other aerosol systems, many different devices have been developed for filtering the fineparticles onto paper or glass fiber filters. Many of the devices used to filter aerosol fineparticles fractionate the airborne fineparticles by means of twists and bends in the equipment and are known as elutriators. Many of the advanced aerosol sampling devices will be discussed in detail in Chapter 5, after we have developed the basic concepts of elutriation and fineparticle fractionation in a moving fluid.

Two special types of filters which can be used to filter aerosols are shown in Figure 1.9(a) and (b). The filter in Figure 1.9(a) is known as a Nuclepore® filter [48]. This type of filter is made by bombarding a thin film of polycarbonate with subatomic particles emerging from a nuclear reactor. The flux of the subatomic particles is ori-

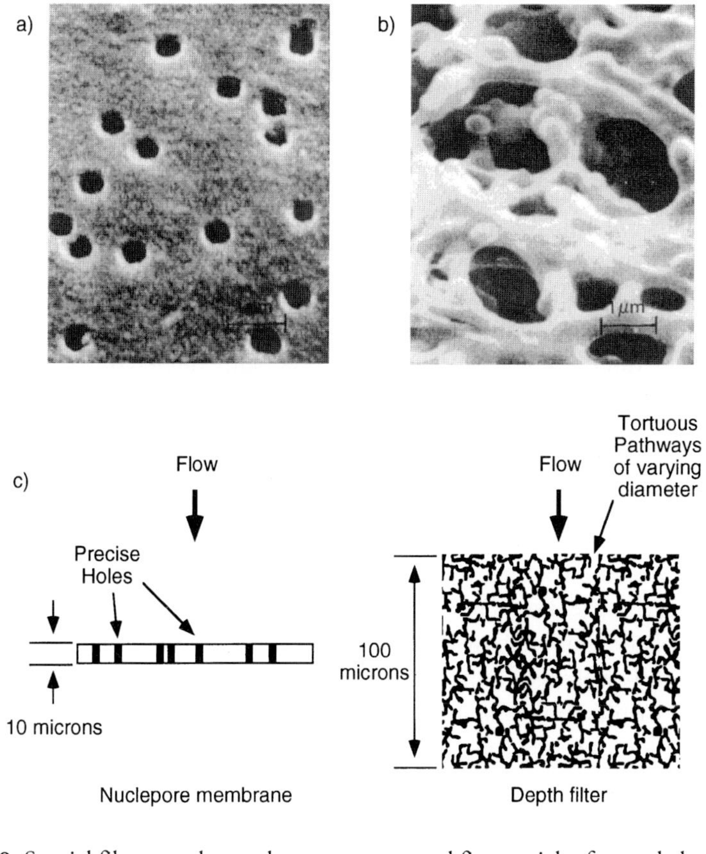

Figure 1.9. Special filters can be used to capture aerosol fineparticles for study by image analysis and other methods of characterization. a) Appearance of the surface of a Nuclepore filter. b) Appearance of the surface of a membrane filter with the same rating as the Nuclepore of (a). c) Structure of the filters of (a) and (b) as they would appear in cross-section. The Nuclepore is a surface filter while the membrane filter is a depth filter.

ented at right angles to the film. The passage of the particles creates weakened paths in the filter which can then be etched chemically. As a consequence the etched holes in the filter are very cylindrical and virtually at right angles to the surface of the film. One cannot control the actual location of the holes and this places a limitation on the number of holes one can permit in the filter since randomly adjacent holes can sometimes fuse to create a larger hole. Thus in Figure 1.9(a) there is a pair almost fused at the lower left and one that is fused at the upper right. However this disadvantage is balanced with the advantages that the holes are very symmetrical and all of the same size. Moreover when fineparticles are filtered onto the surface they remain on the surface and are very visible. A similar type of filter is marketed by Poretic Corporation [49]. The stopping power, of a Nuclepore® type filter, i. e., the size of fineparticles which cannot pass through the filter, is directly visible in the size of the hole in the filter. Nuclepore® filters are very fragile and they can very easily become electrostatically charged as they are removed from the supply container and placed in their operational position. For this reason the air in the locality of a Nuclepore® type filter is often treated with a radioactive source to discharge the electrostatic charges of the filter.

The other type of filter shown in 1.9(b) is what is known as a membrane filter. Membrane filters are available from several manufacturers [50, 51]. These are more like a sponge in structure and the stopping power of the filter has to be measured using a test aerosol. The difference in the structure of two types of filters are illustrated in Figure 1.9(c). The membrane type filters are stronger than the Nuclepore type of filter but have the disadvantage that the fineparticles being filtered often disappear into the structure of the filter. Many of the filters can however be dissolved to free the filtered fineparticles for subsequent study. Characterizing the size distribution of particles trapped on a filter when examined by image analysis techniques will be discussed in Chapter 2.

When sampling an aerosol onto a filter using a probe into a cloud of fineparticles, one must be very careful with regard to the suction rate used to remove the fineparticles for study. Essentially one should not draw air into the probe at a faster speed than wind in the vicinity of the probe. Drawing at the rate of the wind is called isokinetic sampling, however truly isokinetic sampling cannot be achieved and there has to be a compromise in the rate of sampling of an aerosol cloud. A full discussion of the many techniques available for sampling aerosol clouds is beyond the scope of this book and has been adequately described elsewhere [52, 53].

Recently some very precise filters made from a bundle of fibres in a glass matrix has been developed by Collimated Holes Incorporated [54]. To make these filters a bundle of fiber optics is assembled in a glass matrix with the core of a fiber perpendicular to the surface of the matrix. The assembled plate is then treated with a fluid to dissolve the central core of the fibre optics which are chemically different from the glass matrix. The resultant porous surface has very accurate holes for filtration purposes.

These can be cleaned and used over and over again because of the robust construction. As in the case of the Nuclepore® type filters the fineparticles trapped by the filter remain on the surface and can be readily examined by image analysis.

1.6 Dispersing Powder Samples for Size Characterization Studies

In an earlier section I gave the reader the advice that they should never spend time and effort on characterizing a powder sample until they had discussed, with the group requesting the characterization study, the way in which the powder had been sampled and that only representative samples should be studied in depth. In the same way, when I was running a service laboratory, I always used to make the statement "You disperse the powder, and I will characterize it." This is because in many methods of size characterization the powder sample has to be dispersed in a liquid or in a small drop of oil on a glass slide. The way in which the dispersion is carried out can drastically effect the size measurement. (Dispersing aerosols for characterization is discussed in Chapter 6.)

When using a dispersing agent, such as detergent, the effect of dispersion technology can create major shifts in the measured particle size distribution. It is difficult to give general advice on dispersion technology since each powder's constitution determines the effective method of dispersion. Readers should consult standard texts on dispersion such as that written by Parfitt [55]. Advice for particular industries is given in other books such as that by Washington [56, 57].

When studying techniques for dispersing a powder, the advice given earlier should be remembered, that one must always recognize the operational integrity of the fineparticles being studied. For example ultrasonic treatment of a suspension of fineparticles can very often break up agglomerates which, under normal circumstances, will persist in the operational process from which the powder has been taken. In particular one should also be aware of the problems of diluting a suspension prior to a characterization study since the suspension can undergo what the chemist calls dilution shock in the addition of pure fluids. Pure fluids, when added to a natural suspension, may interact with dissolved constituents of natural suspensions to cause drastic changes in the state of dispersion of the fineparticles such as flocculation (the creation of large low strength agglomerates) and/or precipitation of fineparticles out of suspension. Diluting aerosols prior to characterization is a particularly difficult task which will be discussed in detail at the appropriate point in the text.

References

[1] See chapter for example G. Deiter, "Powder Fabrication", 465–473 in *Modern Science and Technology*, R. Colborn (ed), E. VanNostrand Company Incorporated, Princeton, NJ., 1965.

[2] M. E. Fayed, L. Otten (eds.), *Handbook of Powder Science and Technology*, Second Edition, Chapman and Hall, New York, 1997.

[3] J. C. M. Marijnissen, L. Gradon. *Aerosol Inhalors: Recent Research*, Kluwer Academic Publishers, 1996.

[4] R. P. Perera and A. Karim Ahmed, *Respirable Particles; Impact of Airborne Fineparticulates on Health and the Environment*, Ballinger Publishing Company, Cambridge, Massachusetts, 1979.

[5] T. T. Mercer, P. E. Morrow, and W. Stöber eds, *Assessment of Airborne Particles – Fundamentals, applications, and implications to inhalation toxicity*, Charles C. Thomas Publishers Illinois, 1972.

[6] A. J. Medalia, "Dynamic Shape Factors of Particles," *Powder Technology*, 4 (1970–1971) 117–138.

[7] H. H. Heywood, "Size and Shape Distribution of Lunar Fines Sample 12057, 72," in *Proceedings of Second Lunar Science Conference,* Vol. 13 (1971) 1989–2001.

[8] H. Heywood, "Numerical Definitions of Particle Size and Shape," *Chem. Ind.*, 15 (1937) 149–154.

[9] H. Heywood, "Particle Shape Coefficients," *J. Imp. Coll. Eng. Soc.*, 8 (1954) 25–33.

[10] H. H. Hausner, "Characterization of the Powder Particle Shape," in *Proceedings of the Symposium on Particle Size Analysis,* Loughborough, England; published by the Society for Analytical Chemistry, London, England, (1967) 20–77.

[11] B. H. Kaye, *Direct Characterization of Fineparticles*, John Wiley & Son, New York (1981).

[12] H. J. Schafer and H. J. Pfeifer, "Sizing of Submicron Aerosol Particles by the Whisker Particle Collector Method," *Part. Part. Syst. Charact.*, 5 (1988) 174–178.

[13] B. H. Kaye, *A Randomwalk Through Fractal Dimensions*, Second Edition, VCH Publishers, Weinheim, Germany, 1994.

[14] B. H. Kaye, "Characterizing the Structure of Fumed Pigments Using the Concepts of Fractal Geometry." *Part. Part. Syst. Charact.,* 9 (1991) 63–71.

[15] B. H. Kaye, *Chaos and Complexity: Discovering the Surprising Patterns of Science and Technology*, VCH Publishers, Weinheim, 1994.

[16] B. B. Mandelbrot, *Fractals, Form, Chance, and Dimension*, Freeman, San Francisco, 1977.

[17] B. H. Kaye and G. G. Clark, "Formation Dynamics Information; Can It be Derived from the Fractal Structure of Fumed Fineparticles?", Chapter 24 in *Particle Size Distribution II; Assessment and Characterization.* T. Provder(ed.), American Chemical Society, Washington (1991).

[18] See "Reference Material" by Rasberry in *American Laboratory*, (March 1987) 128–129.

[19] R. Wilson, "Reference Materials of Defined Particle Size Certified Recently by the Community Bureau of Reference of the European Economic Community", *Powder Technology*, 27 (1980) 37–43.
[20] Duke Standards Company, 445 Sherman Avenue, Palo Alto, California 94306.
[21] Dow Diagnostics, The Dow Chemical Company, P. O. Box 68511G, Indianapolis, Indiana 46268.
[22] D. R. Alliet, "A Study of Available Particle Size Standards for Calibrating Electrical Sensing Zone Methods," *Powder Technology.* 13 (1976) 3–7.
[23] Dyno Particles A. S., P. O. Box 160, N-2001 Lillestrom, Norway.
[24] Interfacial Dynamics Corporation, 4814 N. E., 107th Avenue, Sweet Bee, Portland, Oregon, U.S.A., 97220.
[25] ASTM C.136–67 (1967), "Sieve or Screen Analysis of Fine and Coarse Aggregates" (contains description of Sampling Protocols).
[26] ASTM C.311–68 (1968), "Sampling and Testing Fly Ash for Use as an Admixture in Portland Cement".
[27] ASTM D.345–48, (1948), "Sampling and Testing Calcium Chloride for Roads and Structural Applications".
[28] BS 1017: 1960: Part 1, "Sampling from Bulk (Coal); Part 2 (Coke)".
[29] BS 616: 1963, "Methods for Sampling Coal Tar and its Products".
[30] British Standards Methods for the Determination of Particle Size Powders, Part 1, "Subdivision of Gross Sampler Down to 0.2 ml.," Part 1 (1961).
[31] B. H. Kaye, "An Investigation into the Relative Efficiency of Different Sampling Procedures." *Powder Metal.* 9 (1962) 213–234.
[32] B. H. Kaye, "Efficient Sample Reduction of Powders by Means of a Riffler Sampler." *Soc. Chem. Ind. Monographs* 18 (1964) 159–163.
[33] Sampling Equipment literature is available from Gilson Screen Company, P. O. Box 99, Malinta, Oh 43535.
[34] B. H. Kaye, *Direct Characterization of Fineparticles*, J. Wiley & Son, New York, 1981.
[35] M. Pearce, "Solids sampling techniques", Proc. Inst. Chem. Engrs, Annual Symposium on Solids Handling (Sept. 1960).
[36] R. E. Cordell, *Automatic Sampling System,* U.S. Patent 3, 472, 079 (Oct. 1969).
[37] J. C. Fooks, "Sample splitting devices", *Br. Chem. Engr.,* 15:6 (1970) 799.
[38] R. W. M. Hawes, L. D. Muller, "A small rotary sampler and preliminary study of its use" A. E. R. E. – R3051 (Harwell) (1960).
[39] M. W. G. Burt, "The accuracy and precision of the centrigugal-disc photosedimentometer method of particle-size analysis," Powder Technology, 1 (1967) 103.
[40] M. W. G. Burt, C. A. Fewtrell, R. A. Wharton, "A suspension sampler for particle size analysis work," *Powder. Technology,* 7:6 (1973) 327–30.
[41] T. Allen, A. A. Khan, *Chemical Engineering,* 77 (1970) 108–112.
[42] Information on the spinning riffler system is available from Microscal Ltd., 20 Mattock Lane, Ealing, London, W5 5BH.
[43] B. H. Kaye, *Powder Mixing*, Chapman and Hall, London, England, 1997.

[44] B. H. Kaye, "Sampling and Characterization Research; Developing Two Tools for Powder Testing," *Powder and Bulk Engineering*, 10:2 (February 1996) 44–54.
[45] Amherst Process Instruments Inc., The Pomeroy Building, 7 Pomeroy Lane, Amherst, MA, USA, 01002-2942
[46] Sampling equipment literature is available from Gustafson, 6340 LBJ Freeway, Suite 180, Dallas, TX 75240.
[47] Isolock samplers are available from Bristol Engineering Company, 204 South Bridge St., Box 696, Yorkville, IL 60560.
[48] Nuclepore® is the registered trademark of the Costar Corporation, 7035 Commerce Circle, Pleasanton, CA 94566. Comprehensive literature on the structure and properties of Nuclepore® filters is available from the manufacturer who kindly provided the photograph reproduced in Figure 1.9(a).
[49] The Poretics Corporation, 151 Lindbergh Avenue, Livermore, CA.
[50] Gelman Sciences, 600 Southway Road, Ann Arbour MI, 98106.
[51] Millipore Corporation, Ashby Road Bedford, Mass., 10730.
[52] Allen, T., *Particle Size Analysis*, 5th Edition, Chapman and Hall, London, 1996.
[53] T. T. Mercer, P. Morrow, W. Stöber (Eds.), *Assessment of Airborne Particles*, Proceedings of the Third Rochester International Conference on Environmental Toxicity, (1972) Charles C. Thomas, Springfield, IL, 1972.
[54] Collimated Holes Incorporated, 460 Division St., Campbell, CA 95008.
[55] G. D. Parfitt, *Dispersion of Powders in Liquids*, 2nd ed. John Wiley & Sons, New York (1973).
[56] C. Washington, *Particle Size Analysis in Pharmaceutics and other Industries, Theory and Practise*. Ellis Heywood, A Division of Simon & Shuster International, Markey Cross House, Cooper Street, Chicester, West Sussex, England, 1992.
[57] I. S. O. Committee is work a standard for dispersive powders prior to size characterization.

2 Direct Measurement of Larger Fineparticles and the Use of Image Analysis Systems to Characterize Fineparticles

2.1 Measurements on Larger Fineparticles

The powder technologist can be concerned with a range of objects, as small as several nanometres as large as fragments of rock and coal which can be several centimeters in size. It is useful to consider some of the measurements that can be made in a direct manner on large fragments of rock since a study of such measurements can form a basis for the understanding of techniques which are also applied to much smaller systems when they have been imaged with the appropriate instrument. For example, consider the field of view shown in Figure 2.1(a). The profiles shown are a set of sharp edged rocks produced by crushing slag from the nickel mines of Sudbury whereas those of 2.1(b) are images of diamond dust polishing powder where the actual grains are of the order of fifty microns. In Figure 2.1(c) a plastic moulding powder is shown. For the fineparticles shown in Figures 2.1(a) and (b) the profiles are clearly separate. When one comes to look at the plastic powder array however, one has to make a value judgment as to which profiles are physically joined and which are forming pseudo agglomerates by chance juxtaposition on the slide. Thus the particles A and B are judged to be separate as are C and D but it is harder to decide whether E is made up of two or is one profile. One can sometimes reduce the guess work in deciding which fineparticles are real agglomerates by reducing the number of fineparticles in the field of view when taking the photograph or looking down a microscope. However, it can be shown that to rely entirely on dilution of the field of view to eliminate chance agglomerates requires a very low density of fineparticles and it is usually uneconomic to work at such very low field of view concentrations [1, 2].

When reporting size characterization data from an image analysis procedure any decisions that were made with regard to separating fineparticles into separate entities should be stated and illustrated with a sketch. When looking down a microscope at a field of view such as that of Figure 2.1(c) a decision as to which profiles will be measured in a characterization study has to be made. Usually several fields of view need to be inspected to complete a study and to aid in the selection of profiles it is usual to place a grid in the eye piece of the microscope or to have a boundary for selection purposes in the field of view such as that shown by the dotted line in Figure 2.1(c).

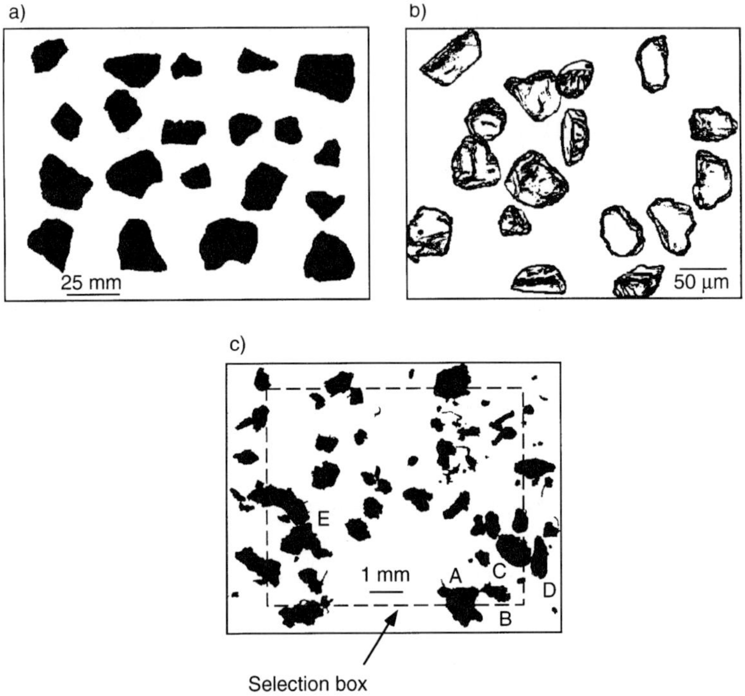

Figure 2.1. The size and shape of fineparticles can vary enormously and be very complex in structure. Selecting the profiles to be characterized from an array of fineparticles can lead to bias in the resulting size distribution [3]. a) Profiles of a group of crushed slag fragments. b) Electronmicrograph of a diamond abrasive powder. c) Micrograph of a plastic moulding powder demonstrating the problems presented by random juxtaposition of individual fineparticles. The selection grid helps establish which fineparticles will be characterized.

A selection difficulty arises because larger fineparticles have a higher probability of intersecting the selection grid to be included in the analysis. This bias toward the larger fineparticles can be avoided in several ways. One of them is to ignore profiles on the top and right side of the selection grid, the other procedure is to include every other profile which intersects the boundary.

2.2 Measuring the Shape Distribution of Fineparticles Using the Concept of Chunkiness

As discussed briefly in Chapter 1, when considering profiles such as those shown in Figure 2.1, one has to specify what one means by size. One of the equivalent size parameters used to describe such profiles is the square root of the area of the profile. In some of the older classical methods of size analysis one measured the area by comparing the area of the irregular profiles with a set of standard circles in a reticule (a grid) placed in the eyepiece of a microscope. In Figure 2.2 one of the widely used reticules for size characterization is shown. This reticule, or graticule, as it is sometimes known, is specified in a British Standard for measuring the size of fineparticles viewed

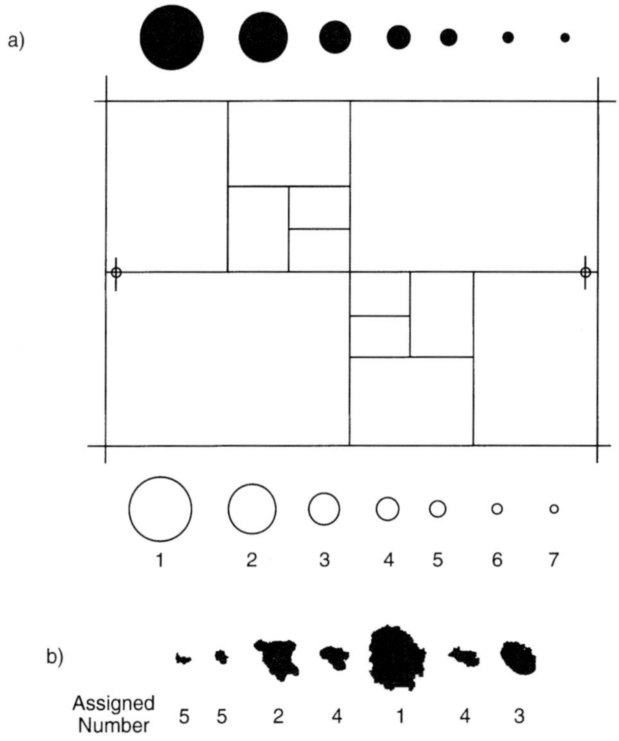

Figure 2.2. One method of selecting and sizing fineparticles viewed through a microscope is to use a grid, known as a reticule, placed in the eyepiece of the microscope. In one technique the fineparticle is allocated the number of the standard circle on a reticule array nearest in size to the fineparticle. a) The appearance of one type of reticule array. b) A sample set of fineparticles with sizes assigned from the reticule of (a).

through a microscope [3]. It has been shown that different operators obtain different results with such graticules depending whether they use the open circles or the closed circles for estimation areas while using this grid [4]. The use of reticules of this type to characterize fineparticles has been widely replaced by low cost television interfaces to computers using the commercially availabile low cost programs for processing size and shape information [5]. Much of the data presented in this chapter was obtained using a computer aided analysis system in which the size of the profiles is specified by the

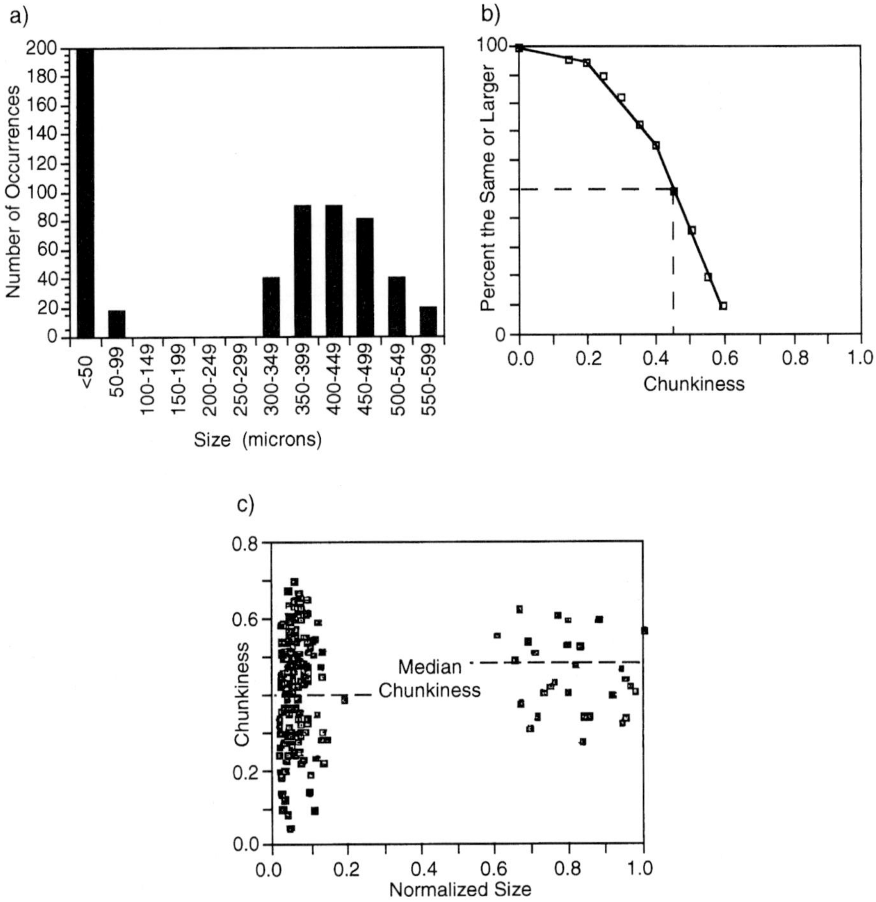

Figure 2.3. There are several ways to graphically summarize the size and shape information from Image analysis of an array of fineparticles. a) Classical size distribution histogram for the powder of Figure 2.1(c). b) Chunkiness distribution for the powder of Figure 2.1(c). c) The Chunkiness-Size domain for the powder of Figure 2.1(c) clearly demonstrates the difference in the range of shape for the fines as compared to the coarse fraction of the powder. This information is lost in a plot such as that of (b).

square root of the area of the profiles [6]. In Figure 2.3(a) the size distribution of the profiles of Figure 2.1(c) is presented as a histogram. This sample had been made by sieving a powder which had been hammer milled. It is interesting to see that there is a population of very small fineparticles which apparently remained by electrostatic attraction on the surface of the larger particles when they were sieved. It should be noted that if this data is converted to a distribution by weight the fines are a very small proportion of the overall powder and this illustrates the power of the microscope method to see a double distribution which is often suppressed by some of the automated methods of characterization such as the diffractometers to be described in Chapter 7.

As mentioned earlier the Aspect Ratio of profiles such as those of Figure 2.1(c) has been widely used to describe the shape, but for reasons that will become apparent later in this section a more useful factor of shape is the Chunkiness. The Aspect Ratio is defined as the length of a profile divided by its width. The Chunkiness is the reciprocal of the Aspect Ratio meaning that now the width is divided by the length. For a discussion of the earlier use of the Aspect Ratio see references 4, 5, and 7. The measurements of the Chunkiness of the profiles of Figure 2.1(c) are summarized in two different ways in Figure 2.3(b) and (c).

In the first technique for summarizing the shape information on the powder population the percentage of the profiles having a Chunkiness the same or larger than the stated Chunkiness on the abscissa, is given. Thus this distribution function tells us that 50 % of the fineparticles had Chunkiness greater than 0.43. It can be seen that the distribution of Chunkiness has three linear regions but the physical significance of the three regions are not readily apparent from this type of plot.

The other way of displaying the data, which was developed by Kaye, is defined as the Chunkiness-Size domain, was not usually feasible until the availability of automated plotting routines with modern data processing equipment [8]. The normalized size is the size of the profile divided by the largest profile present. The utility of the Chunkiness when domain plotting stems from the fact that Chunkiness ranges from 0 to 1 allowing all the information to be plotted on a single domain graph conversely, Aspect Ratio can run from l to infinity and is not readily plotted in a domain type of format. In Chunkiness-Size domain plotting, the information generates one point per fineparticle in data space as shown in Figure 2.3(c). The domain type format shows immediately that the larger fineparticles have a smaller range of Chunkiness which is not dependent on the size of the profile. The median Chunkiness is a line drawn through the data parallel to the abscissa. On the other hand the fine materials present in the powder have much greater range of Chunkiness ratios and are actually more string like. This indicates that the string like fragments, probably created by the hammer mill, have clung to the coarser profiles.

The Chunkiness factor is one in a class of shape factors which are properly described as dimensionless shape factors. In many discussions of the shape of profiles, the discussion often quickly moves to the question of three dimensional shape factors. One can measure three dimensional shape factors but the technology is expensive and labor intensive. One technique for describing three the dimensional structure of fragments such as those shown in Figure 2.1 is to measure three dimensions of the profile seperately and use triaxial graph paper to summarize the data. In Figure 2.4(a) the measurements of length, width and thickness that can be made on a large fineparticle are illustrated. When these three measurements are made one can summarize the three dimensional shape information on triangular graph paper. The graph paper and illustrative shape data are given in Figure 2.4(b) and (c). To be able to use this graph paper one first normalizes the length, width and height data in the way shown in the figure. First add up the three measurements and then normalize the values by dividing the actual measurements by the total. Thus for the cube all three measurements have normalized values of 0.33. To plot the data point for the cube, one moves up the length axis to the

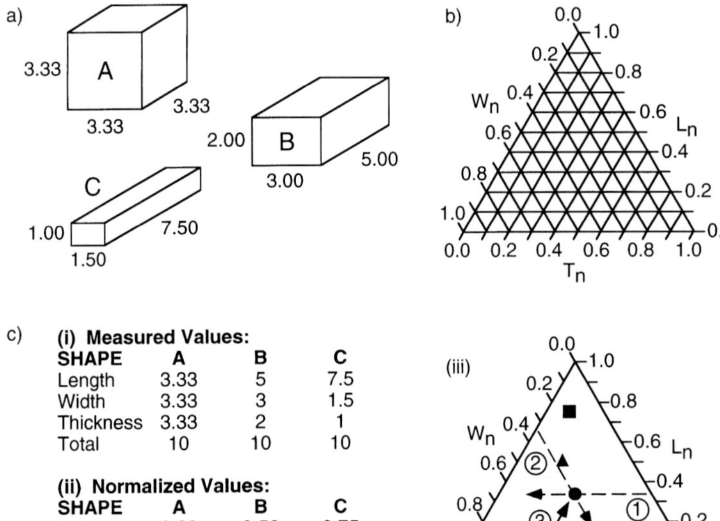

c)

(i) Measured Values:

SHAPE	A	B	C
Length	3.33	5	7.5
Width	3.33	3	1.5
Thickness	3.33	2	1
Total	10	10	10

(ii) Normalized Values:

SHAPE	A	B	C
L_n	0.33	0.50	0.75
W_n	0.33	0.30	0.15
T_n	0.33	0.20	0.10
	●	▲	■

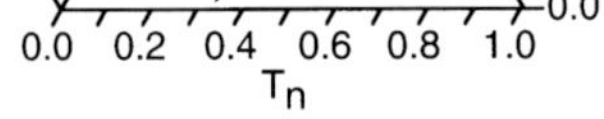

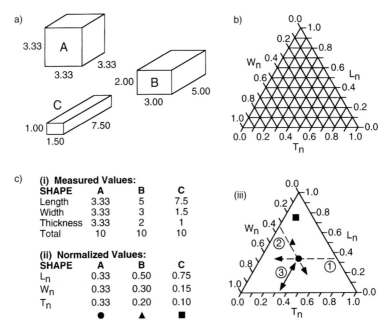

Figure 2.4. Information from the three dimensional measurements performed on powder grains can be summarized on triangular graph paper. a) Measurements of length, width and thickness of some basic shapes. b) Triangular grid graph paper for plotting normalized data from three measured dimensions of a fineparticle. c) (i) Measurements of the three shapes of (a) are first normalized, (ii), and then plotted on the triangular grid graph paper (iii) by first finding the length (1) then finding the intersection of that line with the width (2). When the data is normalized the third dimension, the thickness, is automatically found (3).

value and then moves along the line parallel to the base line as shown by the dotted lines and arrow until it meets the similar line from the width axis, the two lines meet at the point shown. The graph of the Figure 2.4(c) shows the 3 data points for the 2 rectangles and cube of Figure 2.4(a) [9, 10].

The rock fragments of Figure 2.1(a) were big enough to carry out the three dimensional measures of the type illustrated in Figure 2.4(a). The normalized information for direct measurements is plotted in Figure 2.5(a). When the fragments are smaller than can be handled directly, shadow casting can be used to obtain the information on the third dimension of the fineparticle. Shadow casting with gold vapor at an angle to the surface the fineparticles are resting on produces a shadow which can be used to measure the third dimension. To illustrate this technique we show the rock profiles of Figure 2.1(a) with a shadow cast by illumination at an angle in Figure 2.5(b). To compare the information available from shadow casting with those made by direct

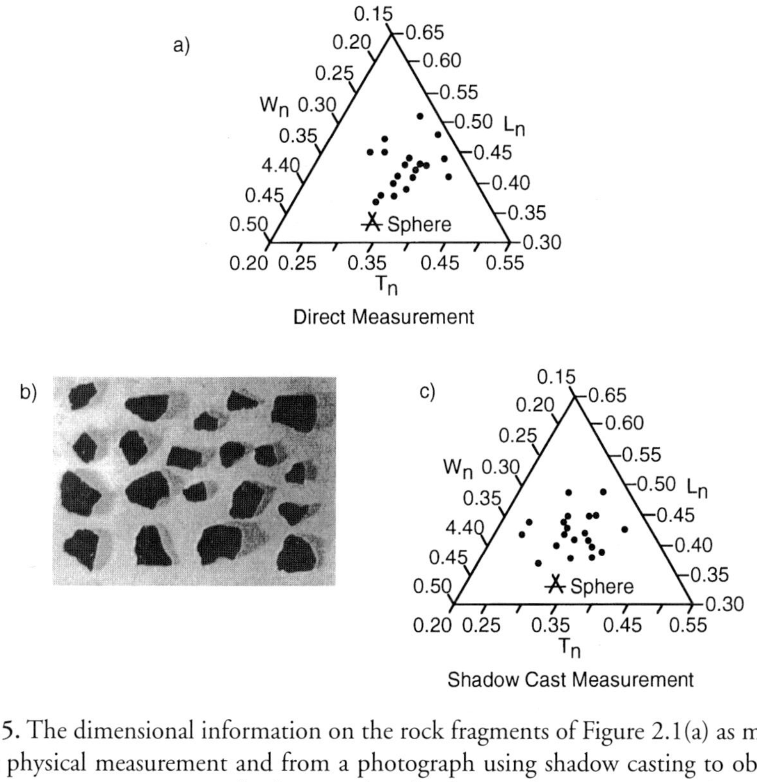

Figure 2.5. The dimensional information on the rock fragments of Figure 2.1(a) as measured by direct physical measurement and from a photograph using shadow casting to obtain the third dimension, generates virtually the same data for the population. a) Direct physical measurements of the rock fragments displayed on triangular grid graph paper. b) Shadow cast image of the rock fragments of Figure 2.1(a). c) Measurements by image analysis of the shadow cast photograph of (b) plotted on triangular grid graph paper.

measurements, from Figure 2.5(a), the shadow cast measurements from Figure 2.5(b) are illustrated in the triangular graph paper summary of the population of shapes shown in Figure 2.5(c). It can be seen that the data is essentially the same for direct measurement and shadow cast data.

The utility of plotting three dimensional shape information on triangular graph paper is illustrated by the data of Figure 2.6. In a laboratory project, J. Gratton took a piece of soft brick (a sand-lime brick which is similar to sedimentary rock in constitution) and studied how it eroded when it was tumbled in a small ball mill filled with sand and water. This simulated what actually happens to a piece of sedimentary rock transported by the flow of a river. Geologists call such erosion fluvial erosion from the Latin word for a river. It can be seen that, as the length of time spent tumbling in the ball mill increased, the rock fragment had its promontories eroded until it began to approach a spherical shape. The changes in the shape of the eroding rock are plotted in Figure 2.6(a). By measuring the shape distribution of rock fragments in a river, a

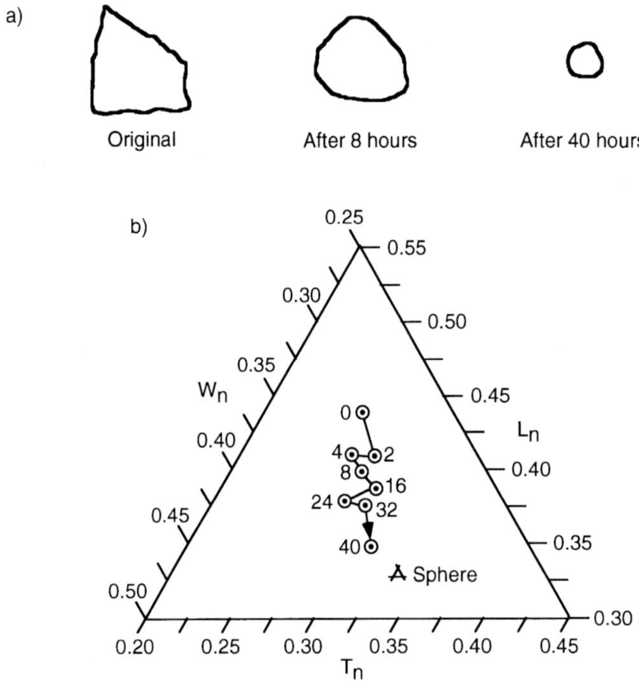

Figure 2.6. Changes in the shape of a brick fragment subjected to erosion in a tumbling sand slurry, simulates a rock fragment tumbling in a river bed over a long period of time. a) Change in shape of a brick fragment with time as it was eroded in a sand slurry. b) Plotting the dimensions of the brick fragment as it was eroded shows the utility of using triangular grid graph paper to summarize shape data.

geologist can learn how long the rock has been in the river and/or how fast it has been tumbling en route to the mouth of the river. For a discussion of other granular shape factors which have been used in earlier work see references 4, 11, 12, 13, and 14.

Other useful graphical displays of the type of information which can be generated by characterizing profiles of powder grains are demonstrated by the data displays presented in Figures 2.7 – 2.9. In Figure 2.7(b) a histogram generated by characterizing a sample of an iron powder, shown in Figure 2.7(a), is presented [15]. This histogram has a shape typical of a distribution function known as a log-Gaussian distribution. If one transforms the histogram of Figure 2.7(b) into a cumulative percentage oversize distribution by weight, the data can be plotted on a special graph paper known as log-Gaussian graph paper [16]. The axes on this graph paper are designed so that if the

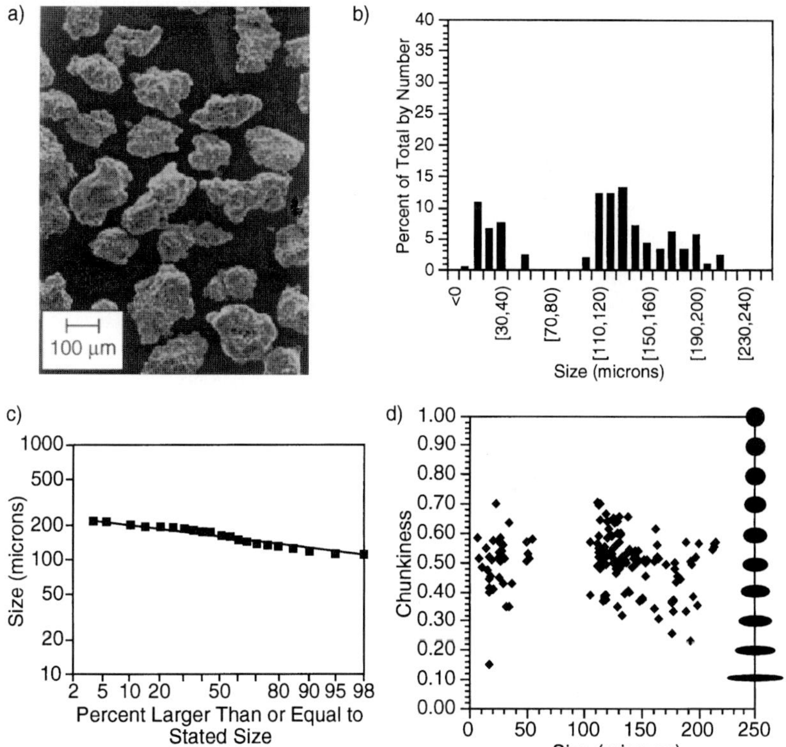

Figure 2.7. When characterizing powders there are several methods of displaying size distribution data [15, 16]. a) Micrograph of typical grains of coarse iron powder. (Metal powders supplied by D. Alliet of Xerox Corp.) b) Size distribution obtained from by image analysis for the coarse iron powder of (a) displayed as a histogram. c) Size distribution of (b) plotted on log-Gaussian graph paper. d) Chunkiness-Size domain for the powder of (a).

appropriate distribution function is log-Gaussian then the cumulative distribution will generate a straight line as shown in Figure 2.7(c). In Figure 2.7(d) the Chunkiness-Size domain generated from an image analysis of the powder is shown. Note that the population is bi-modal with an isolated cloud of fines that were probably clinging to the larger fineparticles after the product was sieved. When converted to a weight distribution, the fines become an insignificant percentage of the powder weight. Note that for the large fineparticles the shape becomes more elongated with larger size [15].

In Figure 2.8 similar data is presented for a spray dried copper-zinc ferrite powder. For this powder the histogram, of Figure 2.8(b) has the familiar bell shaped structure associated with the Gaussian distribution function [16]. For this powder the cumulative percentage oversize by weight generates a straight line on Gaussian graph paper.

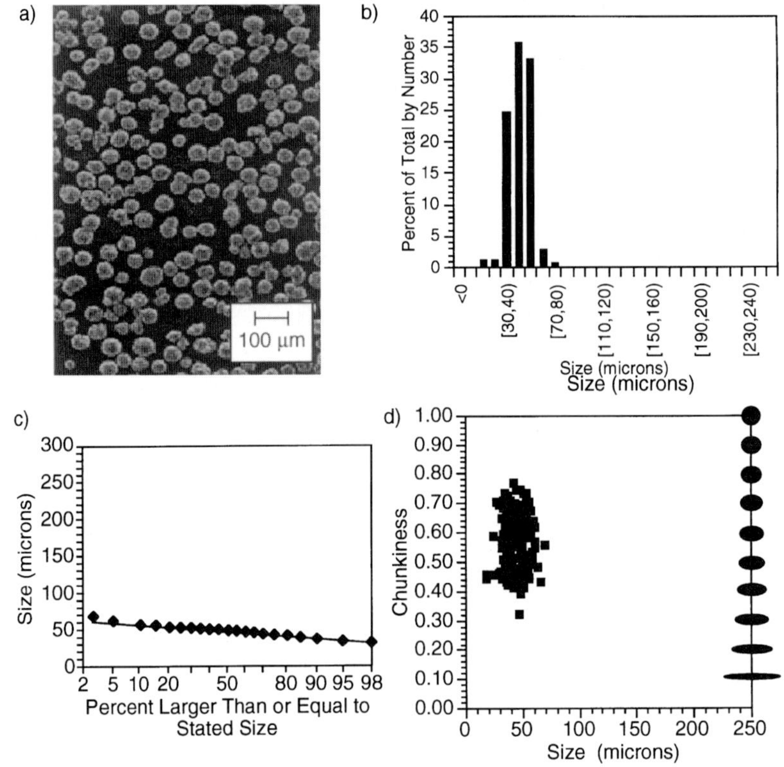

Figure 2.8. Size distribution data for a spray dried copper-zinc ferrite powder. a) Micrograph of typical copper-zinc ferrite powder grains. (Metal powders supplied by D. Alliet of Xerox Corp.) b) Size distribution of the powder of (a) as measured by image analysis displayed as a histogram. c) Size distribution of (b) plotted on Gaussian graph paper. d) Chunkiness-Size domain for the powder of (a).

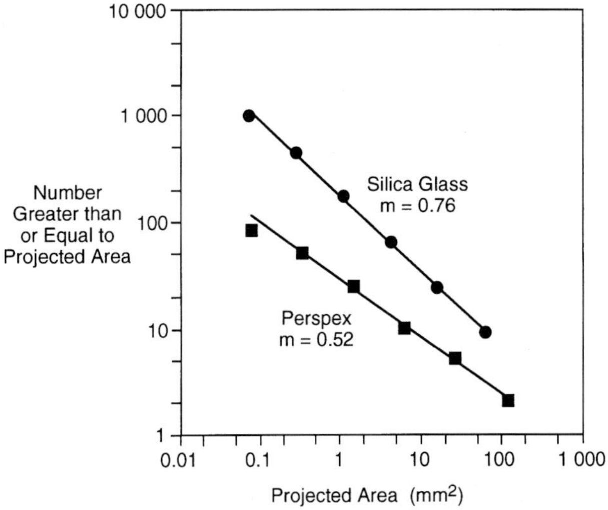

Figure 2.9. The cumulative undersize distribution of fineparticle size is an important way of displaying size distribution data. Shown above, plotted on log-log scales, are the size distributions of the fragments produced when two different amorphous materials were shattered by impact after being cooled to low temperatures [18]. From the perspective of chaos theory and applied fractal geometry explained in more detail in a later chapter, the slope of this type of data line is described as a fractal dimension in data space.

Note that in the Chunkiness-Size domain of Figure 2.8(d) that the range of size is much smaller than for the powder of Figure 2.7 which was produced by sieve fractionation reaffirming it it's Gaussian distribution. Note also that the shape range is closer to spherical, as expected for a spray dried powder [15].

In Figure 2.9 the cumulative oversize distribution for fragmented materials studied by Miles and Brown is shown plotted on log-log graph paper [17, 18]. In general powder studies, this distribution function has not been widely used, but because the significance of such a straight line on this type of graph paper can be related to fragmentation dynamics, one can anticipate that it will be more widely used in the future.

2.3 Characterizing the Presence of Edges On a Fineparticle Profile

In several industries, particularly the abrasives manufacturing and usage industries, not only is the general shape of a profile of interest but the number and sharpness of the edges on the profile. These create the polishing or cutting power of the abrasives and are of great interest to the technologist. In other industries where material may be precipitated, the resultant crystalline fineparticles may have a number of sharp edges the characterization of which is useful information on formation dynamics. To measure the number of edges, and their sharpness, on a profile one can use the procedure illustrated in Figure 2.10 The first step in the characterization process is to digitize

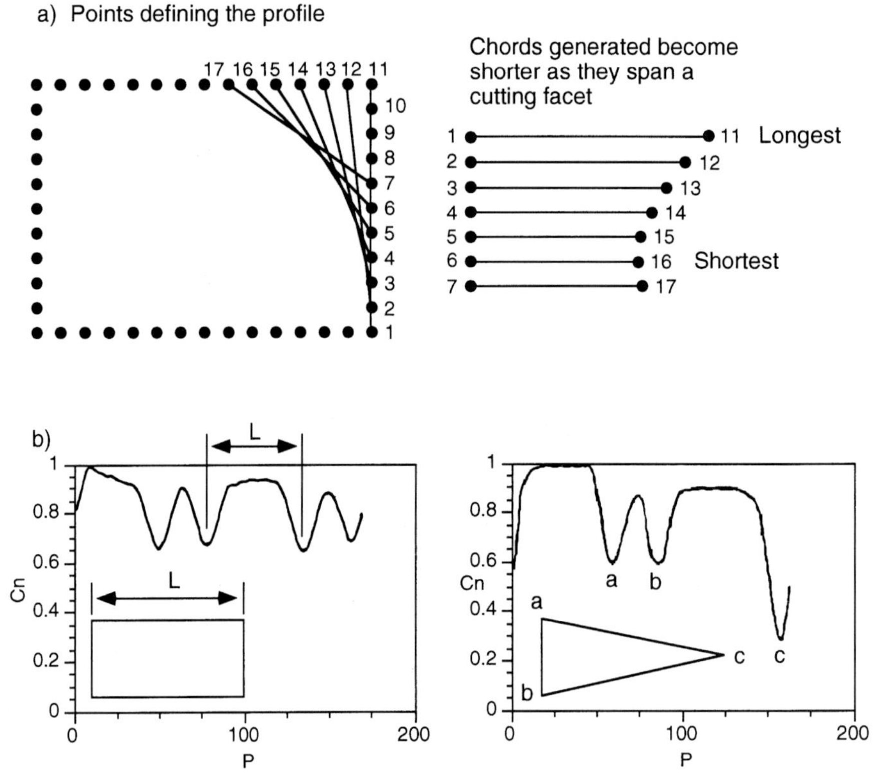

Figure 2.10. A process known as slip-chording can be used to characterize a fineparticle profile for the presence of facets and edges. The chord length between sets of points forming a digitized version of a profile are used to generate a signature waveform of the profile. Dips in the waveform indicate the presence of facets and their depth indicates the sharpness of the point. Flat portions of the waveform indicate edges and the distance between dips represent the length of the edges.

the profile as shown for the rectangular profile of Figure 2.10(a). Then start to draw chords between a specified number of steps around the digitized profile. For instance if starting at the point 1 draw a chord to the position of the 11th step. Then "step" forward 1 step to draw a chord between 2 and 12 and so on as shown. As one moves around a corner the chords obviously shorten as shown. By plotting a graph of the chord length against the position on the perimeter one generates what is known as the facet signature waveform [19]. Dips in the facet signature waveform indicate the presence of a facet on the profile. The distance between any two sequential facets can be read off the facet signature waveform as shown by the distance L in Figure 2.10(b) in which the signature waveform for the rectangle is shown. When using this technique one can tune the exploration technique to demonstrate corners of interest in a

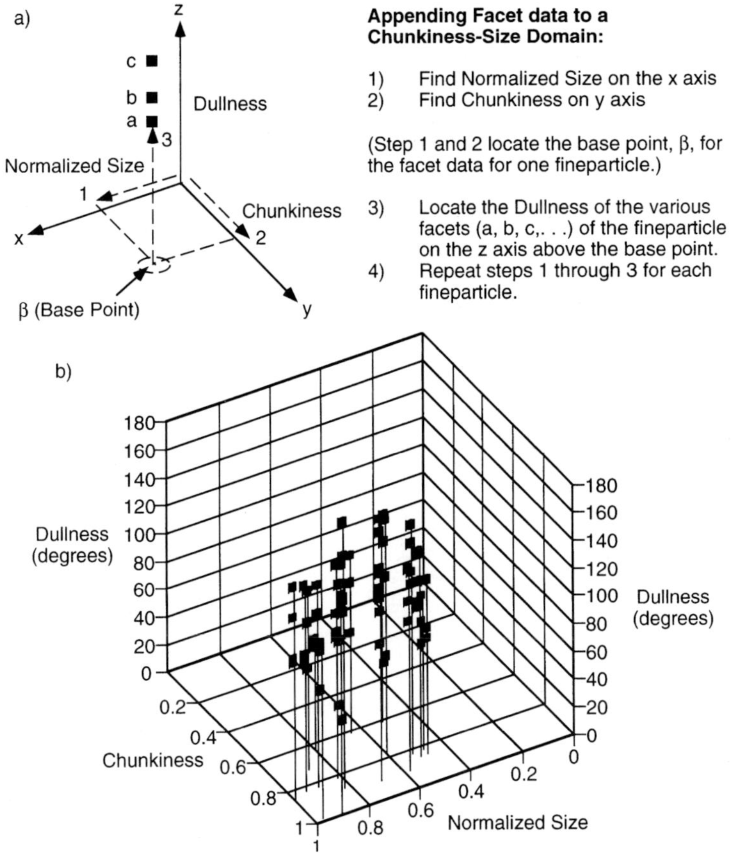

Figure 2.11. Information on the facets present in a population of fragments can be displayed in a three dimensional domain plot

specific study. For example by altering the number of steps taken before drawing a chord may emphasize some feature of the structure.

To summarize graphically the data on the size, shape and number of facets present in a set of abrasive or crystalline profiles one can use the three dimensional domain plotting strategy illustrated in Figure 2.11. The three axis of the graph represent normalized size, chunkiness and what is defined as dullness. For a given abrasive fineparticle plot a point such as β in the Chunkiness-Size domain of the graph. Then erect an ordinate at this point. On this ordinate locate the number of facets on a profile of a

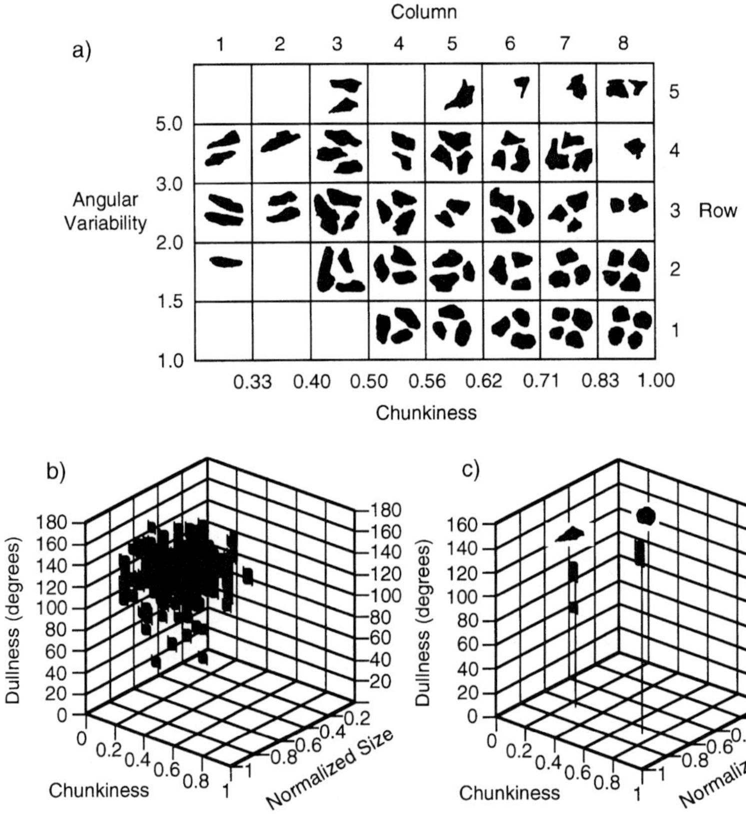

Figure 2.12. Facet signature characterization of a profile in conjunction with size and shape measurements of fineparticle profiles can be used to automate allocation of the fineparticle to a location within a reference grid used by geologists to describe the shape and size of rock fragments [20]. a) Reworked standard grid replacing elongation with the more easily handled Chunkiness. b) Three dimensional summary of the array of (a) using facet data generated for each profile. c) Location of two significantly different profiles in the three dimensional space showing their facets distributed along the vertical axis by order of dullness. (Sharper facets are placed lower while dull facets are nearer the top of the space).

given angularity. A point such as **c** is approaching an angle which is considered dull whereas **b** and **a** are sharper. In Figure 2.11(b) the information of the edges from the facet signatures of the profiles shown in Figure 2.1(a) are shown. The cloud of data points representing the number of facets on the different profiles can be treated as a set of points in a three dimensional data space. They can then be characterized by their centroid and first and second moments about the centroid.

The three dimensional plotting technique just described can be useful in geology. Geologists and scientists concerned with the crushing and processing of fragmented rock sometimes make use of an array of shapes to allocate rock fragments to a descriptive shape group. In Figure 2.12(a) a reference array of variously shaped fineparticles originally described by Austin et al has been reformulated using the concept of Chunkiness [20]. When the facet signatures of their profiles are measured, the reference array of Figure 2.12(a) becomes the data display of Figure 2.12(b). The physical significance of any particular profile could be plotted as shown in Figure 2.12(c). A part of the reference array such as column 6, row 3 can become a spatial region in the three dimensional graph. Computer data processing of the facet signature waveform of a profile can be tied to a location in the measured profile shape and facet richness category automatically as illustrated in Figure 2.12(c).

2.4 Geometric Signature Waveforms for Describing the Shape of Fineparticles

In the early 1970's, as the cost of data processing started to fall drastically, several scientists independently developed the concept of using what are known as geometric signature waveforms which could then be subjected to Fourier analysis to describe the shape of fineparticle profiles [21, 22, 23, 24]. To generate information on the structure of the perimeter of the profile the geometric signature waveform can be used. Flook used the centroid of a profile, as if the profile were a laminar shape, as a reference point as illustrated in Figure 2.13(a) [23]. To generate the signature waveform, move a rotating vector around the perimeter with respect to some fixed direction. Thus, in the case of the profile shown in Figure 2.13(a), the vector $\mathbf{R}_1$ drawn from the centroid reference point to the most distant point serves as the reference direction. Then the vector **R** is rotated to various angles, $\boldsymbol{\theta}$, noting the magnitude of the vector at each angle as shown. When the magnitude of the vector is plotted against the angle, a graph is generated that looks like a waveform as shown in Figure 2.13(a). When this is done in greater detail the highly resolved graph of Figure 2.13(b) is generated. Strictly speaking one can only apply Fourier Analysis to such waveforms if it is part of a continuous

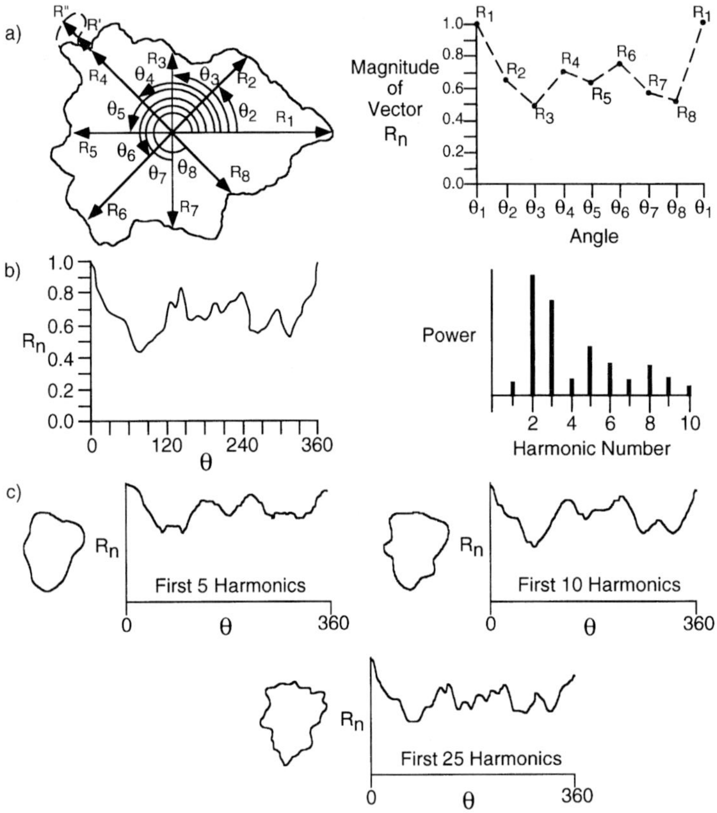

Figure 2.13. A geometric signature waveform, of the type reported by Flook, can be Fourier analyzed and used to characterize the fineparticle profile. a) Method for generating a geometric signature waveform. b) Performing a Fourier analysis of a high resolution signature waveform yields a harmonic spectrum showing the various component waves and their power which would be required to produce the original waveform. c) Profiles regenerated using the number of harmonics stated on the graphs. (R_n is the magnitude of the radius vector from the centroid to the perimeter of the profile at the angle θ.)

waveform but this is assumed when the wave is analyzed. Readers unfamiliar with Fourier analysis can find an introduction to the topic in reference 25. For the profile of Figure 2.13(a) the relative power levels of the various harmonics which contribute to the complex waveform are shown as a part of Figure 2.13(b). It has been shown by several workers that the basic structure of the profile is described by the information in the first five harmonics contributing to the signature waveform. The next 20 harmonics generate information on the texture of the profile. In Figure 2.13(c) the reconstituted signature waveform using the information on the strength of various numbers of harmonics are shown. It can be seen that a reconstruction using only the first five harmonics summarizes the basic shape of the profile. In the subsequent parts of the figure the waveform for the first ten harmonics and then twenty five harmonics are shown and it can be seen that the texture of the profile increases with the addition of higher harmonics. The use of this type of waveform is limited if the profile contains any indentations or convolutions. The value of the vector from the reference point to the perimeter does not have a unique value since it crosses the profile more than once. This fact is not usually a limitation for most of the profiles studied by geologists and mineral processing engineers.

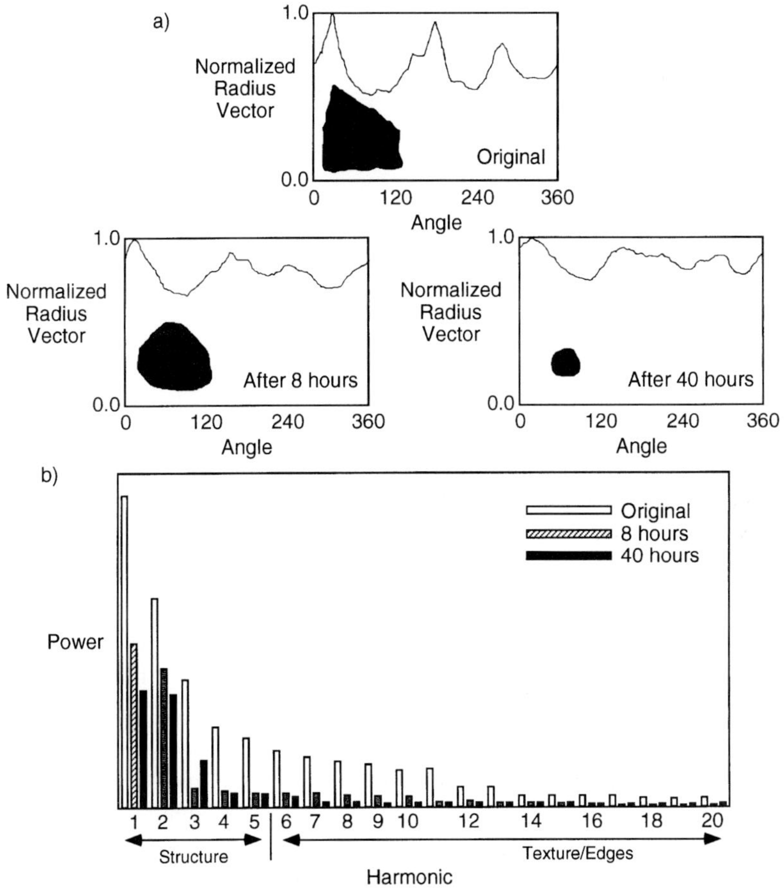

Figure 2.14. The usefulness of Fourier analysis of the geometric signature waveform for characterizing a profile is demonstrated by the application of the technique to the study of erosion of a fragment of material as discussed in Figure 2.6. a) Geometric signature waveforms of the eroding fragment. b) Harmonic spectra of the signature waveforms of (a).

In Figure 2.14 the construction of the geometric signature waveform followed by Fourier Analysis is shown for a rock undergoing simulated fluvial erosion as described in the experiment summarized in Figure 2.6. The shape and signature waveform of the fragment as it originally appeared, at 8 hours erosion, and 40 hours erosion are shown. The power spectra for the three signature waveforms shown in Figure 2.14(a) summarized in Figure 2.14(b). The original profile had some sharp edges which resulted in higher frequencies in the harmonic spectrum being relatively strong. As the rock moved towards a more spherical shape the power of the higher frequencies diminished drastically.

With modern computers one can generate the Fourier transform of an actual profile which is technically the same as the Fraunhoffer diffraction pattern of the profile. This aspect of shape analysis of profiles will be discussed in more detail in Chapter 7 when we discuss the use of diffractometers to characterize fineparticles.

2.5 Using Automated Image Analysis Systems to Size Fineparticle Populations

One of the difficult tasks presented to the fineparticle technologists involves characterizing an array such as that shown in Figure 2.15(a). This is a highly magnified view of a cross section through a toner bead used in xerographic copiers. The word toner is used in the xerographic industry to describe the dry powdered ink used to develop the picture created in the electrostatic copier. The xerographic specialists are interested in several aspects of this array. First of all they are interested in the structure and size of the individual carbonblack profiles which are dispersed in the transparent plastic. Secondly they are interested in whether the dispersion of the profiles is random and if one can arrange for a non-random dispersion in certain circumstances. Such a non-random dispersion would be known as a structured mix [26].

Geologists are interested in systems similar to that of Figure 2.15(a) when they study sections through a sample of mineral ore. A study of ore sections is known to the geologist as petrographic modal analysis [27]. Thus, to the geologist studying an ore sample, the dispersed fineparticles of Figure 2.15(a) would represent the valuable mineral in the ore. When assessing the appropriate crushing and grinding procedure the geologist needs to estimate the richness of the ore (the percentage of valuable material in the ore) and the average size of the dispersed ore fineparticles. To evaluate these two parameters interesting techniques have been developed by geologists which exploit some of the concepts of a subject known as geometrical probability and fractal geometry. In fact, the method we now know as an application of fractal geometry, was developed

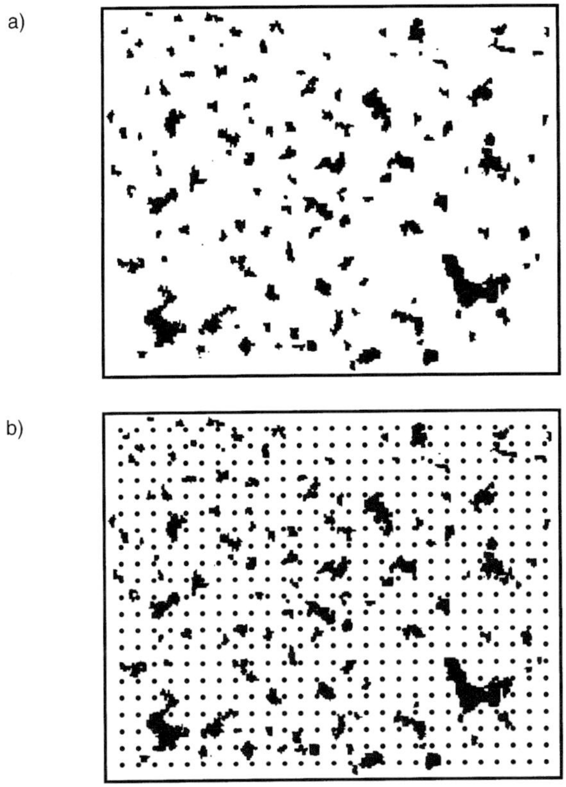

Figure 2.15. Statistical counting procedures can simplify such tasks as estimating the area of a field of view covered by a dispersion of fineparticles. a) Image derived from an electron micrograph of a section through a xerographic toner bead. b) A regular array of dots superimposed on the image of (a) can be used to estimate the area of the field of view occupied by the dispersed fineparticles. The fraction of dots falling within the profiles of an array is an estimate of the fraction of the area of the field of view covered by the fineparticles.

prior to the actual appearance of the theories of fractal geometry. This method is known as the Rosiwal intercept method and will be discussed later in this section.

The first technique for evaluating the richness of the ore depends upon the fact that if one makes a section through a piece of ore the volume fraction of the valuable material in the ore is the same as the area fraction exposed in the section. Analysis of fields of view such as those of 2.15(a) also correspond to the problems of assessing the porosity of a piece of sandstone or for the structure of a section through a filter. This will be discussed in greater detail in Chapter 9. To assess the percentage area of the field of view such as that of Figure 2.15(a), the geologists would make use of what is known as dot counting procedures, (also known as Chayes dot counting procedure) [27]. The principles of this technique are illustrated in Figure 2.15(b). If a regular grid of dots is placed over the random array of profiles it can be shown that the percentage number of dots falling within the profiles is the same as the percentage of the field of view covered by the fineparticle profiles. Experimental measures of the field of view shown in Figure 2.15(a) have established that the carbon black occupies 7.8 % of the field of view. The reader can check how good this estimate is by counting the dots falling within the profiles in the system of Figure 2.15(b) [28]. Because it can be assumed that the carbon black profiles are themselves randomly distributed there is no need to randomize the dots. Double randomness does not have any advantage for the measurement procedure. The theory of randomly interacting dispersed systems in two dimensional space and three dimensional space is part of a subject known as geometrical probability [29]. The size of an irregular profile can also be measured by using enlarged images and a more intense array of dots. Although such a method could prove to be very efficient it has not found favor with technologists in the field of powder technology who seem to prefer making direct measurements of area and perimeter than relying on statistical theorems for a fluctuating dot population [28, 30].

To understand the basis of another procedure for estimating the area fraction of a field of view occupied by dispersed fineparticles consider the simpler system shown in Figure 2.16(a). This figure was constructed by measuring the area of each of the carbon black profiles of Figure 2.15(a) and then a circle of the same size as the irregular profile was drawn at the point where each profile was located. (Note that the array of circles has been rotated 90° clockwise with respect to the array of Figure 2.15(a) in order to accommodate the space available for the diagram.)

It can be shown from the theorems of fractal geometry that the sets of intercepts created in a line search examination of a dispersed field of view constitutes a Cantorian set in one dimensional space. If limited to the points where the search tracks, enter and leave the profiles, revealed is a Cantorian dust that can be linked to the richness of the ore [31, 32]. It can be shown that the fraction of the random lines falling within the profiles is the same as area fraction occupied by the dispersed fineparticles in the field of view. It can also be shown that the average size of the chord tracking across the dispersed fineparticles is a function of the average size of the dispersed fineparticle. These relationships were first developed by Rosiwal [33, 34, 35].

Ceramicists and powder metallurgists have also made extensive studies of techniques for studying the structure of holes and dispersed species in items made from powder compacts and these technologists refer to the measurement technology either by the term quantitative stereology or quantitative microscopy [30]. It has been shown that if a dispersion such as that shown in Figure 2.16(a), is randomly dispersed then the sets of chords drawn in a parallel line scan search such as that of Figure 2.16(c)(i) has a log-normal distribution. Whereas an agglomerated field of view, or a non randomized

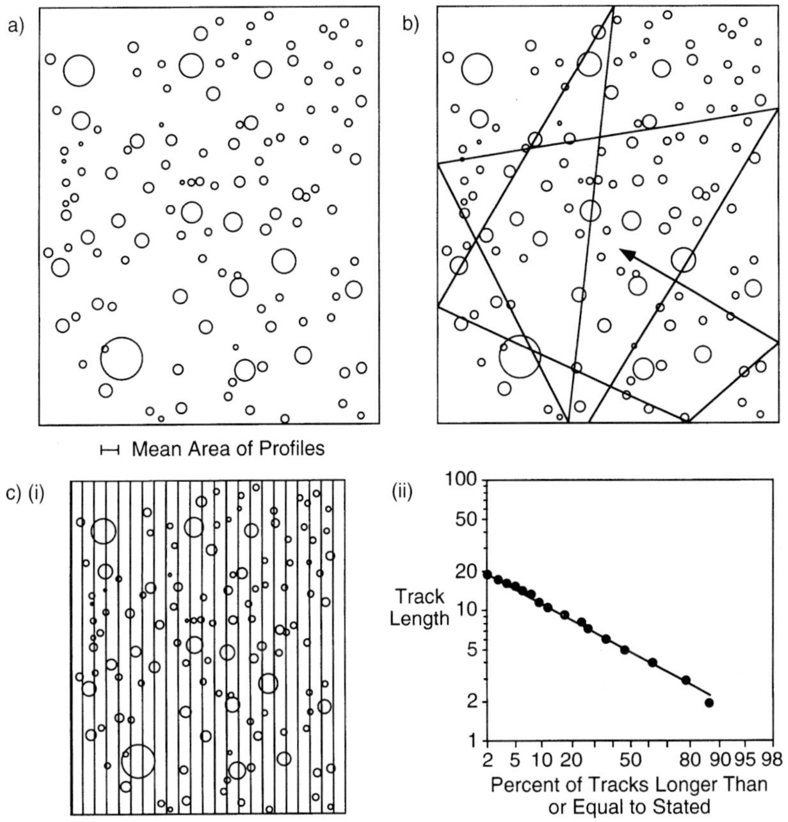

Figure 2.16. Random tracks drawn across a dispersed set of fineparticle profiles can be used to characterize the important features of the array. This type of evaluation is known as the Rosiwal intercept method. a) Array of Figure 2.15(a) converted to circles of equal area. b) Random lines can be used to generate a set of intercept tracks. c) Random lines used to inspect a random distribution are no better than using a set of parallel lines. (i) Parallel lines on the dispersion. (ii) Data generated by the set of parallel lines.

field of view, gives a different population of chords [34]. This randomness contained in Figure 2.17(b), is represented by the data of Figure 2.17(a). Consider a digit such as 9 dispersed in a random number table of the type shown in Figure 2.17(a). That digit constitutes an ideally randomized dispersion of fineparticle if one considers the digit to represent a fineparticle. Thusly the random number table in 2.17(a) has been turned into a simulated dispersion where each 9 in the original random number table has been turned into a small black square. If one now tracks between each black pixel in a consistent direction as shown by the tracking lines in Figure 2.17(b) one genera)

3	0	3	9	6	9	8	4	4	7	8	3	1	6	4	4	9	1	5	0
1	6	0	7	0	7	5	7	2	4	0	6	3	1	7	4	3	5	5	0
4	2	7	4	1	5	1	2	9	6	0	0	1	0	6	8	9	2	0	5
6	5	0	5	4	1	7	7	7	8	5	3	9	7	3	1	9	8	6	0
2	2	6	2	7	9	1	5	7	2	2	4	0	8	6	2	8	0	2	7
8	2	7	1	7	7	8	6	7	4	2	1	9	5	8	0	0	1	8	6
7	5	7	1	2	5	6	3	5	2	6	1	8	5	9	3	5	8	1	4
6	5	4	8	4	8	5	1	0	8	4	9	7	6	5	3	4	5	7	4
5	2	1	3	1	0	4	9	1	9	2	0	6	0	0	4	4	3	0	3
0	6	5	0	4	9	4	0	1	5	9	1	1	0	2	1	7	2	9	5
4	1	3	1	6	1	7	1	8	4	6	2	1	1	3	5	5	7	5	9
4	9	3	1	9	6	6	3	6	9	5	4	4	6	3	3	9	2	4	0
7	5	6	3	0	0	2	1	7	7	9	9	6	7	0	3	9	8	1	5
9	7	4	1	2	8	0	3	7	9	2	2	1	8	1	0	7	7	9	1
3	6	9	3	5	1	5	1	2	0	5	2	8	1	4	2	2	2	0	0
0	0	2	8	8	7	3	4	3	3	1	8	5	0	2	6	0	6	3	1
6	1	1	1	3	3	3	0	2	8	3	9	7	7	9	7	1	2	6	9
5	7	8	4	3	5	5	3	1	7	5	6	3	5	2	2	2	3	2	5
8	6	7	8	3	9	1	4	4	0	6	2	0	7	7	6	8	2	6	0
7	0	6	8	3	2	4	6	6	3	6	2	5	0	2	0	7	0	5	0

b)

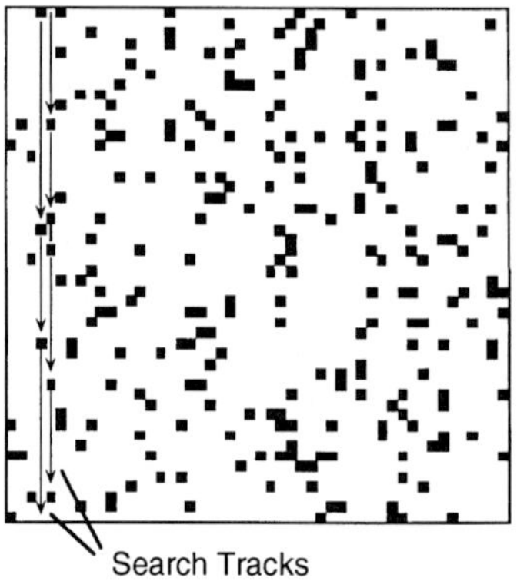

c)

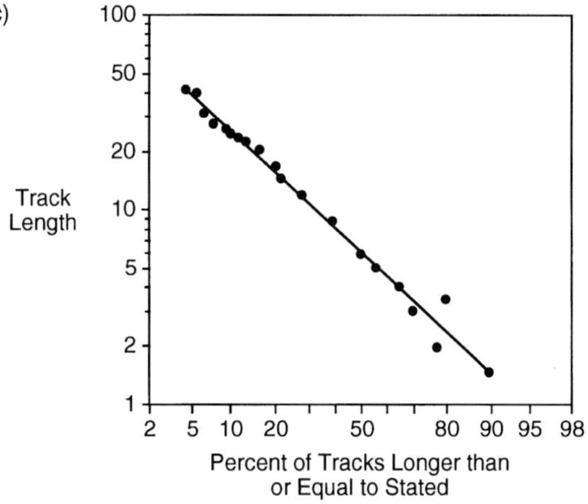

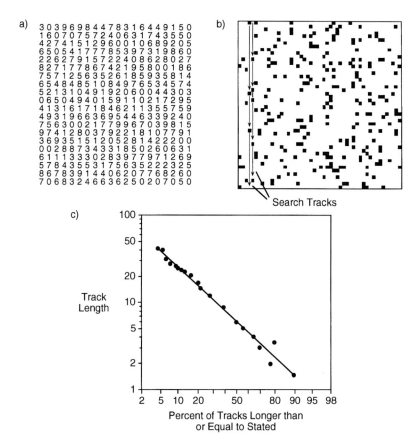

Figure 2.17. It has been shown that track lengths between fineparticles in a random array are log-normally distributed. a) A random number table consists of the digits 0 through 9 chosen in random order to form a table. b) A simulated random field of view can be generated by turning every 9 in the random number table of (a) into a black square. The distance between black squares then generates various track lengths when examined with a series of parallel lines. c) Track length distribution generated from the array of (b).

ates the set of track lengths. This data plots on log-Gaussian probability paper as cumulative distribution demonstrating a straight line relationship as shown in Figure 2.17(c). For a full discussion of using this technique to assess the structure of a pigment dispersion see reference 34.

In the days when measurements in a field of view of fineparticles had to be carried out by manual inspection through a microscope several so called statistical diameter measures of a fineparticle system were developed. One of these was described as the Feret's diameter and was the projected length of the profile with respect to fixed direction. Another statistical diameter uses the Martin's diameter which is defined as the chord length parallel to a fixed tracking direction which divided the profile into two equal sections. When tracking across a field of view these diameters could be measured easily without adjusting the microscope. These measures were sometimes more objective than the measurements made by comparing areas using the reticules described earlier. Such statistical measures are no longer in wide spread use but reader who wishes to familiarize themselves with such measurements, if they encounter them in earlier research literature, can consult references 4 and 5.

Line scan logic of the type illustrated in Figure 2.18 formed the basis of the logic used in a generation of image analyzers based on line scan television cameras. During the 1970s and 1980s these instruments were widely used. They made measurements

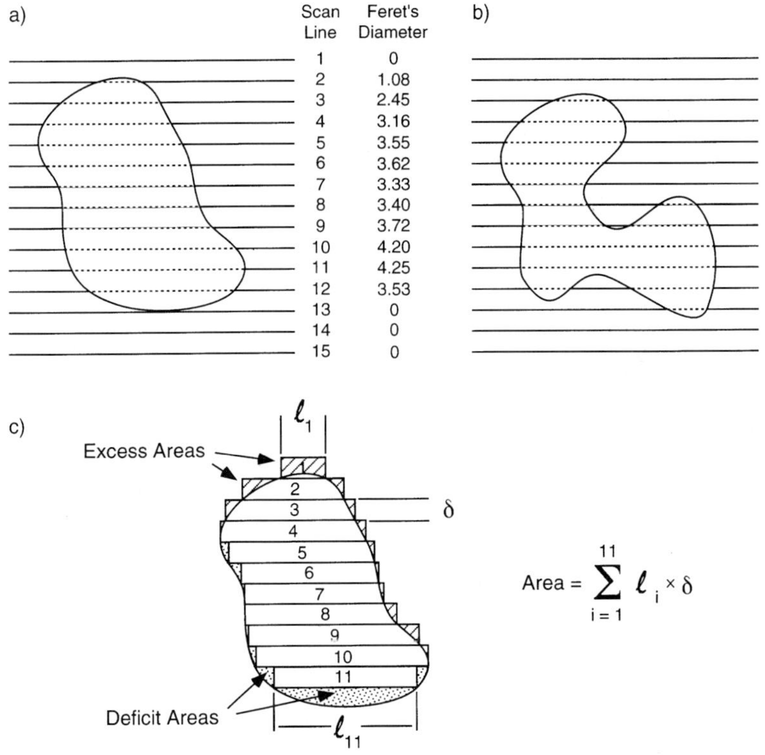

Figure 2.18. In line scan logic the area of a profile was estimated by adding up the length of the chords generated by the intersection of the parallel line scans with the profile, and multiplying by the line scan separation, δ. Excess areas generated by this method tend to be compensated by deficit areas resulting in a quite efficient area estimate [4]. a) Profile with line scans showing the measurement of Feret's diameters on successive lines. b) Convoluted profiles present a problem for line scan logic. c) Method of determining area using line scan logic.

on profiles by logic based on the fact that, if a fineparticle was constituted from a series of line scans, then the area of that profile and the perimeter could be deduced from the line scan information. The Galai and Lasentec size analyzers also line scan image analysis to characterize fineparticles in a suspension but the discussion of these instruments is more appropriately dealt with in Chapter 6. This type of line scan integration worked beautifully with sets of circles but if one were interested in rugged profiles then multiple crossings of the profile by the search scan led to uncertainty in the significance of the data. To overcome such multiple interception problems and also to separate profiles in the field of view which were judged to be just touching, erosion-dilation logic was developed.

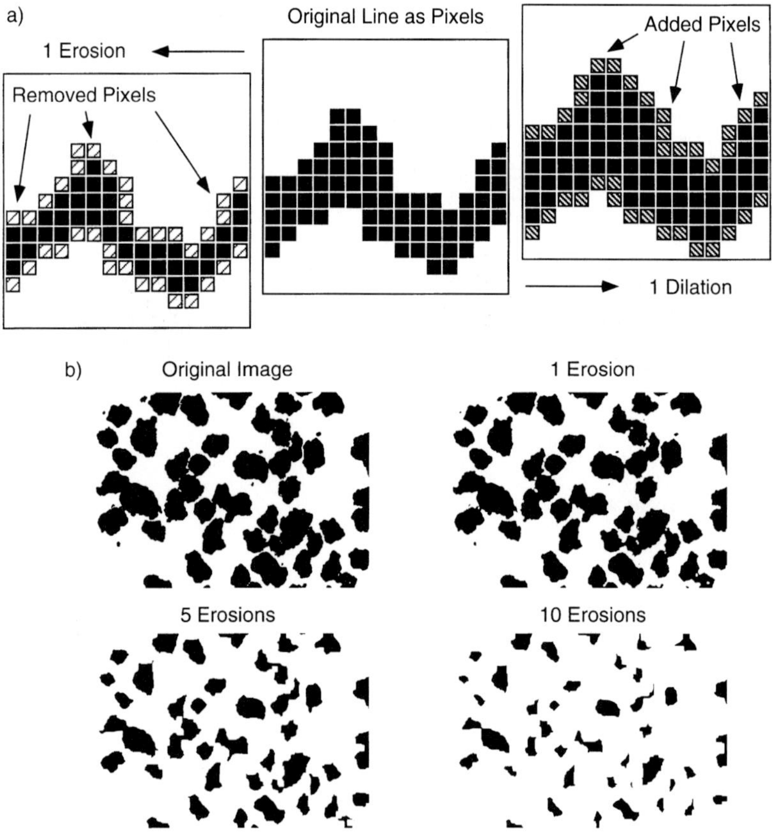

Figure 2.19. Procedures for automated counting of fineparticles using images captured by a computer may make use of erosion/dilation logic in order to separate then count and size individual fineparticles. a) Basic principles of erosion and dilation of a digitized image. b) Erosion logic was originally developed to separate adjacent profiles in a field of view for automated counting.

To implement-erosion dilation logic the information from the scan of a profile was stored in the memory of the computer as a mosaic of the type shown in Figure 2.19(a). The small squares of such a data mosaic are known in the language of image analysis as pixels. The word pixel is short for picture element. The processes of pixel dilation and erosion are illustrated in Figure 2.19(a). Erosion logic was first developed to separate adjacent profiles when counting with automated logic the number of profiles in a field of view. This aspect of the use of erosion logic is shown in Figure 2.19(b) The way in which successive dilation followed by erosion back to the original size could smooth out a complicated profile to avoid multiple intercepts of the profile by scan logic as illustrated in figure 2.20(a). In Figure 2.20(b) the process by which erosion of the profile can be used to explore the significance of its structure is shown. For example a carbonblack of the type shown in Figure 2.20 was probably formed by the colli-

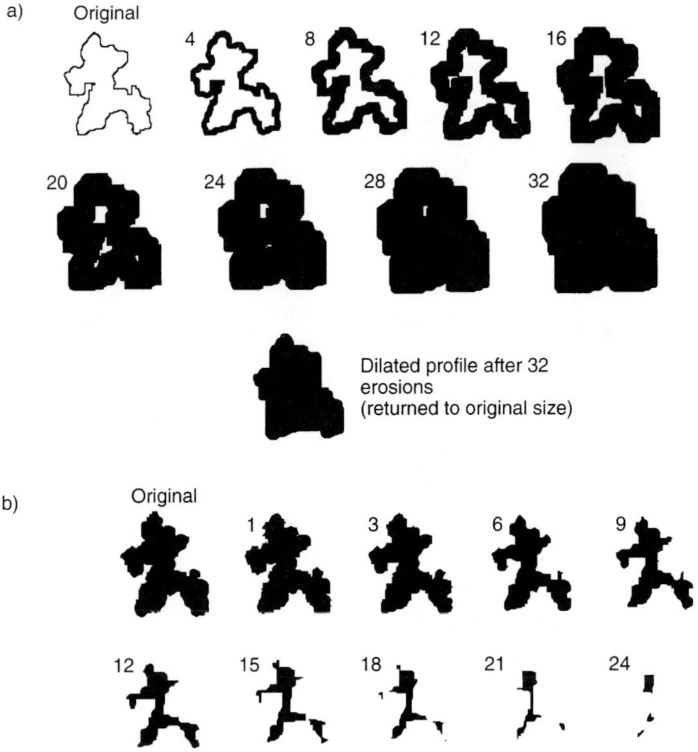

Figure 2.20. Erosion dilation logic is often used as a processing stage in the characterization of complex fineparticle profiles. a) Dilation followed by erosion can smooth out a rugged profile so that it can be used with linescan logic procedures. b) Erosion of a complex agglomerate can be used to study it's probable formation dynamics. In this case the agglomerate appears to consist of at least three primary agglomerates.

sion of several sub-agglomerates in the flame producing the carbon black. One can erode the structure of the profile to see how many subunits were probably combined to form the agglomerate.

A full discussion of the sophisticated logic used in modern image analysis systems is beyond the scope of this text and the reader should consult the literature of the manufacturers of such instruments.

2.6 Fractal Characterization of Rugged Boundaries

The detailed description of boundaries such as those of the carbonblack profiles of Figure 2.15(a) has been revolutionized by the development of a subject known as fractal geometry. In 1977 Mandelbrot published a book entitled "Fractals, Form, Chance and Dimension" in which he set out the basic theorems and concepts of a new form of geometry to be used to describe the structure of rugged boundaries such as coastlines [31]. In his book Mandelbrot tells us that one of the key ideas underlying the development of fractal geometry was the fact that there is no answer to the simple question; "How long is the coastline of Great Britain?" He points out that, in fact, the coastline of Great Britain is infinite and that all one can do is estimate the perimeter from a given operational perspective. Thus he pointed out that if one attempted to estimate the coastline of Great Britain by striding around it with a given step size, that the estimate of the perimeter of the coastline achieved in this way increased as the size of the step decreased. Eventually if one would use a trained ant to estimate the coastline it would be very very large as compared to the estimate made by a human being walking around the coast. This fact is illustrated by the sketches of Figure 2.21 in which the coastline of Great Britain is estimated using decreasing step size exploration of the perimeter.

One of the key ideas suggested by Mandelbrot is that to characterize indeterminate boundaries such as the coastline of Great Britain one should switch their attention from attempts to find a finite answer to the question of how great the perimeter is to measure how fast the estimates of the perimeter increase as the inspection resolution is increased. Thus, by looking at an attempt to measure the coastline of Great Britain the perimeter of the polygon created by striding around the coastline with a series of decreasing step size exploration generates a series of polygons of increasing length. As a mathematical strategy one can normalize these estimates of a rugged profile using the maximum projected length of the profile. For historic reasons this projected maximum length is known as the maximum Ferret's diameter. Mandelbrot showed that, for a structure describable as being fractal in nature, by plottig the normalized value of the perimeter estimate against the step size on log-log graph paper would obtain a

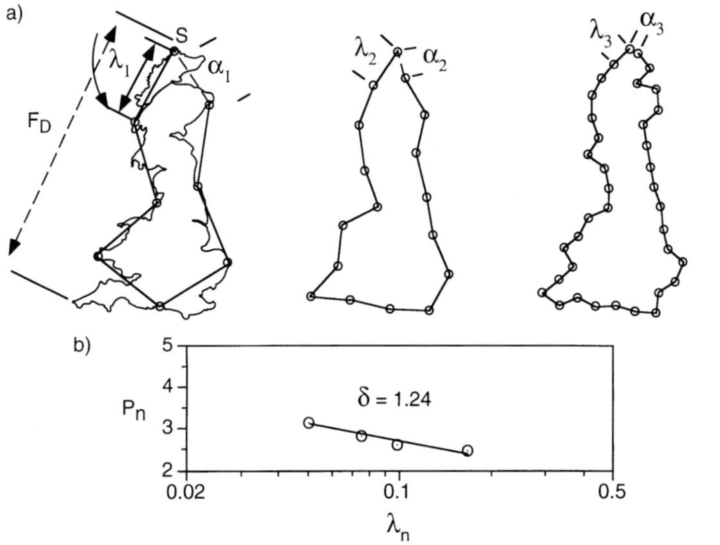

The Perimeter Estimate, P, of the profile found by walking around the profile using stride length λ is given by the equation:

$$P = n\lambda + \alpha$$

where where n is the number of steps (λ) needed to walk around the profile and α is the small remaining distance ($< \lambda$) needed to close the polygon back to the starting point, S, of the walk.

In order to plot the data it is first normalized by dividing by the maximum Feret's Diameter, F_D

$$\lambda_n = \frac{\lambda}{F_D} \qquad P_n = \frac{P}{F_D}$$

The normalized data is then plotted on log-log scales and the fractal dimension, d, is determined by adding 1 to the absolute value of the slope, m, of the best fit line through the data

$$\delta = 1 + |m|$$

Figure 2.21. A simple example of the application of the structured walk method of determining the fractal dimension of a rugged profile.

straight line relationship. Mandelbrot showed that a useful way of describing an indeterminate structure, such as the coastline of Great Britain, was to add the magnitude of the absolute value of the slope of the straight line relationship generated by the perimeter estimates to the topological dimension of a structure. Therefore the coastline of Great Britain topologically has a dimension of 1. From the experimental data of Figure 2.21(b) the slope of the line is –0.24. (Note that if the physical length of one cycle on the logarithmic scales on the two axis are not the same, the slope of the best fit line to the data measured directly on the graph is not the same as the actual mathematical value of the slope.) Mandelbrot described the combination of numbers such as 1.24 as the fractal dimension of the structure. It should be noted that the range of explorations used in the construction of Figure 2.21 is somewhat limited and that one can only say that over the ranges of inspection used to construct the graph that the system is describable as a fractal structure. The graphical display of the perimeter estimates against the step size of the exploration on log-log graph paper is known for historic reasons as a Richardson plot. The name comes from the name of a scientist who pioneered the exploration of the structure of coastlines [36, 37].

Fineparticle specialists had come up against the same dilemma of indeterminism in their exploration of carbonblack profiles as they were able to employ more and more powerful microscopes in the exploration of the structure of fineparticles. In Figure 2.22 the appearance of a nickel pigment at 4 different levels of magnification are shown. What is interesting about these series of pictures is that without information on the magnification of the profiles they all look very similar as the magnification is increased.

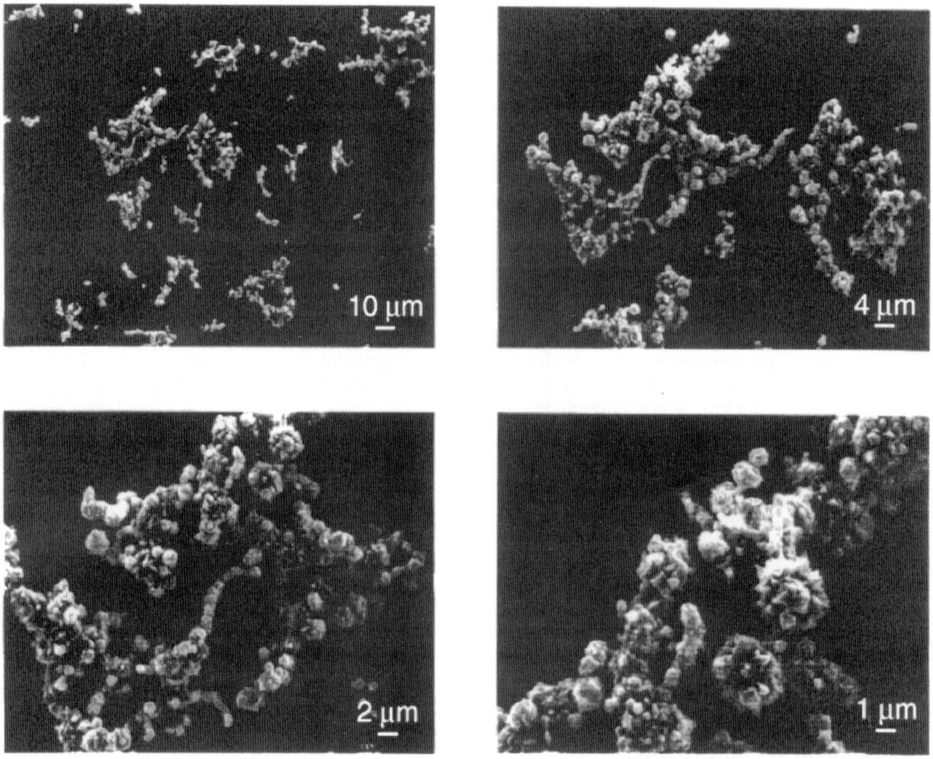

Figure 2.22. The appearance of a nickel pigment viewed through a scanning electron microscope at various magnifications demonstrates the concept of self-similarity. As the resolution is increased, the complexity of the boundary remains the same. (Micrographs provided by and used with the permission of INCO Limited, Sudbury.)

This is an important property of systems which can be described as fractal structures; it is known as self similarity [36, 37].

In his book Mandelbrot discussed curves such as the Koch triadic island which have truly infinite boundaries and which have Richardson plots manifesting a linear relationship at all levels of exploration. In the real world structures manifest fractal dimensions only over a range of inspection values. Reporting the range of inspection values is part of the information content of a fractal description of a profile. The explorations of the outlines of the coastline of Great Britain shown in Figure 2.21 were carried out using a pair of dividers to simulate a walk around the profile to estimate the perimeter of the coastline. Several techniques have been reported by various workers for characterizing the structure of fractal boundaries. It has been shown recently that probably the most efficient technique for characterizing fractal boundaries involves a procedure known as a equipaced structured walk [37, 38]. To illustrate the basic concepts involved in this procedure consider the carbon black profile shown in Figure 2.23.

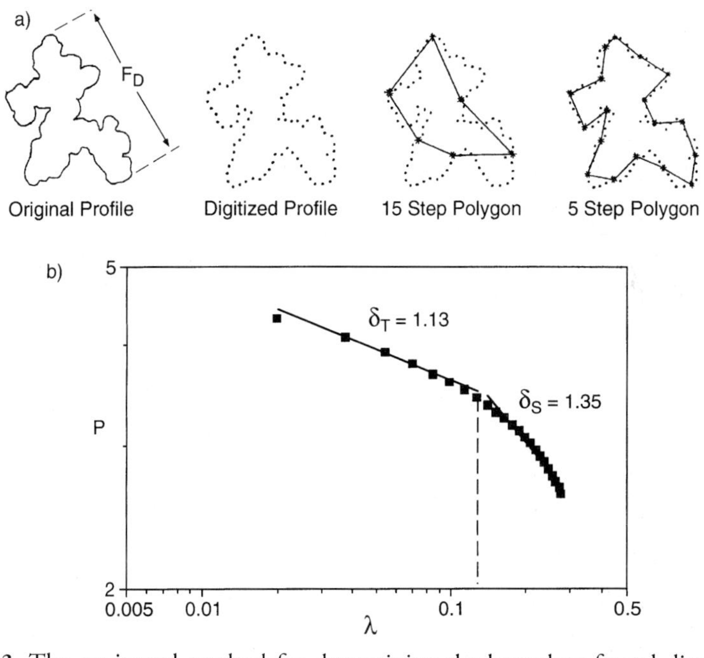

Figure 2.23. The equipaced method for determining the boundary fractal dimension of a profile involves digitizing the profile into small equal "paces", then using chords of a given number of paces to form a polygon which is a perimeter estimate of the profile. The length of each chord is calculated using Pothagoras' theorem and the coordinates of the two end points. a) A carbonblack profile, a simple digitized version of the profile and polygons formed using chords of two different pace numbers. b) Richardson plot of the data generated by equipaced exploration of the carbonblack profile of (a).

The first step in this procedure for evaluating the fractal structure is to digitize the profile. To create a polygon estimate for estimating the perimeter as one steps around the profile a specified number of steps. Polygons for 15 and 5 step chords are shown in Figure 2.23(a). As one decreases the number of steps involved in creating the side of the polygon obviously a better and better estimate of the structure of the profile is had. When the data for such a series of experiments is plotted on a Richardson plot as shown in Figure 2.23(b) it can be seen that the data manifest two different approximately linear regions. It can be shown that the linear region at large step exploration is characteristic of the gross structure of the profile. In fact a fractal dimension such as 1.35 is typical of a system which has grown by agglomeration of subunits. When smaller

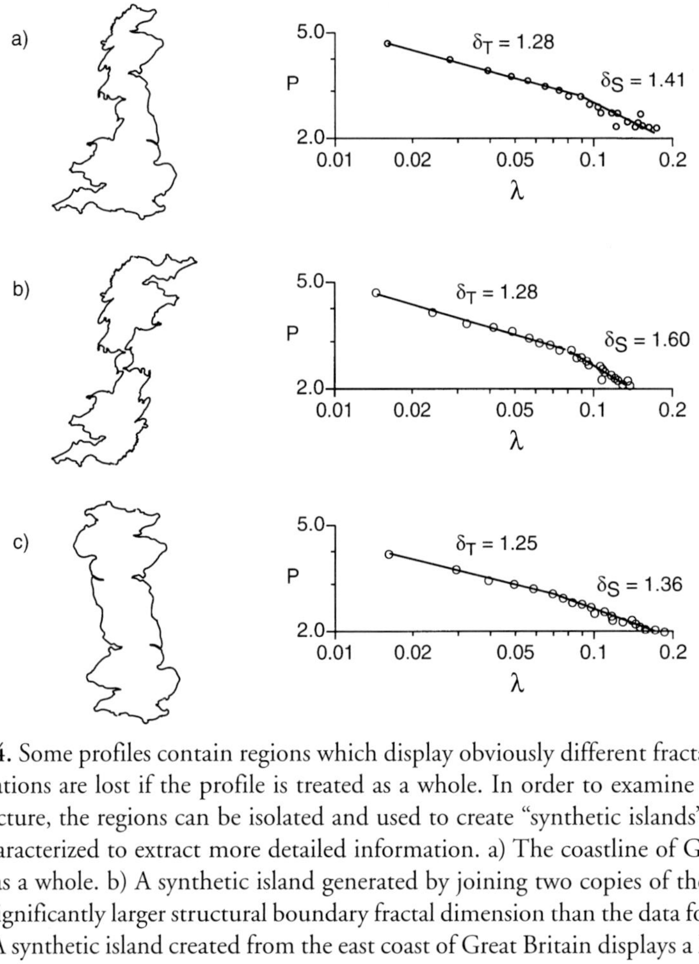

Figure 2.24. Some profiles contain regions which display obviously different fractal structure. These variations are lost if the profile is treated as a whole. In order to examine the varying fractal structure, the regions can be isolated and used to create "synthetic islands" which can then be characterized to extract more detailed information. a) The coastline of Great Britain examined as a whole. b) A synthetic island generated by joining two copies of the west coast displays a significantly larger structural boundary fractal dimension than the data for the whole profile. c) A synthetic island created from the east coast of Great Britain displays a lower structural boundary fractal than the island as a whole.

steps of exploration are used one is exploring the texture of the profile so that the break point between the two linear regions of the graph is actually indicative of the size of the subunits creating the texture of the profile.

When carrying out an exploration intended to arrive at a fractal description of a rugged boundary one must be aware of the fact that different portions of a profile may exhibit different fractal structure. This is illustrated dramatically for the profile of the coastline of Great Britain. If one visually inspects the profile it can be seen that the west coast is much more rugged than the east coast. In Figure 2.24(a) the coastline as a whole has a structural dimension, from a low level inspection, of 1.41 whereas a textural fractal dimension of 1.28 emerges from the high level inspection. If one creates a synthetic island by adding the west coast to its mirror image, a profile that is much more rugged than the overall coastline is created. This is demonstrated by the fact that such a synthetic island has a structural fractal dimension of 1.60 as shown in Figure 2.24(b). On the other hand a synthetic island created by combining the east coast with its mirror image has a much less rugged profile as shown by the data of 2.24(c).

It is rather interesting to note that overall the textural fractal dimensions stays virtually the same. The different structural fractal dimensions have physical significance in that the east and west coasts of Great Britain consist of different rock types and are exposed to different formative processes. The west coast, which has a rugged profile, is made up of more mountainous material and faces into the storms of the Atlantic.

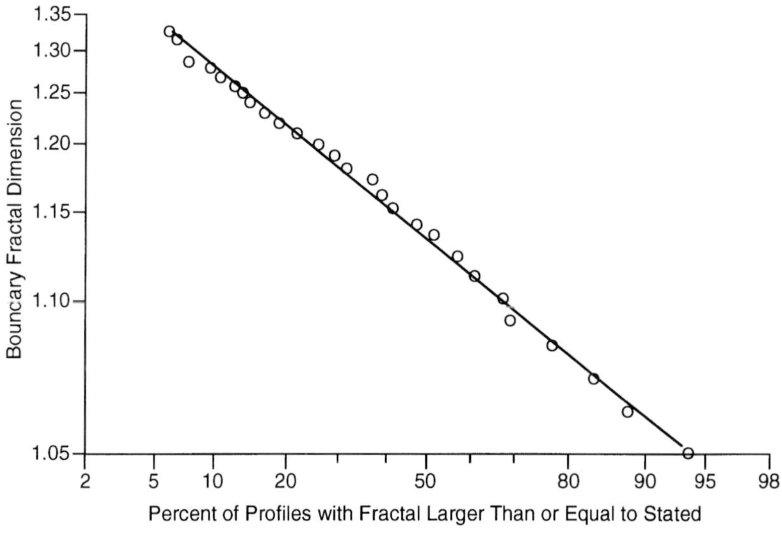

Figure 2.25. Distribution of the boundary fractal dimensions of the carbon black profiles of Figure 2.15(a).

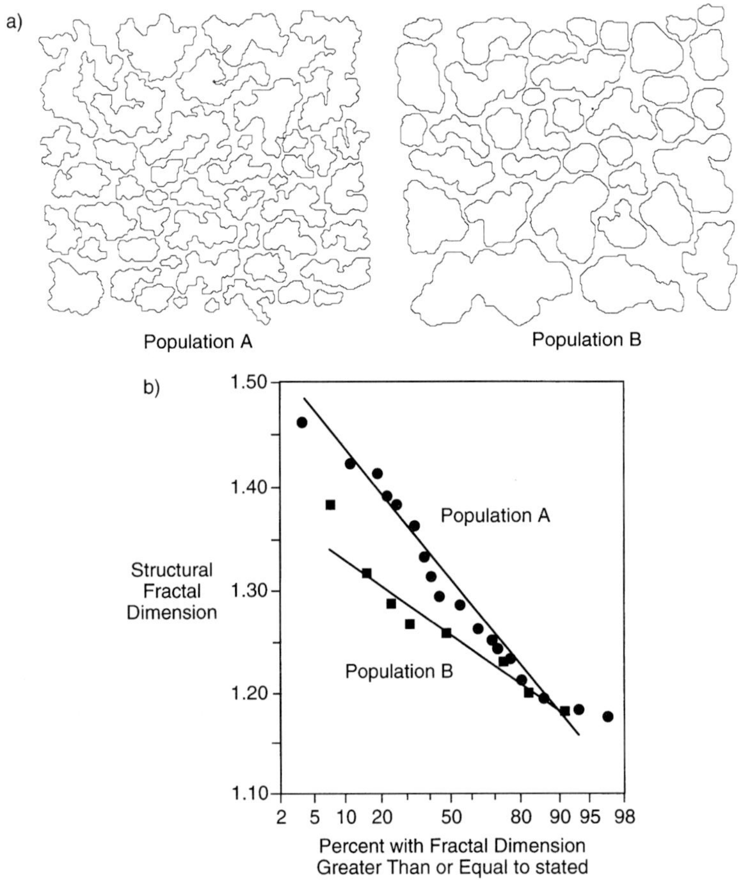

Figure 2.26. Summary of variations in the structural boundary fractal of two populations of carbonblack profiles [38]. a) Tracings of profiles of two carbonblack populations produced by different methods. b) Distribution functions of the structural boundary fractal dimensions for the two populations of (a).

The east coast has a different structure and does not face the same type of formative erosive forces. Similar types of differences in part of the structure of a profile has been found useful in other areas of fineparticle science [37, 38].

When one attempts to study a set of profiles such as those shown in Figure 2.15(a) it is obvious that there is a range of fractal dimensions manifest by the various profiles. In Figure 2.25 the distribution of fractal dimensions for the profiles of Figure 2.15(a) are shown. The fact that the range of fractal dimensions manifest by the carbonblack profiles in the toner section is not surprising since the creation of a high fractal dimension requires co-operative interaction of many random variables; see discussion of log-gaussian distributions in reference 36.

Another example of the utility of fractal description of fineparticle profiles is demonstrated by the data summarized in Figure 2.26. The profiles of this diagram are an assembly collected from high magnification electron micrographs for two different carbonblack products of the Cabot Corporation. In Figure 2.26(b) the distribution of the structural boundary fractals of the two sets of carbon black profiles are shown. It can be seen that for these two sets of data the distributions of fractal dimensions are Gaussianly distributed. From the distributions one can deduce that the carbonblack profiles described as being from Population A were formed by allowing the smoke from a flame to agglomerate for a longer period than for the Population B profiles.

2.7 Stratified Count Logic for Assessing an Array of Fineparticle Profiles

When the neophyte analyst looks at an array of fineparticles, either in a set of pictures or through a microscope the basic instinct is to size everything in sight. If there is a large range of sizes present in the population to be characterized this often results in a distorted set of data for calculating size distribution functions. Thus I have seen reported data in which there were three profiles of one thousand microns in size in a set of data which included fineparticles down to one micron, with 10,000 fineparticles counted in the smallest size group. When it is appreciated that one 1,000 micron profile is equivalent in volume to a billion one micron profiles, the lack of balance in the data count becomes very obvious. In such situations one should adopt a stratified count logic [39]. In this process one sets a microscope for example to a magnification such that the largest fineparticles are located. One then examines areas of the microscope slide until one has found ten of these largest profiles. One then proceeds to change the magnification and look for ten of the next largest profiles and so on down to the smallest profiles. In this way one establishes the population per unit area of each size. Often the size distribution based on as few as two hundred profiles using this search procedure can yield very accurate data. For a detailed discussion of a stratified logic count see page 411 of reference 36.

2.8 Special Imaging Procedures for Studying Fineparticles

One of the special imaging procedures which have been successfully used to study fineparticle systems is far-field holography. The basic system of this technique is illustrated in Figure 2.27. In the example shown the technique was used to size fog droplets. The laser beam was shone through the fog system and light scattered from the droplets interacted with the forward traveling beam to produce the holographic record

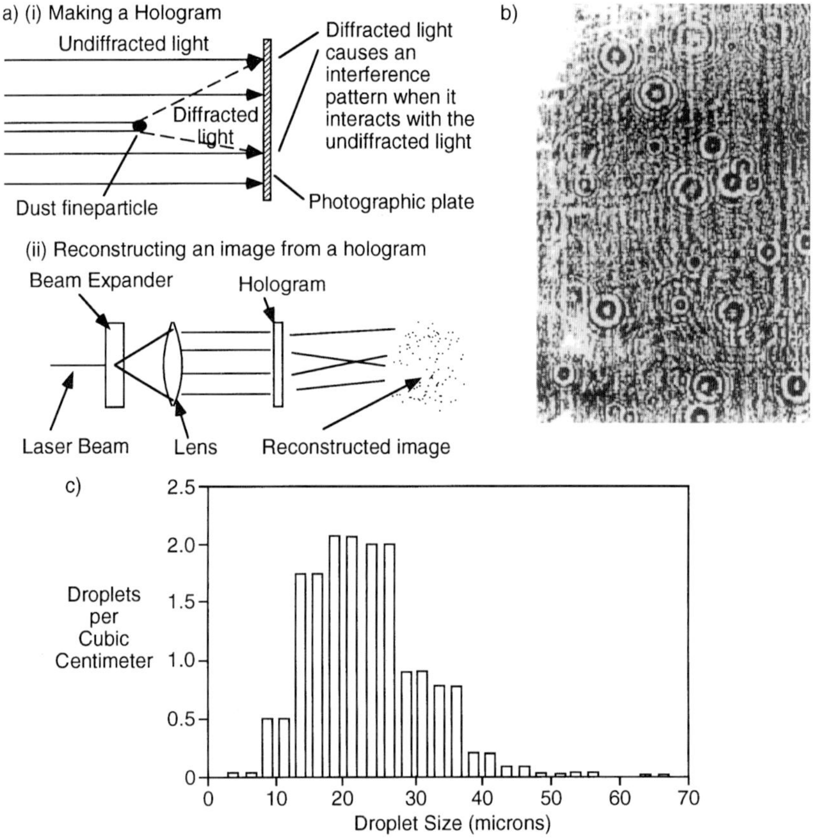

Figure 2.27. Far-field holography can be used to create a three-dimensional image of a cloud of fineparticles which can then be studied to provide a size distribution from the three dimensional image. a) Method of capturing and reconstructing a far-field holograph. b) The appearance of a holographic plate used to reconstruct the image. c) Size distribution of fog droplets determined from analysis of the holograph. (The Optical Society of Amaerica is the copyright holder of the material cited herein. [40])

on the photographic plate. The photographic plate was then used to reconstruct the three dimensional image of the cloud in the laboratory and the droplets in the holographic record were sized using standard procedures. The size distribution of fog droplets measured by B. J. Thompson and colleagues using this technique is shown in Figure 2.27(c). Holographic imaging and subsequent reconstruction is particularly useful when looking at fineparticle systems in a hostile environment such as the unburnt fuel in a space rocket exhaust [40, 41].

Tomography is a new imaging process which is being increasingly applied to the study of fineparticle systems [42]. In tomography multiple images using penetrating radiation taken from various angles around a body are used to synthesize on a computer screen a simulated section through the object. This simulated section can then be processed using the general image analysis procedures described earlier.

References

[1] B. H. Kaye and G. G. Clark, "Monte Carlo Studies of the Effect of Spatial Coincidence Errors on the Accuracy of the Size Characterization of Respirable Dust." *Part. Part. Syst. Charact.*, 9 (1992) 83–93.

[2] B. H. Kaye, "Operational Protocols for Efficient Characterization of Arrays of Deposited Fineparticles by Robotic Image Analysis Systems." Chapter 23 in *Particle Size Distribution II: Assessment and Characterization.* T. Provder (ed.), ACS Symposium Series 472. Published by the American Chemical Society, Boston (1991).

[3] British Standards 3406 1961, "Methods for the Determination of Particle Size of Powders," information available from the British Standards Institute. 2 Park Street, London W1.

[4] See discussion in B. H. Kaye *Direct Characterization of Fineparticles,* John Wiley, New York, 1981.

[5] See discussion of automated image analysis in T. Allen, *Particle Size Analysis,* Chapman & Hall, 5th Edition, 1996.

[6] Ultimage Pro 2.5 by GrafTek France, is available from GTFS Inc., 2455 Bennett Valley Road, #100C, Santa Rosa CA 95404.

[7] T. Mercer, P. Morrow, W. Stöber, *Assessment of Airborne Particles,* Charles C. Thomas, Il, 1972.

[8] See for example B. H. Kaye, J. Junkala & G. G. Clark, "Domain Plotting as a Technique for Summarizing Fineparticle Shape, Texture and Size Information," *Part. & Part. System Charact.* in press.

[9] For a full discussion of the use of three dimensional shape data see R. Davies, "A Simple Feature Based Representation of Particle Shape." *Powder Technol.* 12 (1975) 111–124.

[10] B. H. Kaye, G. G. Clark, Y. Liu, "Characterizing the Structure of Abrasive Fineparticles." *Part. Part. Syst. Charact.,* 9 (1992) 1–8.

[11] H. H. Heywood, "Size and Shape Distribution of Lunar Fines Sample 12057, 72," in *Proceedings of Second Lunar Science Conference,* Vol. 13, (1971) 1989–2001.
[12] H. Heywood, "Numerical Definitions of Particle Size and Shape." *Chem. Ind.,* 15 (1937) 149–154.
[13] H. Heywood, "Particle Shape Coefficients," *J. Imp. Coll. Eng. Soc.,* 8 (1954) 25–33.
[14] H. H. Hausner, "Characterization of the Powder Particle Shape," in *Proceeding of the Symposium on Particle Size Analysis,* Loughborough, England; published by the Society for Analytical Chemistry, London, England, (1967) 20–77.
[15] B. H. Kaye, D. Alliet, L. Switzer, C. Turbitt, "The Effect of Shape on Intermethod Correlation of Techniques for Characterizing the Size Distribution of a Powder. Part 1: Correlating the Size Distribution measured by Sieving, Image Analysis, and Diffractometer Methods" *Part. Part. Syst. Charact.,* 14 (1997) in press.
[16] For those unfamiliar with the specialty graph paper such as the Gaussian Probability paper used in this chapter, their use is described in detail in B. H. Kaye, *Chaos and Complexity; Discovering the Surprising Patterns of Science and Technology,* VCH, Weiheim, Germany (1993).
[17] B. H. Kaye, "Fractal Dimensions in Data Space; New Descriptors for Fineparticle System," *Part. Part. Syst. Charact.,* 10 (1993) 191–200.
[18] G. J. Brown, N. J. Miles, "The Fragmentation Fractal – A Useful Descriptor of Broken Rock", *Part. Part. Syst. Charact.,* 12(3) (1995) 166–169.
[19] B. H. Kaye, G. G. Clark, Y. Liu, "Characterizing the Structure of Abrasive Fineparticles," *Part. Part. Syst. Charact.,* 9 (1992) 1–8.
[20] L. G. Austin, M. Yerkeler, T. F. Dumm, R. Hogg, "The Kinetics and Shape Factors of Ultrafine Dry Grinding in a Laboratory Tumbling Ball Mill", *Part. Part. Syst. Charact.,* 7(4) (1990) 242–247.
[21] H. P. Schwartz and K. C. Shane, "Measurement of Particle Shape by Fourier Analysis." *Sedimentology* 13 (1969) 213–231.
[22] R. Ehrlich and B. Weinberg, "An Exact Method for Characterization of Grain Shape." *J. Sediment. Petrol.* 40, 1 (March 1970) 205–212.
[23] A. G. Flook, "A Comparison of Quantitative Methods of Shape Characterization." *Acta Stereol.* 3 (1984) 159–164.
[24] See Chapter 15 of B. H. Kaye, *Chaos and Complexity; Discovering the Surprising Patterns of Science and Technology,* VCH, Weiheim, Germany (1993).
[25] I. Stewart, Concepts of Modern Mathematics, Penguin Books, Harmondsworth, Middlesex, England; also, 41 Steelcase Road, West Markham, Ontario, Canada 91975).
[26] See discussion of Random and Structured mixtures in B. H. Kaye *Powder Mixing,* Chapman & Hall, London, England, 1997.
[27] Many of the appropriate statistical relationships involved in the assessment of rock structure by techniques such as the Rosiwal intercept method are to be found in F. Chayes, *Petrographic Model Analysis,* Wiley, New York, 1956.
[28] See the discussion of dot counting procedures for evaluating the area of a profile in Chapter 8 of B. H. Kaye, *Chaos and Complexity; Discovering the Surprising Patterns of Science and Technology,* VCH, Weiheim, Germany (1993).

[29] M. J. Kendall and P. A. Moran, *Geometric Probability,* Griffiths Statistical Monographs. Charles Griffith, London, 1963.

[30] The science of studying two-dimensional sections using statistical search methods, with consequent description of three dimensional properties of the system, is known as quantitative microscopy or quantitative stereology. Comprehensive texts on these two subjects are available in R. T. DeHoff and F. N. Rhines, *Quantitative Microscopy,* McGraw-Hill, New York, 1968, and E. E. Underwood, *Quantitative Stereology,* Addison Wesley, Reading,, MA., 1970.

[31] B. B. Mandelbrot, *Fractals; Form, Chance, and Dimension,* Freeman, San Francisco (1977).

[32] B. H. Kaye, *A Randomwalk Through Fractal Dimensions,* VCH, Weinheim, Germany (1989).

[33] For a readily accessible discussion of the Rosiwal intercept method, see G. Herdan, *Small Particle Statistics,* 2nd ed., Butterworths, London, 1960.

[34] B. H. Kaye, G. G. Clark, "Computer Aided Image Analysis Procedures for Characterizing a Stochastic Structure of Chaotically Assembled Pigmented Coatings," *Part. & Part. Syst. Charact.* 9 (1992), 157–170.

[35] Techniques for using the Rosiwal intercept technique are discussed in B. H. Kaye, *A Randomwalk Through Fractal Dimensions,* 2nd ed., VCH, Weinheim, 1995.

[36] B. H. Kaye, *Chaos and Complexity, Discovering the Surprising Patterns of Size and Technology,* VCH Publishers, Weinheim, 1993.

[37] B. H. Kaye, G. G. Clark, Y. Kydar, "Strategies for Evaluating Boundary Fractal Dimensions by Computer Aided Image Analysis." *Part. Part. Syst. Charact.* 11 (1994) 411–417.

[38] B. H. Kaye and G. G. Clark, "Formation Dynamics Information; Can It be Derived from the Fractal Structure of Fumed Fineparticles?" Chapter 24 in *Particle Size Distribution II; Assessment and Characterization,* T. Provder (ed.), American Chemical Society, Washington (1991).

[39] B. H. Kaye, "Operational Protocols for Efficient Characterization of Arrays of Deposited Fineparticles by Robotic Image Analysis Systems." Chapter 23 in *Particle Size Distribution II: Assessment and Characterization,* T. Provder (ed.), ACS Symposium Series 472. Published by the American Chemical Society, Boston (1991).

[40] B. J. Thompson, J. H. Ward, W. R. Zinky, "Applications of Hologram Techniques for particle Size Analysis", *Applied Optics*, 6 (1967) 519–526.

[41] Techniques for studying fog droplets etc. with holography are reviewed in B. H. Kaye, *Soot Dust and Mist,* (in preparation).

[42] See special issue of Particle and Particle Systems Characterization devoted to Tomography in powder science and Technology Vol 12, no 2 April 1995.

3 Characterizing Powders Using Sieves

3.1 Sieving Surfaces

A sieve is defined as a container the base of which has a series of holes. Ideally the holes are all of the same size and act as go/no-go gauges for the powder grains placed on the sieve. By shaking the sieve, the powder able to pass through the sieve apertures falls through the holes so that, at the end of the shaking of the sieve, in theory all of the fineparticles remaining on the sieve are larger than the holes in the sieving surface. The two fractions of powder are referred to as the oversize and undersize fraction of the powder. To characterize a powder one uses a whole series of sieves, the apertures of which are smaller as one moves down through the nest. For historic reasons the apertures in commercially available sieves are often arranged in the root of 2 progression [1, 2, 3]. In practice difficulties arise because it is not possible to create all of the apertures of the sieving surface exactly the same. Furthermore, during the use of the sieve, the apertures can become distorted or change through wear. The second practical problem is that except for spheres, the probability of passage through the holes is not easy to predict and indeed a method of shape analysis uses a series of sieves of the same apertures to sort fineparticles for shape [4]. This technique known as cascadography will be discussed in more detail later in this chapter [5]. (Note that in the discussion of sieving machines we will restrict the discussion to sieves used to characterize powders. In the processing industry, large machines are used to produce powders of a given size fraction, such processes are known as "screening". A good reference article on industrial screening has been written by Beddows [6].)

To allow for these practical difficulties it is usually necessary to rigidly specify and adhere to sieve fractionation study protocols. One also may fix limitations on the permitted variations in sieve apertures and sieving times according to empirical limits. In the manufacturing of sieve surfaces using wire woven cloth most major industrial nations have established specifications for sieves used in test studies [3, 7–10]. See Table 3.1.

In this section we will concern ourselves with the manufacture and characterization of sieving surfaces. The first type of sieving surface to be widely used in powder technology was woven wire. In the manufacture of sieves for analytical studies, it is usual to have the wire used to weave the cloth forming the surface of the sieve the same diameter as the aperture created by the weaving process. The sieve apertures created by such a process have the appearance shown in Figure 3.1(a). The aperture defined

Table 3.1. [a] For sieves from the 1000 mm (No. 18) to the 37 mm (No. 400) size, inclusive, not more than 5 % of the openings shall exceed the nominal opening by more than one-half of the permissible variations in maximum opening.

Permitted Variations in Aperture Dimensions for Wire-Woven Sieves Conforming to U.S. Sieve Series (Fine Series)

		Permissible Variation, %		
Width of Opening um	Average Screen No.	Maximum Opening (±)	Wire Opening (+)[a]	Diameter mm
5660	3.5	3	10	1.28–1.90
4760	4	3	10	1.14–1.68
4000	5	3	10	1.00–1.47
3360	6	3	10	0.87–1.32
2830	7	3	10	0.80–1.20
2380	8	3	10	0.74–1.10
2000	10	3	10	0.68–1.00
1680	12	3	10	0.62–0.90
1410	14	3	10	0.56–0.80
1190	16	3	10	0.50–0.70
1000	18	5	15	0.43–0.62
840	20	5	15	0.38–0.55
710	25	5	15	0.33–0.48
590	30	5	15	0.29–0.42
500	35	5	15	0.26–0.37
420	40	5	25	0.23–0.33
350	45	5	25	0.20–0.29
297	50	5	25	0.170–0.253
250	60	5	25	0.149–0.220
210	70	5	25	0.130–0.187
177	80	6	40	0.114–0.154
149	100	6	40	0.096–0.125
125	120	6	40	0.079–0.103
105	140	6	40	0.063–0.087
88	170	6	40	0.054–0.073
74	200	7	60	0.045–0.061
62	230	7	90	0.039–0.052
53	270	7	90	0.035–0.046
44	325	7	90	0.031–0.040
37	400	7	90	0.023–0.035

by the wires in the weave varies according to slight changes in the positions of the wire in the weaving process as illustrated by the array of apertures shown in Figure 3.1(b). Because of the geometric structure the aperture in a wire woven sieve is actually a trapezium, as shown in Figure 3.1. One way of measuring the size distribution of apertures present in the sieving surface is to examine the surface directly through a microscope. The aperature size can be measured by either taking the square root of the area or considering for a spherical particle the mid point of the trapezium, shown as **m** in the picture, is the effective aperture. To measure the size distribution of apertures in a

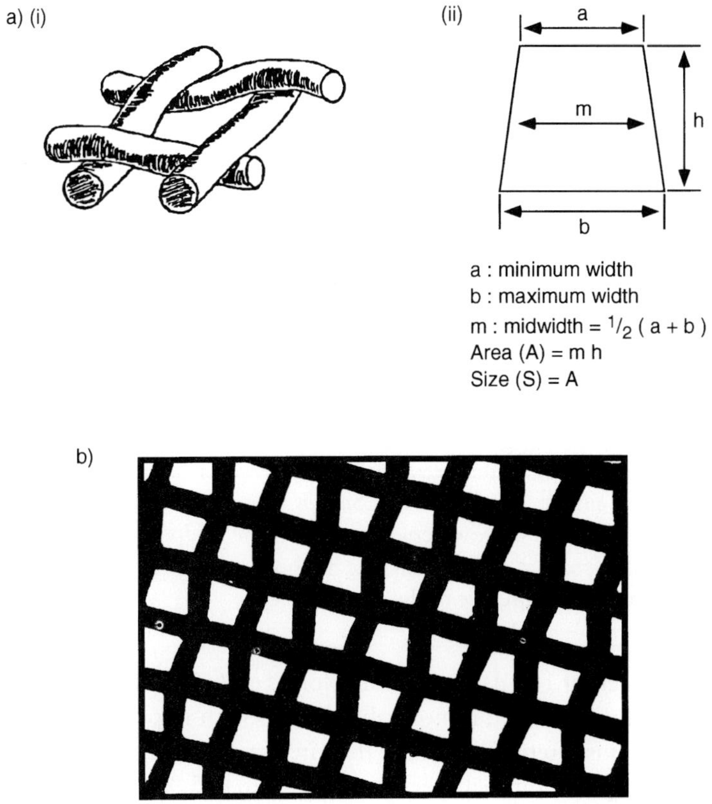

Figure 3.1. Various measurements used to define the operative size of an aperture in a woven wire sieve cloth. a) (i) The appearance of an aperture formed by weaving wires. (ii) Measurements possible on an aperture in a wire woven sieve. b) Appearance of a wire woven sieve as viewed with a microscope.

sieving surface of the type shown in Figure 3.1(b) first choose an area of the mesh at random and then measure ten adjacent apertures in a straight line. Then move the viewing area until another randowm field is chosen and measure a gurther ten apertures. Carrying this out for one hundred apertures would give enough data to construct a size distribution of the apertures. In Figure 3.2 the results of carrying out this kind of measurement for two sets of 100 apertures for a 65 mesh ASTM standard sieve are shown. In a description of a sieving surface the mesh number designates the number of wires per linear inch. In this case the 65 mesh sieve corresponds to a two hundred micron opening. Because the aperture range is the result of the interaction of many small causes, it is reasonable to anticipate that the size distribution function of the apertures is a Gaussian probability function [11]. In Figure 3.2 the experimental meas-

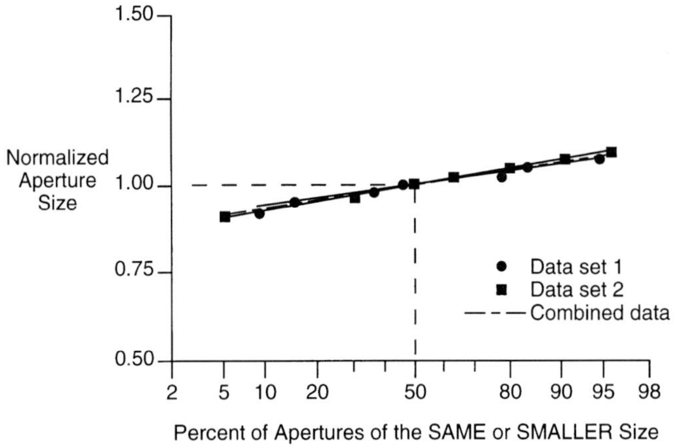

Figure 3.2. The range of apertures permitted in a wire woven sieving surface are strictly limited by ASTM and similar standards [10]. Shown above are the aperture size distributions derived from two sets of 100 measurements made at the midpoint of the apertures of a 65 mesh ASTM sieve. Note that the aperture size has been normalized by dividing the measurements by the nominal aperture size of the 65 mesh sieve.

urements of two sets of 100 data points are plotted on Gaussian probability paper with the apertures being expressed in normalized terms with respect to the nominal aperture of the sieve. It is interesting to note that, in fact, 50 % of the apertures were equal to, or smaller than, the nominal aperture and that up to 5 % of the apertures exceeded the nominal size by more than 5 %. These are the limits of variables set in the ASTM standards [1, 10].

Detailed investigation of new sieve surfaces can generate some surprising data. In Figure 3.3(a) mesh aperture distributions measured on two sets of sieves, which were nominally within the British standard sieve specification, but bought at different times are shown. In fact only one sieve mesh of the B series has a mean aperture size which failed to meet the permitted variations in a sieve mesh which are shown by the data of Table 3.2. To test the operational effect of the mesh aperture distribution variations of the two sets of 4 sieves, a sample of a bronze powder, having spherical particles, was fractionated on each of the sieves in turn. The residue on the sieve was calculated from the rate method to be described later in the chapter. The oversize and undersize fraction from one experiment were recombined and used on the next sieve. By using the same sample and amount of powder on each sieve, and by using spherical grains, effects due to shape and bed loading of the sieve were minimized. As can be seen from a summary of the results presented in Figure 3.3(b) the measured residues varied from 76.4 to 72.5. In practice in industry a standard powder is used to check on the performance of a new sieve to see how its values need to be adjusted with respect to the

Table 3.2. [a] Manufactured by W. S. Tyler Company, Cleveland, Ohio.
[b] National Bureau of Standards L.C.-584 and ASTM-E. II.
[c] BS-410 1943.
[d] DIN 1171 (German Standard Specification).

Comparison of Various Types of Standard Sieve							
Tyler[a]	U.S.[b]		British[c] Standard		German DIN[d]		
Equivalent Mesh	Mesh No.	Openings mm	Mesh No.	Openings mm	DIN No.	Mesh/cm^2	Openings mm
3½	3½	5.66	...	...	1	1	6.000
4	4	4.76	...	...	...	...	...
5	5	4.00	...	...	...	...	...
6	6	3.36	5	3.353	2	4	3.000
7	7	2.83	6	2.812	...	...	...
...	...	...	...	...	...	...	...
8	8	2.38	7	2.411	2½	6.25	2.400
9	10	2.00	8	2.057	3	9	2.000
10	12	1.68	10	1.676	4	16	1.500
...	...	...	...	...	...	...	...
12	14	1.41	12	1.405	...	...	...
...	...	...	...	...	...	...	...
14	16	1.19	14	1.204	5	25	1.200
...	...	...	...	...	...	...	...
16	18	1.00	16	1.003	6	36	1.020
20	20	0.84	18	0.853	...	...	...
...	...	...	...	...	8	64	0.750
24	25	0.71	22	0.699	...	...	...
...	...	...	...	...	...	...	...
28	30	0.59	25	0.599	10	100	0.600
...	...	...	...	...	11	121	0.540
32	35	0.50	30	0.500	12	144	0.490
35	40	0.42	36	0.422	14	196	0.430
42	45	0.35	44	0.353	16	256	0.385
...	...	...	...	...	...	...	...
48	50	0.297	52	0.295	20	400	0.300
60	60	0.250	60	0.251	24	576	0.250
65	70	0.210	72	0.211	30	900	0.200
80	80	0.177	85	0.178	...	...	...
100	100	0.149	100	0.152	40	1600	0.150
...	...	...	...	...	...	...	...
115	120	0.125	120	0.124	50	2500	0.120
150	140	0.105	150	0.104	60	3600	0.102
170	170	0.088	170	0.089	70	4900	0.088
...	...	...	...	...	...	...	...
200	200	0.074	200	0.076	80	6400	0.075
250	230	0.062	240	0.066	100	10000	0.060
270	270	0.053	300	0.053	...	...	...
325	325	0.044	...	...	...	...	...
400	400	0.037	...	...	...	...	...

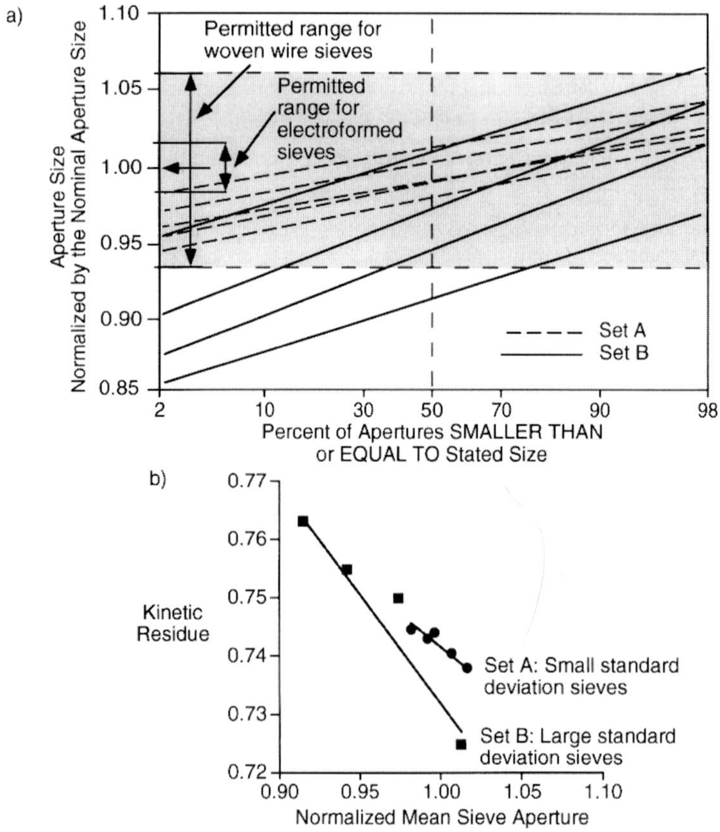

Figure 3.3. Sieves purchased as being within a standard specification can differ significantly in their aperture distribution and generate significantly different measured sieve residues [1, 2]. a) Aperture size distributions from two sets of sieves of the same nominal aperture purchased at different times. Note that only one sieve in Set B has a mean aperture size that falls outside the standard specification. b) Sieve residues from the sieves of (a) cover a wide range. Note that the largest residue is 5.4 % larger than the smallest residue.

sieve it is replacing. The data of Figure 3.3(b) demonstrates why this type of calibration is necessary [12]. A more detailed discussion of the data of Figure 3.3 is to be found in reference [1]. Some sieve manufacturers offer a selection service to select special sieves within their stock to achive a closer match for the sieve being replaced than a stock standard. [13].

It is not always possible or convenient to examine the mesh of a sieve through a microscope to make direct measurements on the aperture variations. An alternative procedure is to use the powder of interest to calibrate the apertures. To calibrate the sieve apertures in terms of spherical particles that will pass through the sieve place a small amount of glass spheres on the sieve mesh. By agitating the sieve for a given period of time then pouring the powder sample off of the sieve surface, there will always be spheres left trapped in the apertures in the sieve. By taking a shallow tray and inverting the sieve over it, after rapping it vigorously up and down the particles trapped in the mesh fall out onto the tray. These spheres which had been trapped in the mesh can then be characterized with image analysis equipment. In Figure 3.4(a) the size distribution of sieve apertures, based on an examination of glass spheres which had been trapped in a 65 mesh sieve, is shown. The size distribution measured this way is virtually indistinguishable from the measurements made by direct examination of the apertures.

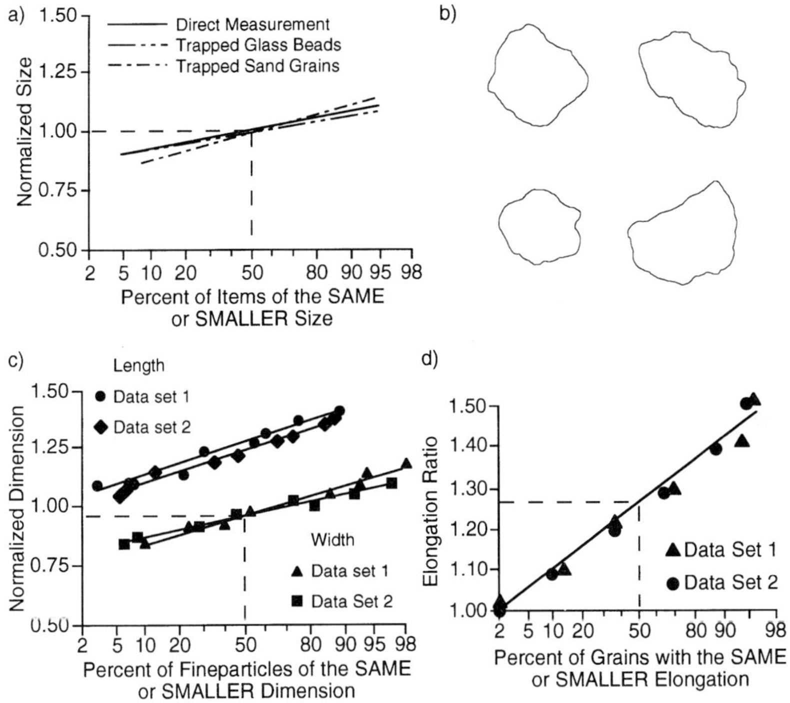

Figure 3.4. The operative aperture size in a sieve can be measured by examining powder grains which have been trapped in the apertures. a) Comparison of the aperture size distribution of a sieve as determined from direct examination of the sieve surface, trapped spherical glass beads, and trapped irregularly shaped sand grains. b) Profiles of typical sand grains trapped in the sieve mesh. c) Length and width distributions of two sets of 100 sand grains trapped in the mesh of the sieve. d) Shape distribution of the sand grains of (c).

When using irregularly shaped fineparticles trapped mesh to calibrate a sieve, a measurment the widths of the fineparticles should be made since this is what determines whether or not the grains of powder will pass the sieve mesh apertures. Consider the sand grains shown in Figure 3.4(b). These are typical grains that were trapped in the mesh of the 65 mesh sieve used in the previous experiments. When the mesh trapped sand grains are examined through a microscope one can measure the length and width distributions. Remember however that the real grains are three dimensional objects and we are only looking at a reduced information set when we look at the sand grains in stable position on a glass slide [14, 15]. In Figure 3.4(c) the length and width distributions of two sets of one hundred sand grains are presented. It can be seen that the width distribution has a mean size approximately equal to that of the nominal aperture of a sieve, this showing that the width distribution of an irregularly shaped powder is the operative distribution with respect to the sieve apertures of the wire woven sieve. It is interesting to note that both the length and width distributions are Gaussian.

For the grains of sand the information on the length and width, shown in Figure 3.4(c), has been measured as separate sets of data. If one measures the length and width of each sand grain and uses them to generate elongation ratios, a shape distribution like that summarized in Figure 3.4(d) is generated. It is interesting to note that again the elongation ratio of the sand fineparticles is Gaussianly distributed. Because of the shape and structure of the sand grains, calibrating the apertures of the sieve surface this way results in slightly different data than that generated by direct inspection or when using glass beads as illustrated by the summary graph in Figure 3.4(a). For a discussion of how to allow for shape and size when calibrating a sieve see reference [16].

The practical effect of the aperture distribution on the way a powder is fractionated by a sieve, is demonstrated by the data of Figure 3.5. The size distribution of a spherical bronze powder sample before sieving, the over and undersize fractions, and the powder trapped in the mesh of the sieve were all sized using an image analysis system. It can be seen that a small number of larger than mesh fineparticles have made their way into the undersized fraction because of the presence of some apertures larger than nominal in the mesh aperture distribution. There are also a small amount of fines (powder able to pass the sieve) which are in the oversize fraction [17]. This cross contamination of the oversize and undersize fractions tends to be a compensating error so that the analytical characterization of the powder by sieve fractionation is more accurate than one would have thought considering the presence of greater then nominal aperatures [18, 19].

More accurate sieving surfaces than the wire woven material have been developed to improve precision of powder fractionation by sieves. One of these more precise surfaces is known as an electroformed sieving surface. These are created by drawing a grid pattern on a surface and then electroplating the pattern using appropriate technologies. This type of sieving surface can have apertures precise to within one micron of the nominal aperture of the sieve but is more expensive than the wire woven sieves. It is very fragile and requires a supporting coarser grid to avoid damage to the aperatures [20].

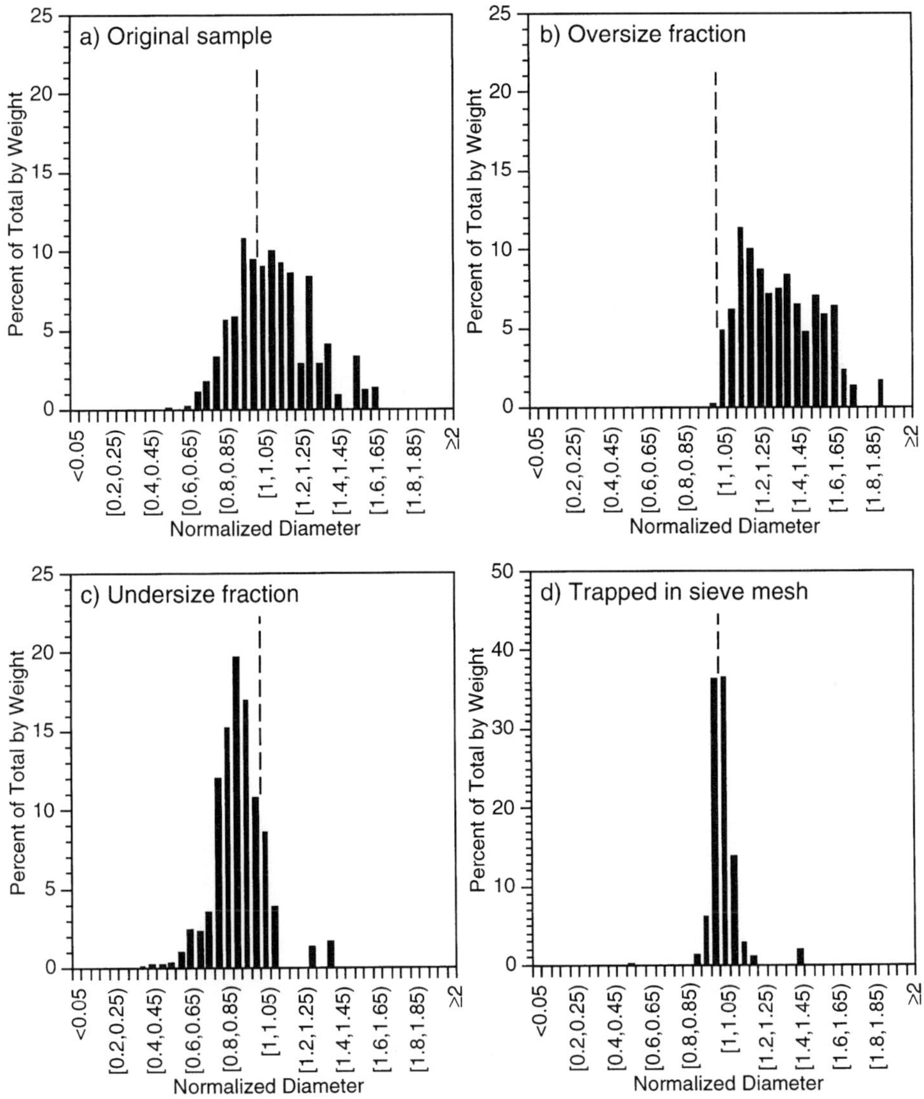

Figure 3.5. Compensating errors (fines in the oversize and vice versa) makes the sieve residue used in a particle size distribution, a better characteristic quantity that would be anticipated from the structure of the sieve surface [17].

Another type of very accurate sieving surface is made by etching holes in a thin steel surface. These etched hole sieving surfaces are available from the Veco Company [21].

In spite of the fact that sieves are very widely used in industry there is limited information on the rate at which apertures in sieves in service deteriorate. Many companies lack any objective rules for time lines by which to replace sieving surfaces. One technique to look for damaged holes in a micromesh sieve surface has been described by Naylor, Kaye and Legault [22–24]. The basic principle of the technique is shown in Figure 3.6. It is based on optical spatial filtering. Light from a laser is used to create a diffraction pattern of the mesh to be examined at the focal point of a lens. In Figure 3.6(a) the diffraction pattern from a virgin piece of mesh is shown. When a negative

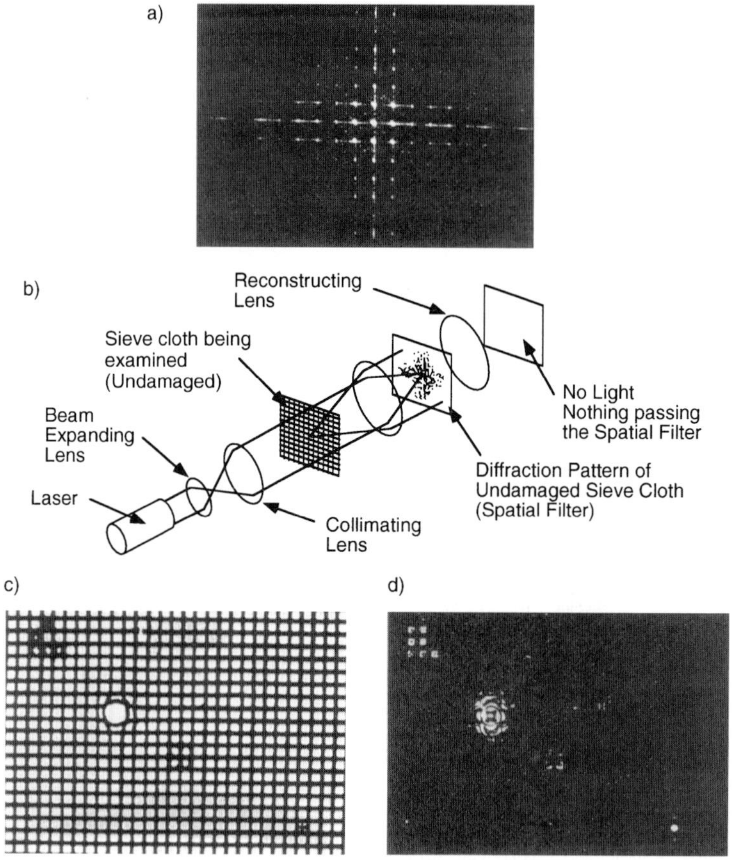

Figure 3.6. Optical methods can be used to discover damaged apertures in a sieving surface [22, 23, 24]. a) Diffraction pattern of an undamaged sieve mesh. b) Optical spatial filtering system used to detect sieve damage. c) Photograph of a damaged electroformed sieve surface. d) Spatially filtered image of the damaged sieve of (c).

of the diffraction pattern of the undamaged sieve is placed in the plane where the diffraction pattern of a sieve to be examined has been created, as shown in Figure 3.6(b), only light from defects in the sieve being examined can pass the negative of the good sieve diffraction pattern. The negative of the good sieve mesh is described in optical technology as an optical filter. When a second lens is used to reconstruct what is left of the filtered light in the image plane, only the defects are made visible in the reconstructed image. Thus for the damaged mesh of Figure 3.6(c), when we look at the filtered, reconstructed image in Figure 3.6(d) we see that the large hole deliberately created in the piece of mesh is clearly imaged and that other defects obviously present in the mesh are clearly identified in the reconstructed image. In the present state of technology, where ordinary wire woven sieves are used, it is probably cheaper to replace the sieve than it is to build an expensive optical filtering device to check on the apertures. The system of Figure 3.6(b) could be used in more critical situations such as to check for holes in a piece of mesh being used to produce diamond abrasive powders where the presence of one large hole would let large pieces of abrasive through the sieve and create havoc in a polishing process. Those interested in a detailed discussion of the technique represented in Figure 3.6 should consult reference [22].

3.2 The Rate of Powder Passage Through a Sieve

The actual rate at which a powder will pass through a sieve depends upon the volume of powder on the sieve and the way in which the sieve is shaken. Intuitively one can anticipate that side to side movement of the sieve would cause the powder grains which are able to pass the sieve to work their way down through the powder bed on the sieve. However, side to side shaking of the sieve has the disadvantage that it causes blinding i. e. blockage of the apertures by powder on the surface of a sieve. To avoid blinding one must adopt a vigorous agitation of the sieve which includes vertical agitation as well as a horizontal agitation. A machine which has been widely adopted which incorporates these movements is the Ro-tap sieve manufactured by the W. S. Tyler Company. The Tyler company has prepared a useful booklet on the use of their sieving machines and sieves to characterize a powder [13].

By taking a sieve with a depth of powder on it and studing the passage of powder through that sieve, by weighing the fractions passing at a series of times, one can establish that passage of the powder takes place in two phases. First, at the beginning of the sieving process, larger amounts of the smaller fineparticles quickly pass through the sieve. Then a point beyond which the rate of the powder passing through the sieve is much slower is reached. The point at which the rate of sieving changes dramatically is known as the Whitby point because this phenomena was first studied by K. T. Whitby [25]. It has been shown that after the Whitby point is reached the size distribution of the powder fractions passing through the sieve remains virtually unchanged over this second phase of sieving. In Figure 3.7 size distributions of a series of fractions of glass beads passing through an 18 mesh sieve are shown. After the Whitby point was reached i. e. from fraction 5 onwards, the size distribution of the fractions did not change significantly [1, 12].

It has been shown that once the Whitby point has reached the final stages, the residue on the sieve decays exponentially. Therefore, if one knew the ultimate residue, $\mathbf{R}_R$, the difference between the residue at time **t**, **R**, minus the predicted ultimate residue, plotted against sieving time, would result in a straight line on semi-log (log-arithmetic) graph paper. If one has a series of $\mathbf{R}_t$ values, an estimate of $\mathbf{R}_R$ can be made and the data series can be plotted on log-arithmetic graph paper. Then adjustments to the estimate of $\mathbf{R}_R$ can be made until the resulting data generates a straight line relationship. In Figure 3.8 a data set from the final phase of a sieving process of a spherical glass powder sieved on a 120 mesh sieve of one and a half inches diameter is shown [26]. To obtain the estimate of the ultimate residue plot the decay of powder on the

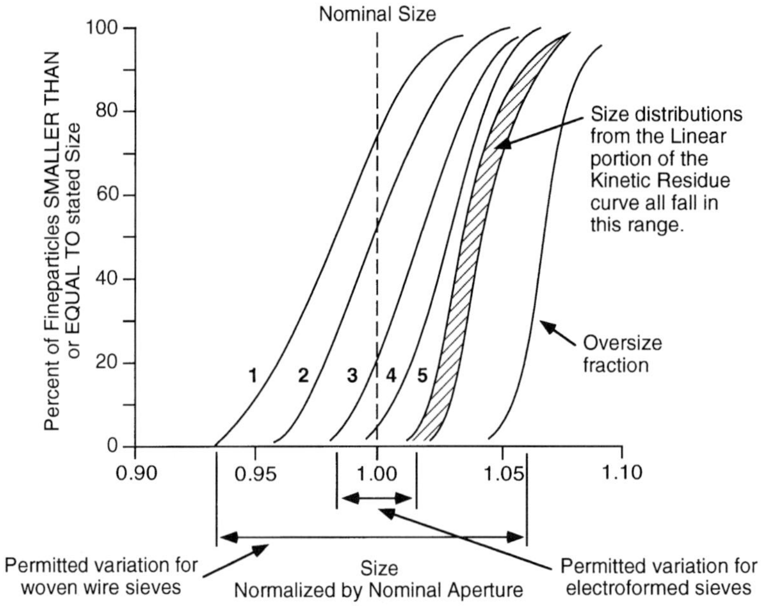

Figure 3.7. In the final stages of a sieving process, the size distribution of the fractions passing the sieve are virtually all the same size. As shown above, all samples after number 5 fall into a very narrow range [1, 12].

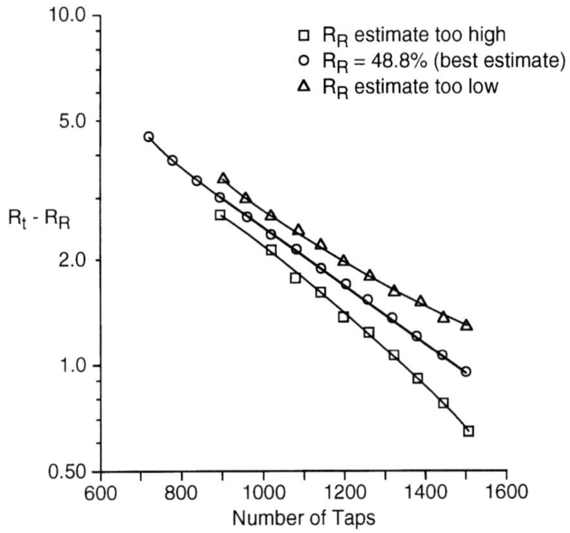

Figure 3.8. Example of a graphical solution for rate data used to find the kinetic residue, R_R, for spherical bronze powder sieved on a one and one half inch, 120 mesh sieve. Various estimates are made of R_R which is then subtracted from the data for the actual sieve residue R_t after a given time (or number of taps), t. The data is then plotted on semi-log graph paper and the best estimate of R_R results in a straight line on the graph.

sieve on log-arithmetic paper using a series of estimates of the ultimate residue. In Figure 3.8 the linear decay relationship is obtained with an estimate of 48.8 % whereas the higher and lower estimates give curved data lines. Therefore the value 48.8 % represents the best estimate of the ultimate residue. Since this type of residue is not obtained by sieving ad nauseam but from the rate of sieving the measurement made is called the kinetic residue. The method avoids variations due to operators and variations in sieving techniques. The method has not however been widely adopted since most sieving machines commercially available are not suitable for the measurements to be made. Furthermore, when the technique was originally described in the early 1960s, the cost of data processing and automatic weighing were high. Currently investigations are underway at Laurentian University to design an integrated automated sieving process in which the kinetic residue is deduced by a computer algorithm and displayed in real time [26, 27].

To speed the processing of a powder on a series of sieves a system known as a nest of sieves is used. A nest of sieves is an interlocking set of sieves equipped with a lid and a base pan which can be placed on a sieving machine. In Figure 3.9(a) a nest of sieves placed in a Ro-tap sieving machine, as available commercially from the Tyler Corporation, is shown. The beam on top of the device moves up and down to vigorously tap the stack of sieves. In the assembled nest of sieves the coarsest mesh opening is placed at the top of the nest. The finer powder works its way down through a series of sieves of decreasing aperture size. Sieving times have to be prolonged by the fact that small amounts of powder have to make their way down the nest as the sieving process passes the Whitby stage. When using nest sieving one must specify precisely the amount of powder and the sieving time otherwise results will vary from one analysis to another. Sometimes with cohesive powders wet sieving is used, this entails using a liquid to flush the powder through the sieve apertures. A wet testing assembly for sieving powders is shown in Figure 3.9(b). The use of a liquid (usually water) to facilitate powder passage through a sieve results in the added complication of having to dry the residues before they are weighed.

The fact that fineparticles of different shape require longer periods to move through the sieve has been exploited by Meloy and co-workers [5]. In their technique for fractionating powders of the same nominal size but of different shape, a charge of powder is placed in the top of a nest of sieves in which each sieve has the same nominal aperture size. The powder grains with chunky shape are able to pass down through the column much more quickly than grains having elongated shape. By removing the

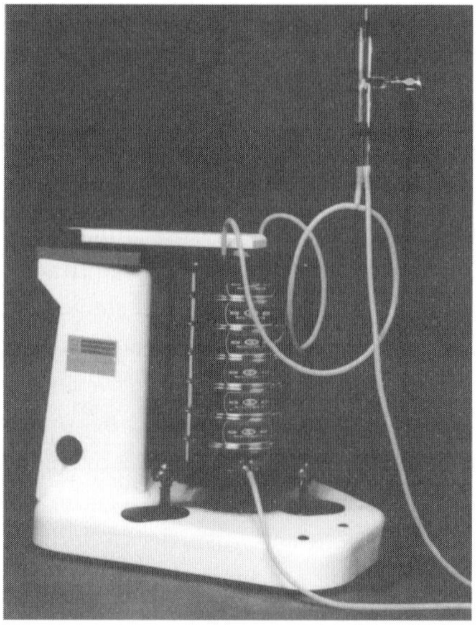

Figure 3.9. A nest of sieves is used to conveniently fractionate powders into fractions of different size range. The equipment shown above is manufactured by the W. S. Tyler Company [13]. a) A nest of sieves mounted in a Ro-tap sieving machine. b) Cohesive powders can be fractionated using wet sieving equipment. (Photographs courtesy of W. S. Tyler Particle Analysis and Fine Screening Group.)

powder from the bottom of the column periodically, a series of powders of different shape but of the same width results [5]. Turbitt-Daoust has recently used this technique to separate an iron powder into different shaped fractions. Typical data generated by Turbitt-Daoust is shown in Figure 3.10 [28].

As mentioned briefly in the first chapter when we considered the filtration of aerosol samples, a recent innovation in the construction of sieves using a matrix of fiberoptic elements has been described for the construction of sieves made of glass. Information on these sieves are available from Collimated Holes Incorporated [29].

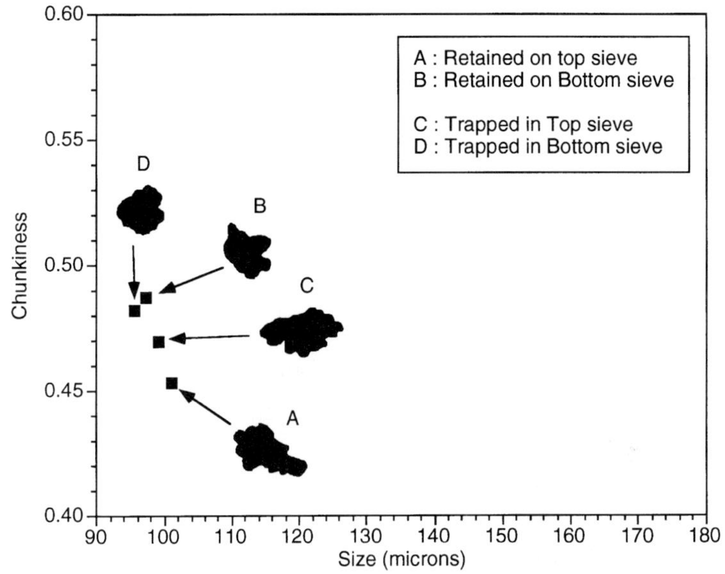

Figure 3.10. The Mean Chunkiness-Size graph of a sample of iron powder grains found retained on the top sieve compared to grains found on the bottom sieve of the nest of four 90 micron sieves after extended agitation. (Typical grains of each group are shown.). Also shown are the mean chunkiness of grains found trapped in the mesh of the top an bottom sieve. Note that the grains which were retained on the top sieve are of lower chunkiness (more elongated) than those on the bottom sieve.

3.3 Sieving Machines

Several attempts have been made to automate sieving procedures but many of the machines described in the research literature have failed to materialize as commercial equipment [30].

The basic nest sieving procedure of the type used in the Ro-tap machine has been automated by the Rotex Company in a machine known as the Gradex 2000 Particle Size Analyzer [31].

In Figure 3.11 the basics of an innovation in nest sieving in which an air column is vibrated to enhance the passage of fineparticles through the sieve is shown [32]. This basic technique was introduced by scientists at the Allen Bradley Corporation in Wisconsin and has been developed commercially by various organizations offering slightly different instrument formats [33, 34]. The information in Figure 3.11 has been reproduced from the literature of the ATM Corporation [33].

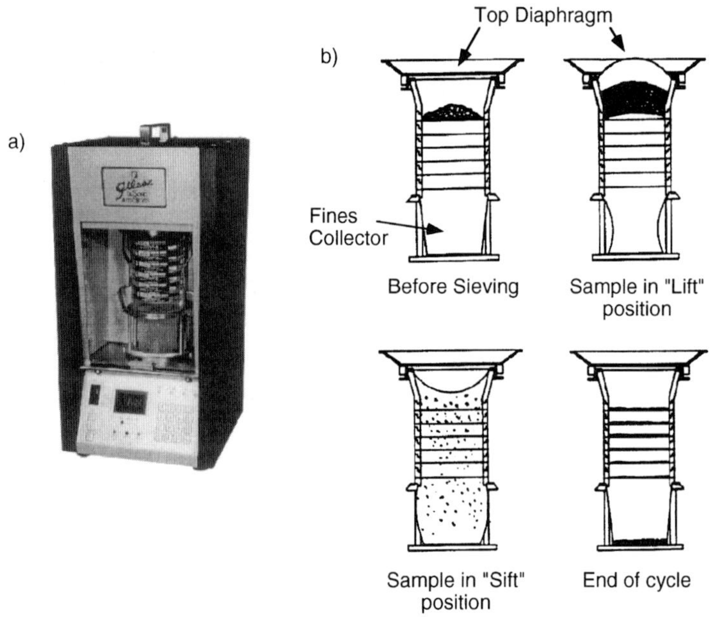

Figure 3.11. The ATM Sonic Sifter uses a pulsating air column to assist the passage of powder down through a nest of sieves [33]. a) Appearance of the Sonic Sifter. b) A single cycle in the operation of the Sonic Sifter. (Images courtesy of, and used with the permission of ATM Corp.)

The Sonic Sifter has been designed to combine two motions to provide a precise particle separation. A vertical oscillating column of air and a repetitive mechanical pulse is used to move the material to be sifted through a single sieve or set of sieves. The name Sonic Sifter appears to come

> *"from the fact that historically, in the original prototype, the moving membrane of a loud speaker device was used to create the oscillating column of air."*

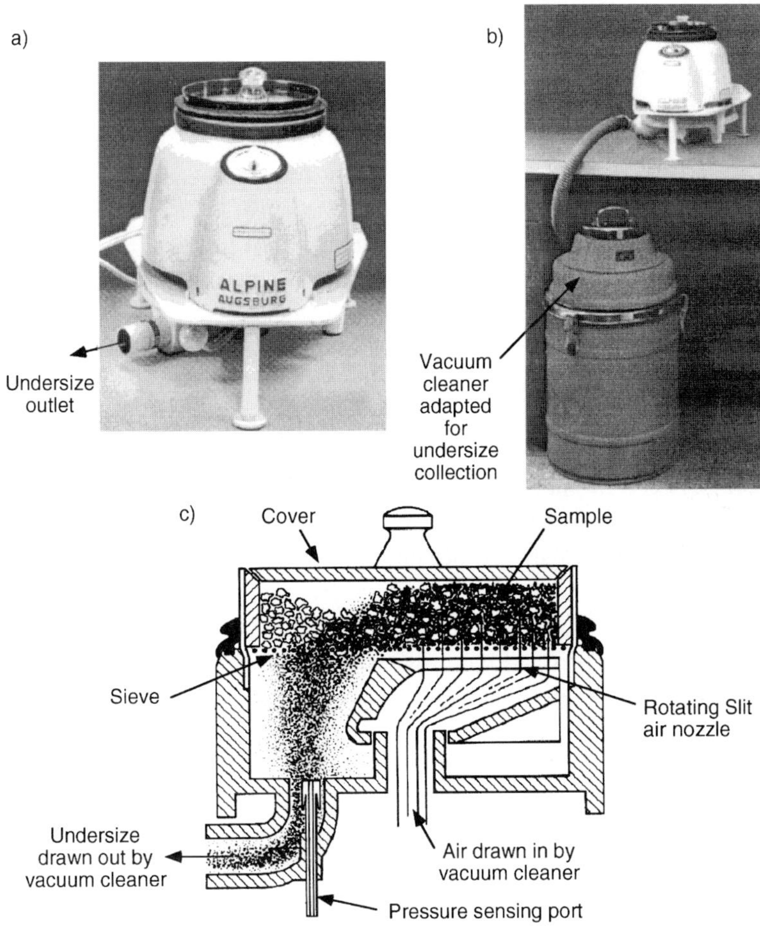

Figure 3.12. In the Alpine Air-Jet sieve, a rotating nozzle feeds air up through the sieve surface. The upward moving air clears blinded apertures and the downward moving air helps fines pass through the sieve [36, 37]. (Used with the permission of Hosokawa Micron International Inc.)

The technical literature of ATM continues,

> *"the oscillating air sets the sample in a vertical periodical motion. It provides a unblinding pulse for every pulse, thus repeatedly giving the particles an opportunity to pass through the sieve."*

The manufactures of this type of sieving machine claim that particle attrition is virtually eliminated and the movement of the powder through the nest of sieves is very rapid. The Gilson Company call their similar device the Gilsonic Autosiever [34]. The Seishin Corporation have described a robotic sieving equipment based upon the use of the nest of sieves operating with the vibrating air column in reference 35.

In Figure 3.11 a piece of equipment specially designed to overcome problems of blinding of the sieves by fine powders is shown. This was originally developed by the Alpine Company and is known as the Alpine Air-Jet Sieve. The Alpine Company has since been taken over by the Hosokawa Company. Inquiries regarding the equipment should therefore be directed to the Hosokawa Company [36, 37].

In the Air-Jet Sieve the sieve is sealed with an air tight plastic lid and fitted onto a cylindrical stand. Air is sucked through the device by means of a specially equipped vacuum cleaner. Air enters the base of the sieve through a set of nozzles in a pipe that rotates just underneath the surface of the sieve. When the powder is placed on the sieving surface and the unit sealed, the rotating air jets clean the sieve apertures immediately above them and fluidize the powder. The return flow of air over portions of the sieve helps the passage of the fines through this sieve. The fines are collected by a filter inserted between the sieve stand and the vacuum pump. General turbulence at the sieving surface jets help to break down agglomerates in the powder. This equipment can be used to fractionate powder down to 10 microns by using electroformed sieves. When very fine powders are sieved with dry air, electrostatic charge can cause problems. It may be necessary either to suspend a Beta-Ray generating anti static capsule inside the sieve or to treat the powder with an electrostatic eliminating type of flow agent such as a silica flow agent [38].

3.4 Possible Future Developments in Sieving

Research projects have suggested that a multisectioned surface of a sieve can minimize the effects due to aperture distributions in the sieve. It has also been suggested that confederated miniature sieves may be able to replace nest sieving [39, 40].

One of the reasons that scientists have been reluctant to move from the standard 8 inch diameter sieves to use more fragile miniature sieves is that many sieving studies are carried out in industrial environments requiring robust equipment. Moreover some companies have up to 50 years of standard data from the Ro-tap devices and are not too sure how the more sophisticated analytical procedures using electroformed sieves and miniature sieves can be compared to the data from older equipment. A further difficulty is that miniature sieves use small amounts of powder of the order of l or 2 grams and scientists know that obtaining the smaller amounts of powder compared to the industrial sampling for several hundred grams is more difficult. However the AeroKaye™ type sampling device described in Chapter 1, along with the falling costs of automatic weighing and data processing equipment, could open up the field for the use of miniature sieves. The fact that the same answer can be obtained using small quantities of powder when one has access to efficient sampling devices has been demonstrated by Humphries at Laurentian University, as part of an M.Sc. project for Loughborough University England. The residue on a sieve was determined by the kinetic residue method for decreasing amounts of powder. The same answer was obtained for 4 grams on a 3 inch sieve as compared to 85 grams of a bronze powder on an eight inch sieve. Furthermore, the total sieving time required for the four gram sample was only ten shakes of the sieve!

References

[1] B. H. Kaye, *Direct Characterization of Fineparticles,* John Wiley and Sons, 1981.

[2] T. Allen, *Particle Size Analysis,* Chapman & Hall, 4th Edition, 1994.

[3] British Standards Institute Publication No. 410 (1943) and 1796 (1952).

[4] B. H. Kaye, D. Alliet, L. Switzer, C. Turbitt-Daoust, "The Effect of Shape on Intermethod Correlation of Techniques for Characterizing the Size Distribution of Powder. Part 1: Correlating the Size Distribution Measured by Sieving, Image Analysis, and Diffractometer Methods", *Part. Part. Syst. Charact.*, 14:5 (1997) 219–224.

[5] T. P. Meloy, K. Makino, "Characterizing Residence Times of Powder Samples on Sieves," *Powder Tech,*, 36 (1983) 253–258.

[6] J. K. Beddows, "Dry Separation Techniques", *Chemical Engineering,* (August 10, 1981) 69–84.

[7] German Standard Specification, DIN 1171 (1934) in Sieving Handbook, W. S. Tyler Inc. 8200 Tyler Blvd. Mentor, Ohio, USA 44060.

[8] International Standards Organization – Sub Committee TL/24-SC1 in Sieving Handbook, W. S. Tyler Inc. 8200 Tyler Blvd. Mentor, Ohio, USA 44060.

[9] Canadian Standard Sive Series 8-GP-1d, in Sieving Handbook, W. S. Tyler Inc. 8200 Tyler Blvd. Mentor, Ohio, USA 44060

[10] ASTM Standard E11-61 in *A.S.T.M. Standards with Related Material,* American Society for Testing and Material, Philadelphia, PA, USA 10 (1967) 528–534.
[11] See discussion of Gaussian probability in B. H. Kaye, *Chaos and Complexity; Discovering the Surprising Patterns of Science and Technology,* VCH, Weinheim, 1994.
[12] B. H. Kaye, *Physical Problems of Particle Size Analysis,* Ph. D. Thesis, London University, England, 1961.
[13] W. S. Tyler Incorporated, 8200 Tyler Boulevard, Mentor, Ohio, 44060.
[14] M. A. K. Yousufzai, *A Study of the Physical Parameters Affecting the Efficiency of Sieve Fractionation of Powders,* M.Sc. Thesis, Laurentian University, Sudbury, Ontario, 1984.
[15] B. H. Kaye, M. A. K. Yousufzai, "How to Calibrate a Wire-Woven Sieve," *Powder and Bulk Engineering,* 6:2 (February 1992) 29–34.
[16] T. Allen, "Sieve Calibration Using Tacky Dots," *Powder Technology.,* 79 (1994) 61–68.
[17] Data for Figure 3.5 was generated by G. G. Clark at Laurentian University as part of a study of sieve blinding and variations in aperture size distribution over a sieve surface.
[18] H. Heywood, "The Origin and Development of Particle Size Analysis," in M. G. Groves, J. L. Wyatt-Sargent, Eds. Proceedings with the Conference on Particle Size Analysis, Society for Analytical Chemistry, 9/10 Savile Row, London, W1X 1A1, (1970).
[19] See discussion of compensating size errors in T. Allen, *Particle Size Analysis,* Chapman & Hall, 4th Edition, 1994.
[20] Electroformed sieves are available from Buckbee-Mears Company, 245 East Sixth Street, St. Paul, Minnesota, 55101.
[21] Photoetched sieves are available from Veco-Stork International, 4925 Silabert Avenue, Charlotte, North Carolina.
[22] H. Konowalchuk, A. G. Naylor and B. H. Kaye, "A Rapid Method for Assessing the Quality of Sieves," *Powder Technology,* 13:1 (1976) 97–101.
[23] H. Konowalchuk, M.Sc. Thesis, Laurentian University, Sudbury, Ontario, Canada.
[24] A. G. Naylor, B. H. Kaye and P. E. Legault, "An Instrument for Assessing Sieve Aperture Damage," *J. Powder Bulk Solids,* 1 (1977), 9–12.
[25] K. T. Whitby, "The Mechanics of Fine Sieving," A.S.T.M. Special Technical Publication No. 234, 1959, p. 3.
[26] B. H. Kaye, "Investigations into the Possibilities of Developing a Rate Method of Sieve Analysis," *Powder Met.* (1962) 199–217.
[27] B. H. Kaye and N. I. Robb, "A New Algorithm for Calculating Kinetic Residues for the Characterization of Powders Using Sieves," *Powder Technology,* 24 (1979) 125–128.
[28] C. Turbitt-Daoust. Laurentian University, Sudbury, Ontario, Personal Communication.
[29] Collimated Holes Incorporated, 460 Divisions Street, Campbell, California, 95008.
[30] See discussion of the Autosieve and the Sorsi system in reference 1.
[31] Rotex Inc., 1230 Knowlton Street, Cincinnatti, Ohio, USA, 45223-1845,
[32] H. O. Sum, "Oscillating Air Column Method for the Dry Separation of Fine and Subsieve Particles Sizes," *Powder Technology,* 2 (1968–69) 356–362.
[33] ATM Corporation, Sonic Safety Division, Post Office Box 10494, Milwaukee, Wisconsin, USA, 532-14.

[34] Information on the Gilsonic Autosiever available from Gilson Company Incorporated, P. O. Box 677, Worthington, Ohio, 4305-0677.
[35] Seishin Enterprise Company Limited, Nipon-Brunswick Building, 5-27-7 Sendagaya, Shibuyo-Ku, Tokyo, Japan.
[36] Alpine American Corporation, 3 Michigan Drive, Natick, Massachusetts, 01760.
[37] See technical literature available from Hosokawa Micron International Inc. 10 Catham Road, Summit, New Jersey, USA 07901.
[38] For a discussion of the use and properties of flowagents see Chapter 3, "Powder Rheology", in B. H. Kaye, *Powder Mixing,* Chapman and Hall, London, 1997.
[39] B. H. Kaye, G. G. Clark, "Confederated Miniature Sieves as an Alternative to Nest Sieving". *Part. Char.*, 3 (1986) 145–150.
[40] B. H. Kaye, "Potential Advantages of Multipartitioned Sieve Surfaces for Characterizing Powders," *Powder Technology*, 19 (1978) 121–123.

4 Size Distribution Characterization Using Sedimentation Methods

4.1 Basic Considerations

Prior to the development of the light diffraction methods and photon correlation spectroscopy to be described in chapters 6 and 8, sedimentation methods were the most widely used methods for characterizing the size distribution of a powder. In this chapter we will discuss the basic concepts involved in sedimentation techniques and review two or three methods still in relatively wide use. We will also describe briefly some of the methods which have been used extensively in past studies so that the reader may be able to assess the value of published data from the archives of powder technology. The reader interested in a detailed discussion of classical sedimentation size characterization techniques should refer to textbooks covering these earlier methods of size distribution in detail [1, 2].

In sedimentation methods for characterizing fineparticles the settling dynamics of a dispersion of the fineparticles under study in a gas or a fluid are studied. The sedimentation conditions are usually adjusted so that the fineparticles can be considered as falling freely at low Reynolds number in a viscous fluid. Under such conditions one can use Stokes law to interpret the size of the settling fineparticles. The Stokes diameter of a fineparticle is defined as the size of the dense smooth sphere of the same density material having the same falling speed as the fineparticle of interest in a viscous fluid at low Reynolds number. The reason that one has to specify low Reynolds number for sedimentation studies is that the falling conditions of the fineparticle must involve laminar flow without turbulence. The formulae for Stokes Law and Reynolds number are

$$R_n = \frac{\upsilon d \rho_L}{\eta} \tag{4.1}$$

where

R_n = Reynolds number
υ = velocity of the fluid
ρ_L = density of the fluid
η = viscosity of the fluid

$$d = \sqrt{\frac{18\eta h}{(\rho_S - \rho_L)gt}} \tag{4.2}$$

where
d = diameter of the sphere
h = height through which the sphere has fallen
t = time for the sphere to fall height h
g = acceleration due to gravity
$\boldsymbol{\rho_S}$ = density of the sphere

Readers unfamiliar with Stokes Law and Reynolds number will find a useful introduction to these concepts in Chapter 6 of Reference 3.

In all sedimentation size studies a key variable in the experimental procedure is the state of dispersion of the fineparticles. If the fineparticles being studied are suspended in a liquid it is common practice to use a suitable dispersing agent and to treat the suspension with some such procedure as ultrasonic agitation to disperse the fineparticles. No absolute rules can be given regarding the dispersion techniques [4, 5]. Recently an ISO standard on dispersion technique has been prepared and it contains much useful information [6]. Usually it is necessary to specify the dispersion technique used in a given procedure and to strictly adhere to that procedure. Sometimes the use of ultrasonics to aid the dispersion of the powder can actually result in degradation of the fineparticle size of the powder being dispersed. Also excessive use of ultrasonic agitation can significantly change the temperature of a liquid suspension and hence the viscosity of the liquid used in the experiment.

A critical factor in the accuracy of fineparticle size determination by sedimentation techniques is the requirement that the fineparticles when dispersed, do not physically interact with each other. In the early days of sedimentation procedures is was not always appreciated that even at relatively low solid concentrations there can be significant interaction between the falling fineparticles to cause errors in the Stokes diameter determination. The interaction of fineparticles in a suspension depends upon the configuration being used to carry out the Stokes diameter determination. Essentially there are four different configurations of suspension used in sedimentation studies. These are shown schematically in Figure 4.1. For reasons that will be discussed later, the interaction of particles is more critical in the line start procedure illustrated in Figure 4.1(b). The interaction of the fineparticles also depends upon the size ratio of adjacent fineparticles interacting in a fluid. Because sedimentation techniques were so important in the 1960's and 1970's the dynamic interactions of sedimenting fineparticles were studied extensively. One of the unexpected problems encountered in these studies was that for concentration ranges between 0.5 % and 5 % volume-solids con-

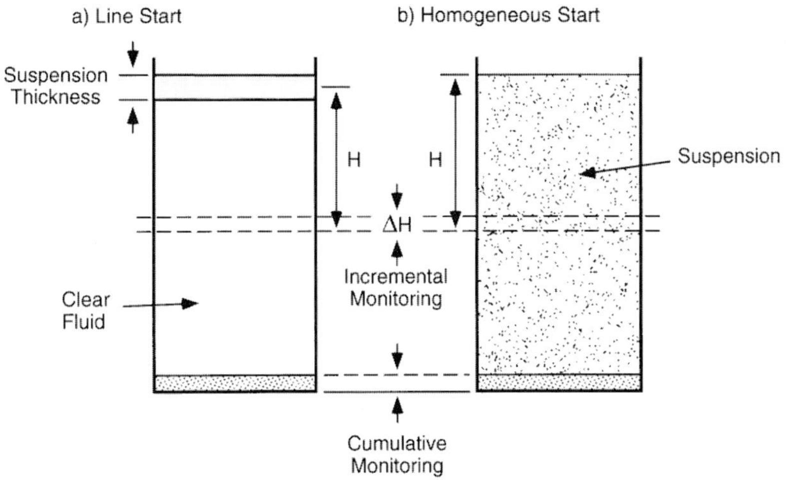

Figure 4.1. Two possible configurations for using suspensions to determine the Stokes diameter distribution of a system of fineparticles. a) Line start system uses a layer of suspension carefully placed on top of a column of clear fluid. b) Homogeneous start system.

centration, the fineparticles in an initially homogeneous suspension actually fell faster then predicted by the Stokes equation. This is because it was easier for the fluid to move around a cluster of fineparticles than in between the fineparticles. This is shown in the data reported by Kaye and Boardman summarized in Figure 4.2 [7, 8, 9].

In their studies Kaye and Boardman studied the dynamics of red marker spheres in suspensions of glass spheres of different size range. The interaction of the spheres was at a maximum when the marker spheres were of the same order of size as the glass spheres in suspension. In Figure 4.2(a) size distributions of the powders used in their studies are shown. In Figure 4.2(b) the higher falling speeds for spheres caused by dynamic cluster interaction of the marker spheres with the spheres of approximately the same size is shown. In the discussion of their results, Kaye and Boardman divided the fineparticle interaction graph into 4 regions. At the low solids concentration the interaction was at a minimum and one could assume for solids concentrations of below 0.05 by volume that the Stokes diameter was being measured. In the concentration region from 0.1 up to 1 % by volume concentration, the formation of clusters which sediment together in the suspension can increase the measured settling velocities by up to 50 %. As the concentration moves above 1 % by volume, the return flow of liquid around one cluster will help to break up another cluster adjacent or above it. By the time the concentration reaches 5 % by volume the measured Stokes diameter is back to the corresponding physical size. The final region of interaction constitutes hindered settling because the upward moving fluid causes fineparticles to sediment

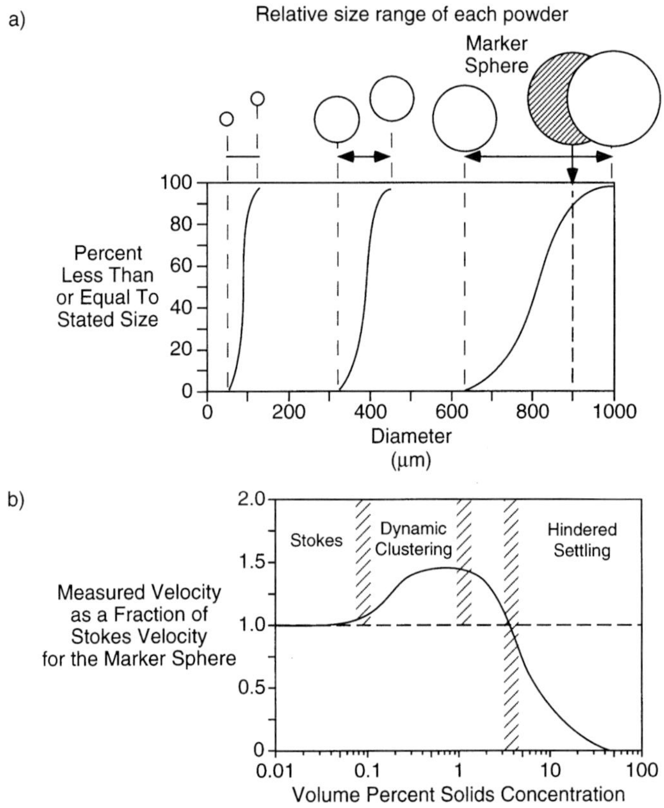

Figure 4.2. Work done by Kaye and Boardman showed that as the percent solids concentration is increased, suspensions behave very differently under gravity sedimentation. a) Size distributions of powders used by Kaye and Boardman and a graphical display of the minimum and maximum size in each powder in relation to the marker spheres used to measure the sedimenting velocity. b) Sedimenting velocity measured for the largest powder of (a) as the percent volume concentration was increased.

more slowly than anticipated [9]. The phenomena summarized in Figure 4.2 has been explored by different workers and are well established facts (see discussion in references 1 and 2).

When choosing the concentration to be used in a given experimental study, scientists face a dilemma. If one chooses to operate at very low concentrations of solid then the analytical procedure for estimating changes in concentration of the fineparticles in suspension becomes difficult. On the other hand, if one uses higher concentrations the interaction error can become significant. The consensus of opinion seems to be that, if studying a powder with a relatively wide range of sizes, operating at 1 % solids concentration, the penalty in the accuracy of the data is of the order of 1–2 %. One must be very careful to specify what concentration is used in a specific study. One must always work at the same concentration in a series of investigations. In later sections in this chapter we will present data illustrating the penalty incurred by working with different concentrations in different experimental procedures.

To maintain the thermal stability of a sedimenting suspension, it is common practice to surround the sedimentation column with an outer jacket through which constant-temperature fluid is circulated. This is not sufficient thermal protection of the suspension if the top surface of the sedimentation column is open to the atmosphere. In such a system there is vapour loss from the surface of the fluid, thus causing surface cooling of the fluid. This heat loss results in convection currents moving down from the surface of the fluid into the body of the suspension [10]. When the sedimenting column fluid is water, one way of preventing evaporative cooling is to float a thin layer of immiscible low vapour pressure mineral oil on top of the column immediately after the last stirring of the suspension. Davies and Kaye described a procedure to prevent evaporation by covering the top surface with a thin layer of microballoons of glass [11]. Because of the ready availability of expanded plastic foams, probably many different kinds of low density plastic beads can be used to achieve the same thermal protection of the open surface.

Thermal control of the outside of the sedimentation column is inadequate when radiant energy is incident on the sedimentation system. For example, the author once visited a laboratory where glass bottles containing the sedimenting suspension were stood in a constant-temperature water bath. The temperature bath was maintained at a steady 25 °C, yet it was obvious that convection currents were occurring within the sedimenting suspensions. The constant-temperature bath was standing on a bench in strong sunlight. Sunlight penetrating the glass cover of the sedimentation column was being absorbed by the fluid. As a consequence, the fluid was considerably warmer than the surroundings. The instability problem was solved by shielding the temperature bath from direct sunlight. Bright laboratory lights adjacent to a temperature-controlled environment, when using transparent containers, can result in the same type of thermal instability. Inexpensive Dewar flasks can sometimes be used to thermally isolate the sedimentation system.

4.2 Size Analysis Procedures Based on Incremental Sampling of an Iinitially Homogeneous Suspension

The basic concepts used in this group of methods are illustrated by considering one of the classical size analysis methods. The simplest analytical procedure in this group of methods is known as the pipette method. In this technique the concentration changes occurring within an initially homogeneous sedimenting suspension at a depth **h** through a time **t** are evaluated by withdrawing a sample of the suspension into a pipette. In the idealized interpretation of the data it is assumed that the sampling zone, when the pipette is placed in the suspension, is a sphere of fluid a around the tip of the pipette. The ideal theory assumes that the concentration measured in the sphere of fluid is a true measure of the solids concentration at the depth **h** at time **t**. Obviously the size of the sampled sphere of fluid alters the veracity of this ideal interpretation. The speed at which the fluid is drawn into the pipette affects the sizes of fineparticles in the suspension that are actually drawn into the pipette. Furthermore, the act of placing the pipette into the suspension disturbs the settling fineparticles and this all has to be taken into account when interpreting the data.

There are several general relationships that are useful in describing the concentration changes occurring in a settling suspension. When developing these equations it is assumed that initial turbulence effects due to the stirring of the suspension are negligible and that at time $\mathbf{t} = \mathbf{0}$ the suspension is homogeneous so that each cubic centimeter of the suspension contains $\mathbf{n}_1$, fineparticles that have a falling speed of $\mathbf{v}_1$, $\mathbf{n}_2$ fineparticles that have falling speeds of $\mathbf{v}_2$, and so on. To develop the necessary equation, consider events occurring at a monitoring zone at depth $\mathbf{h}_1$ and time $\mathbf{t}_1$. A rigorous derivation of the equations would take into account the finite width of the monitoring zone since this effects the resolution of the characterization data. However, except in the case of the photosedimentometer and the x-ray sedimentometer, the exact dimensions of the effective monitoring zone of a real instrument are indeterminate. As a consequence, the effects of the finite dimensions of the monitoring zone often occur as one of many virtually inseparable ingredients in the general resolution of the characterization procedure that is determined empirically by direct comparison data obtained by various characterization procedures. Heywood [12] carried out experimental work on the relationship between the width of an interrogation zone, when monitoring a homogeneous suspension incrementally, and the resolution of the characterization procedure. He concluded that if the zone were less than one-sixth of the mean depth of the zone, error introduced into the characterization procedure would be negligible. This conclusion should not, however, be treated as a general conclusion since modern electronic equipment permits higher levels of scrutiny in characterization procedures and the effective resolution in modern methods can be very different from that in earlier instruments.

At a depth $\mathbf{h_1}$ in the suspension at time $\mathbf{t_1}$ in a sedimenting suspension all fineparticles with velocities $\mathbf{v_1} = \mathbf{h_1}/\mathbf{t_1}$ will have cleared the monitoring zone. The number of fineparticles of settling velocities smaller than $\mathbf{v_1}$ leaving the zone and being replaced by an equal number of fineparticles per unit volume in the monitoring zone is given by the relationship

$$(N)_{h_i t_i} = \sum_{i=s}^{i=h_1 t_1} n_i \tag{4.3}$$

where

$(N)_{h_i t_i}$ = number of fineparticles per unit volume at depth $\mathbf{h_i}$ and time $\mathbf{t_i}$
$\mathbf{n_i}$ = number of fineparticles per unit volume of velocity $\mathbf{v_i}$
$\mathbf{v_s}$ = velocity of smallest fineparticle in the suspension

Consider again at time $\mathbf{t_2}$ depth $\mathbf{h_2}$, where $\mathbf{t_2} \geq \mathbf{t_1}$ and/or $\mathbf{h_1} \geq \mathbf{h_2}$. The number of fineparticles in the zone is

$$(N)_{h_2 t_2} = \sum_{i=s}^{i=h_2 t_2} n_i$$

where

$$V_{h_2 t_2} = \frac{h_2}{t_2}$$

and $\mathbf{d_s}$ is the smallest fineparticle present. Therefore

$$(N)_{h_1 t_1} - (N)_{h_2 t_2} = \sum_{i-h_1 t_1}^{i=h_2 t_2} n_i \tag{4.4}$$

For the situation where concentration gradients are followed by gravimetric assessment, let $\mathbf{w_i}$ = the weight of fineparticles of Stokes diameter $\mathbf{d_i}$. Then, assuming that the shape of the fineparticles does not vary with their size,

$$w_i = \gamma d_i^{\,3} \rho_P \tag{4.5}$$

where γ is a shape factor. The value of γ for dense smooth spheres is $\pi/6$. The initial weight of fineparticles per unit volume $\mathbf{C_0}$ is given by

$$C_0 = \sum_{i=8}^{i=l} n_i \gamma \rho_p d_i^{\,3} \tag{4.6}$$

where $\mathbf{d}_i$ is the largest fineparticle in the suspension. Let $\mathbf{C}_{ht}$ denote the concentration at depth **h** at time **t**, then

$$C_{ht} = \sum_{i=8}^{i(ht)} n_i \gamma \rho_p d_i^{\,3} \tag{4.7}$$

From equations 4.4 and 4.3, $\mathbf{R}_{ht}$ the percentage weight of fineparticles with Stokes diameters smaller than $\mathbf{d}_{ht}$, is given by

$$R_{h_1 t} = \frac{C_{ht}}{C_0} \times 100 \tag{4.8}$$

It follows from this equation that the cumulative undersize distribution function of a fineparticle system can be calculated by measuring the concentration within a settling, initially homogeneous, suspension at a series of specified times and/or depths and by plotting $\mathbf{R}_{ht}$ against $\mathbf{d}_{ht}$.

It should be noted that pipette procedures do not readily yield information on the largest and smallest fineparticle present in the suspension. Sometimes Stokes diameter distributions measured by pipette procedure are extrapolated to what are claimed to be reasonable upper and lower limits on the distribution. It is better to terminate the curve linking the various data points at the first and last data points.

Andreasen pioneered the use of pipette methods for particle size characterization [13, 14, 15]. His equipment is shown in Figure 4.3. Although it is a simple device and prone to several sources of error it is discussed here in some detail because it is embodied in the ISO standard and used as a reference method in the evaluation of the BCR standards mentioned in an earlier chapter.

In the Andreasen procedure the pipette is left in the sedimenting suspension and a sequence of samples extracted through the pipette at a set of predetermined times. Each time a portion of suspension is removed from the sedimentation column, the effective depth of the pipette tip below the suspension surface changes; this is allowed for in the calculation of the Stokes diameter distribution. The Andreasen equipment is provided with several pipettes of varying stem length to facilitate the sampling of suspension at different depths; however, it is not usual to change pipettes during any particular characterization study. A separate sedimentation vessel would normally be used for each stem length selected.

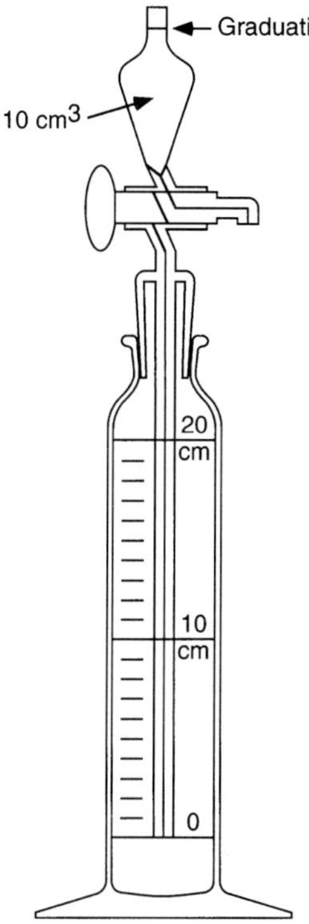

Figure 4.3. Appearance of the Andreasen pipette equipment is 5 cm inside diameter and holds 450 cm^3 of suspension. the bottom of the pipette is positioned 2.5 cm above the base.

An alternative to the fixed pipette procedure is to insert a pipette to the required depth at a specified time whenever it is required to examine a portion of the suspension. In this procedure, termed the free pipette technique, there is an unknown source of variation in the characterization data arising from the disturbance of the suspension caused by the insertion of the pipette. When using the free pipette procedure, it is often necessary to make a mechanical rig to aid in the exact positioning of the pipette tip prior to the withdrawal of a sample of suspension. If ample laboratory space and equipment are available, a useful technique for avoiding uncertainties associated with the free pipette procedure is to make several equally representative suspensions of the fineparticles to be characterized, each suspension should be used only once for a given measurement of the concentration at a given time and depth. This also has the advantage that the measurements are averaged over several samples of the fineparticle system. Statistically, this is a better procedure than carrying out all measurements on one suspension.

The free pipette procedure has an advantage over the fixed pipette procedure in that there is no shadow zone effect. The shadow zone effect arises from the fact that fineparticles leaving the suspension under the column of the pipette are not replaced by fineparticles from above, thus lowering the true concentration of the monitoring zone with respect to its true value in the absence of the pipette. Furthermore, the developing zone of clear fluid immediately under the stem of the pipette is unstable with respect to the adjacent body of the suspension. As a consequence, there will be gravity convection currents in the vicinity of the pipette tip. This causes disturbance to the trajectories of the fineparticles in the immediate vicinity of the pipette stem. When selecting a characterization strategy, the presence of the shadow zone must be traded off against the fact that the free pipette requires a ready access to the sedimentation column, thus increasing the difficulty of protecting the free surface of the sedimentation fluid from evaporative cooling. Leschonski has described equipment in which the pipette stem is continued to the base of shadow zone under the tip of the pipette [16]. The sample of suspension is then aspirated through a set of holes around the periphery of the glass stem of the sampling device. Kaye and Davies described a design for use in pipette procedures for characterizing powders in which the sedimentation vessel is a cylinder constructed out of insulated rigid foam. The glass stem of the sampling system now is able to move along the axis of the cylinder of the sedimentation vessel. This has the advantage over the Leschonski-type pipette in that it is possible to vary the height at which a portion of the suspension is extracted for examination [17].

When sampling from a suspension the actual fineparticles that move into the pipette depends on the flow conditions around the pipette tip. Thus, for a given volume of suspension aspirated into the pipette in a fixed time, different fineparticles will be taken into the pipette for various orifice sizes because of the different fluid flow rates that will exist for the various orifices. Normally it is desirable to extract the suspension into the pipette as fast as possible to reduce the uncertainty in **t**, the time at which the sample concentration is measured. If a large orifice is used on the pipette, the flow rate into the pipette will be relatively low so that larger fineparticles in the bottom hemisphere of the approximate spherical monitoring zone will not be pulled up to the orifice. All too often, experimental data do not quote the speed at which the samples of suspension are extracted from the sedimenting suspension, nor are the dimensions of the orifices used. One of the advantages of using the Andreasen pipette system, which is available commercially, is that in this equipment all dimensions are standardized. When setting up an analytical procedure using a pipette sampling procedure, it is useful to arrange for the suction on the pipette to be from a partial vacuum reservoir to ensure steady withdrawal rates. It should be noted that the interrogation zone for the Leschonski system and the Kaye-Davies equipment is different from that of the simple pipette and consists of hemispheres centered on the orifices located around the stem of the pipette.

When using a fixed pipette to take several samples out of a suspension, an amount of suspension between the tip of the pipette and the operating tap of the pipette remains in the pipette between successive samples. This space between the pipette tip and the tap is known as the dead space. The uncertainty in the characterization study due to this factor is known as the dead-space error. To minimize the effect of the dead-space error it is usual to specify a very narrow channel in the pipette stem. The lower limit for this channel diameter is fixed by the need to aspirate the sample in minimum time and avoid blockage of the tube in the cases where relatively coarse fineparticles exist in the suspension.

Kaye and Davies suggest that dead-space errors can be minimized with their equipment because their continuous-stem pipette permits the flushing out of the dead-space suspension between sampling operations [17].

When choosing the volume of suspension to be extracted in a pipette procedure, one should remember that the resolution of the study is increased by minimizing the volume of suspension. This requirement, however, conflicts with the need to characterize the solid contents of the extracted samples. If the solids content is to be determined chemically by dissolving the fineparticles and reacting them with appropriate reagents, one can often work with smaller volumes of suspension than in techniques in which the fineparticles must be filtered and weighed. Sometimes samples of suspension are evaluated by weighing the residues left after evaporation of the suspension fluid. It should be noted that in such situations the residual dispersing agent is rarely driven off by the drying process. Instead, it forms a hygroscopic ingredient in the residue that causes error in the weighing process because of the absorption of moisture from the atmosphere. For this reason the use of the membrane filters has the advantage that the residual dispersing agents can be washed off the fineparticles by appropriate wash fluids before drying the sample.

The data in Table 4.1 illustrates the magnitude of the type of error that can arise from concentration effects in pipette-type characterization studies. The exact interaction effects in any given suspension will depend on the range of sizes present so that

Table 4.1. Stokes Diameter distributions of a spherical bronze powder, as determined with an Andreasen Pipette at various suspension concentrations.

Solids Concentration by Weight	0.24	0.53	0.89	2.01	2.87	3.67	4.45	13.6
Stokes Diameter µm	**Cumulative Percentage by Weight Less than Stated Stokes Diameter**							
26.0	82.4	83.4	87.1	90.0	89.4	93.1	94.0	100.0
18.2	65.2	66.0	72.0	71.0	72.2	75.0	80.4	95.0
13.0	29.5	33.0	34.0	39.4	38.4	40.0	50.1	76.2
9.0	8.1	8.4	8.6	10.0	9.0	11.0	12.8	23.0

the data in Table 4.1 can be used only as relative indication of the order of magnitude of concentration effects. The concentration effects in any particular situation would have to be investigated empirically by carrying out a series of characterization studies at a range of solids concentration.

A simple method for measuring sedimentation changes within a suspension is based upon the use of hydrometers measuring the density of the suspension. Such hydrometer based methods have been widely used in the ceramics industry and in soil mechanics. The solids concentration required to be able to follow changes of concentration within a sedimenting suspension using a hydrometer, unfortunately, often requires the concentration of the suspension to be of such a level that free falling conditions no longer exist within the suspension. Accordingly it is necessary to specify the operation the experiment or method in great detail and to specify the design of the hydrometers used in the investigation [18, 19].

The density changes within a homogeneous suspension can be related to the Stokes diameter distribution function of the sedimenting fineparticles by using the following relationships [20].

$$\rho_{su} = \rho_L + \alpha(\rho_P - \rho_L) \tag{4.9}$$

where

ρ_{su} = density of suspension
ρ_L = density of dispersing liquid
ρ_P = density of powder fineparticles
α = volume fraction of solids per cubic centimeter

If ρ_1 is the initial density of the suspension, it can be shown that

$$R_{ht} = \frac{(\rho_{su})_{ht} - \rho_L}{\rho_l - \rho_L} \tag{4.10}$$

where

R_{ht} = percentage by weight of fineparticles of Stokes diameter less than $\mathbf{d}_h$, that falls **h** cm in **t** seconds

$(\rho_{su})_{ht}$ = density of suspension at depth **h, t** seconds after settling commenced (and other symbols are as defined)

The idealized hydrometer shown in Figure 4.4(a) consists of two main parts, a symmetrical bulb and a long stem. When placed in a fluid, the bulb descends until the weight of the fluid displaced equals that of the hydrometer. When used to measure the density gradient within a suspension, the finite volume of the bulb of the hydrometer determines the resolution of the experimental measurements. The insertion of the bulb also causes a disturbance in the suspension, and it displaces the suspension in the interrogation zone, thus causing uncertainty in the interpretation of the experimental data. When interpreting data obtained by this technique, it is usual to assume that the volume of the stem is negligible in comparison with that of the bulb and that the density measured is an average value over the depths **h** and $\mathbf{h_l}$ as shown in Figure 4.4(a). The exact nature of this average will depend on the shape of the hydrometer bulb. In the various standard procedures established for hydrometer-sedimentation characterization techniques the shape of the hydrometer is strictly specified, and the

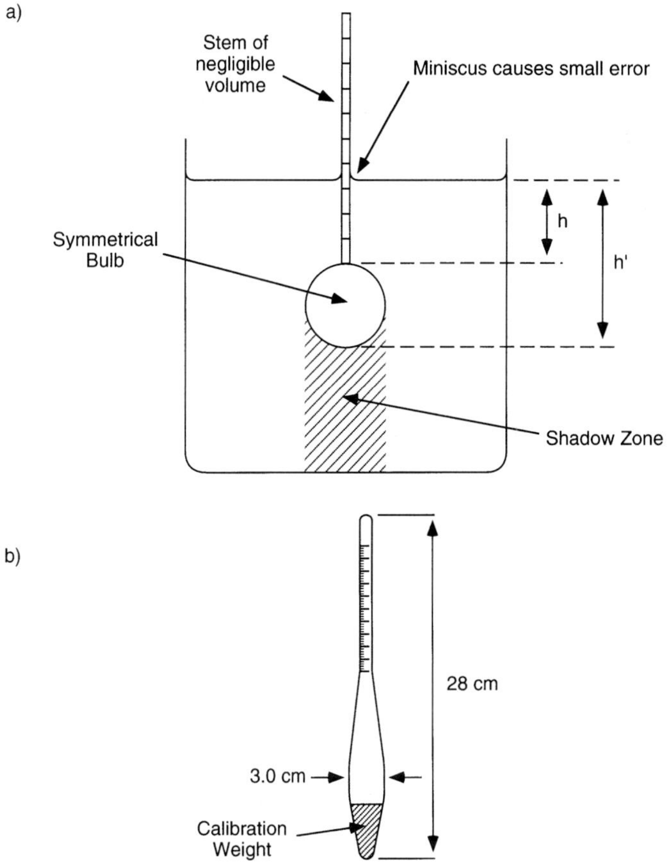

Figure 4.4. Hydrometers can be used to monitor a sedimenting suspension, however, a number of uncertainties are associated with the presence of a hydrometer in a suspension. a) An idealized hydrometer and relevant considerations. b) A streamlined hydrometer specified by British Standard 1377 [18].

appropriate formulas for calculating the effective value of **h** from the dimensions of the hydrometer and the depth to which it sinks are given [18, 19].

When using the hydrometer to follow density shifts within a settling suspension one can follow special procedures. In the first method the hydrometer is placed in the suspension and readings of the depth of immersion taken at a series of times. In the alternate procedure, the hydrometer is placed in the suspension prior to each measurement and then withdrawn immediately.

When the hydrometer is left in the suspension, fineparticles settle from the suspension onto the shoulders of the hydrometer. This changes the effective weight of the hydrometer and causes systematic error in the density readings taken from the stem of the hydrometer, which is usually calibrated to give density readings directly according to the depth to which it sinks in the suspension. The streamlined hydrometer in Figure 4.4(b) has sloping sides designed to help the hydrometer shed any fineparticles falling on it. However, the flow of fineparticles shed by the hydrometer will cause currents around the bulb and further disturb the settling suspension. If the hydrometer is left in position during settlement, another source of error is the shadow zone underneath the bulb, a region denoted **A** in Figure 4.4(a). Within this zone fineparticles settling out are not replaced. For this reason, after a period of time, the density of Zone **A** is less than in the surrounding suspension. This causes turbulent mixing at the boundaries and thus disturbs the settling pattern and causes error of unknown magnitude.

The alternative experimental procedure of withdrawing the hydrometer between readings over a period of time gives rise to two types of error. When the hydrometer is inserted into the suspension, it displaces and generally disturbs the suspension as it sinks through it. The effect of this disturbance is difficult to estimate, and the cumulative result of a series of insertions and withdrawals can be measured only empirically. Factors that influence the magnitude of the error caused by this disturbance of the suspension are; 1) the ratios of the bulb dimensions to those of the sedimentation vessel and 2) the depth to which the hydrometer sinks.

The other main source of error is caused by the fact that it is difficult to decide when the hydrometer is in equilibrium with the surrounding suspension. This problem is usually overcome by specifying arbitrarily some time period between the insertion of the hydrometer and the taking of a reading; again, errors due to this factor are difficult to estimate.

The higher the suspension concentration used in the characterization study, the easier the calibration of the hydrometers because the density changes to be measured are varying over a wider range. The advantage gained in hydrometer calibration by using higher solids concentrations is offset by the fact that dynamic clustering becomes a severe problem at higher concentrations. Usually it is not practical to use concentrations less than 2 % by volume, and unavoidable concentration effects are accepted as a tolerable penalty to be traded off against low cost, portability, and ruggedness of the procedure and equipment.

Surface tension effects on the emergent hydrometer stem are another source of error. A correction factor can be calculated for this from the known dimension of the stem and the surface tension of the fluid, plus dispersing agent.

Provided all experimental conditions and equipment dimensions are carefully specified, the techniques give reproducible results, but it is difficult to relate data obtained by this procedure to those generated by other sedimentation methods.

Berg pioneered the use of miniature hydrometers that can be completely immersed with in the suspension [21]. These miniature hydrometers were described as divers. The diver can only measure the concentration at one zone and moves down with a zone of a constant specific gravity. Therefore in the diver method one requires a series of divers of known density and a technique for locating them within the suspension. Although the divers are not available commercially they could be very useful in following density gradients within high concentration suspensions, a problem encountered in the use of slurries and dense suspensions in various industries. For use in such dense slurries Kaye an James developed a diver that shed powder settling onto the diver in the suspension and could be used to follow a given gradient moving down through a heavy suspension [20]. The history and the modern design of divers for measuring changes in the concentration of a settling suspension is discussed in detail in Reference 1.

The desire to automate sedimentation methods of size characterization, and to use as low a solids concentration as possible in the sedimentation system, led to the use of light beams to monitor concentration changes in the sedimenting system. Instruments using light beams to monitor concentration changes are referred to as photosedimentometers. The basic configuration used in such instruments is illustrated in Figure 4.5. A beam of light passes through a tank of width **L** in order to monitor the settling suspension. The energy removed from the forward beam is dependent on the concentration of the fineparticles, the length of the path of the beam through the tank, and the size and optical properties of the fineparticles. Some of the light scattered by the fineparticles is received by the photocell. The amount of scattered light is governed by the solid angle subtended by the photoelectric device as viewed from the various portions of the suspension through which the light beam is passing. Normally it is assumed that some effective solid angle, created with respect to the center of the sedimentation tank, is useful in describing the properties of a given photosedimentometer. In the early days of the development of photosedimentometers there were many attempts to develop the relationship between the energy removed from the forward beam and the properties of the scattering fineparticles using light scattering theory. When evaluating this early work it should be appreciated that there are several detailed theories of light scattering appropriate to different fineparticle systems. When the fineparticles are of the order of 10 wavelengths of light and above, Fraunhofer diffraction theory can be used [22]. When the fineparticles have diameters of the order of half a wavelength, up to 10 wavelengths, the appropriate light scattering theory has been developed by Mie. For very small fineparticles, less than half a wavelength of light, the appropriate light scattering theory is so-called Rayleigh theory. These theories of light scattering will be discussed in more detail in Chapter 7.

When attempting to derive an equation linking the removal of light energy from a beam to the properties of the fineparticle in a system, such as that in Figure 4.5(a), using light scattering theory, the problem is considerably simplified by considering the situation in which no scattered light is picked up by the photoelectric device; in other words, instruments in which the solid angle α is vanishingly small. To achieve such a situation physically, one has to create a large distance between the sedimentation tank and the photoelectric device. Although this approach to photosedimentometer design facilitates attempts at theoretical calibration, it enormously increases the dimensions of the equipment and the cost of the appropriate optical components necessary to maintain the alignment of the optical system. In addition, because of the low power being received at the photoelectric device, problems associated with signal-to-noise ratio in the recording of the appropriate signals causes difficulties. Early equipment designed by Rose is typical of attempts to exclude scattered light from the receiving device to facilitate subsequent data interpretation [23]. This same worker attempted to derive a so-called universal calibration curve relating the scattering power of a fineparticle to its diameter uniquely for all powders studied with photosedimentometers. This universal calibration curve has, of course, proven to be an elusive dream since the power removed from the forward-traveling beam involves interference effects as well as diffraction effects, and depends on the refractive index of the fineparticle with respect to the fluid, and on the wavelength of the light as well as the geometry of the fineparticles involved. The calibration curve for correcting photosedimentation data described by Rose, was used in some early models of photosedimentation equipment but should now be considered to be of only historic interest.

The heavy theoretical consideration of light-scattering theory and its involvement in the early discussions of the design of photosedimentometers frightened many technologists away from the photosedimentometer. This was unfortunate since the technology of electro-optics is so advanced that it is possible to build efficient low cost monitoring equipment using photosedimentation principles, which are excellent for quality control and rapid generation of data in the research laboratory. Moreover, the problems of calibrating photosedimentometers are no more difficult than many of the problems of calibrating other physical procedures for characterizing fineparticles. In photosedimentation equipment one can work with very low concentration of solids in the suspension, thus circumventing all problems associated with fluid-fineparticle interaction. The photosedimentometer has the advantages that it is easily adapted for hostile environments and gives information on the largest fineparticle present in the suspension. In my opinion, the number of situations in which the relation between the power removed from a beam of light and the size characteristics of the fineparticles intercepting the light beam, can be calculated from theoretical considerations is very low. Therefore, one should adopt an empirical approach to the design of the photosedimentometer and thus should build the equipment to take maximum advantage of electro-optical equipment available and then calibrate the transmission signals of the light beam in terms of the solids concentration of the suspension. One should also use a white light system to integrate the effect of the many different wavelengths of light on the light scattering properties of the system.

The equation used in photosedimentation theory to link the opacity of a suspension to the properties of the suspension is the Lambert-Beer law. The Lambert-Beer law is an energy exponential decay law of the type that occurs in many physical situations. [Note that in some optical textbooks the law is called the Lambert law and in others, the Beer law, since different historians of science variously attribute the law to either the German scientist Johan Heinrich Lambert (1728–1777) or to the German astronomer Wilhelm Beer (1797–1850). The Lambert-Beer law states that

$$I_C = I_O e^{-CLM} \tag{4.11}$$

where
I_C = intensity of the beam after passing through the suspension
I_O = original intensity of the beam
e = the exponential function
C = concentration of the suspension
L = path length through the suspension
M = a constant that incorporates effects related to the solid angle subtended by the photoelectric detector, the spectral response of the detector, the geometry of the sedimentation tank, the nature of the walls and their thickness, and the optical characteristics of the beam, and generally involves the size of the fineparticles involved in the scattering of light from the beam.

The range of suspension concentrations for which the relationship is valid in any given photosedimentometer fineparticle system should always be investigated experimentally.

Equation 4.11 can be written in the form

$$\log_{10} \frac{I_C}{I_O} = -CLM \tag{4.12}$$

where **m** is a modified constant incorporating the essential variables present in the constant **M** of equation 4.12. The quantity $\log_{10} I_C/I_O$, is the logarithm of the transmittance of the suspension, this is replaced by the symbol γ. It follows that

$$\gamma = -CLM = -C\mu \tag{4.13}$$

where μ is a constant for a given instrument.

Equation 4.13 predicts that the measured values of γ for a series of suspensions of a given powder, plotted against concentration, will generate a straight-line relationship such as that illustrated in Figure 4.5(b). In this graph, the line marked d_g is a typical

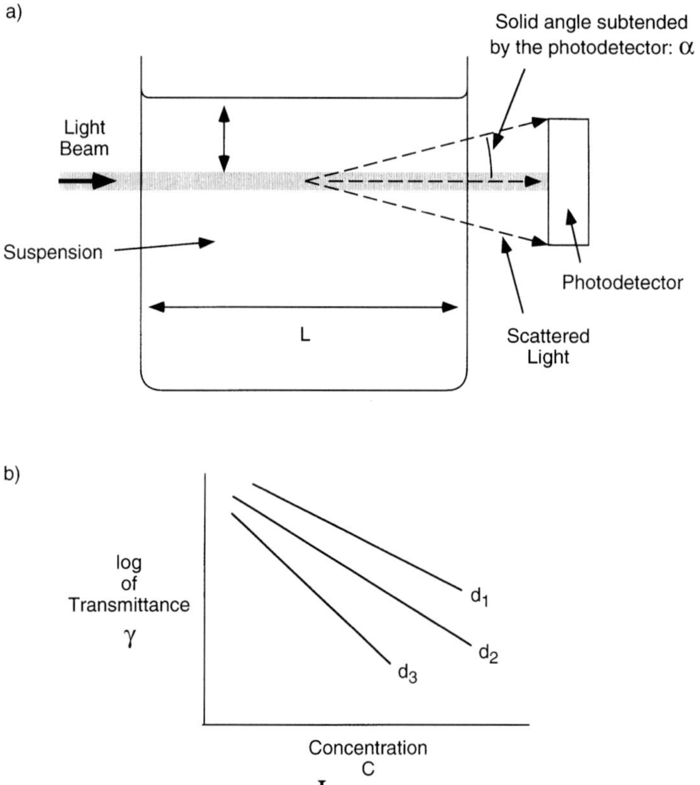

$$\log_{10}\frac{I_C}{I_O} = -CLM = \gamma$$

Figure 4.5. Photosedimentometers monitor the sedimenting of a suspension by monitoring the light intensity as the suspension sediments. a) A simple photosedimentometer showing the solid angle subtended by the photodetector. b) Light energy scattered for various suspension concentrations.

line obtained for a powder that contains fineparticles with a given range of Stokes diameters. If experiments were to be repeated with suspensions of fineparticles that all had the same Stokes diameter, a line of a different slope would be obtained. Thus for fineparticles that generally were coarser than $\mathbf{d}_g$, a line such as that labeled $\mathbf{d}_1$ would be obtained, whereas for fineparticles much smaller than $\mathbf{d}_g$, the slope would be similar to that labeled $\mathbf{d}_3$. Equation 4.13 can be written in the form

$$C = \left|\frac{\gamma}{\mu}\right| \tag{4.14}$$

where the symbol | | denotes the positive value of the enclosed symbols. This equation is used to deduce concentrations from measured changes in transmittance $\boldsymbol{\gamma}$ when we know that the change is due to the light scattering of fineparticles of known magnitude. When monitoring the opacity changes in a sedimenting suspension, it is usual to record the opacity of the suspension on a pen recorder to produce a data display of the type illustrated in Figure 4.6. For historic reasons it has become usual to arrange such transmission records in such a way that high intensity of transmission through the clear fluid before the fineparticles are added corresponds to the bottom position of the pen recorder. Therefore, in the chart shown in Figure 4.6(b) the line indicated as $\mathbf{I}_I$, is the light transmitted through the sedimentation tank full of clear fluid to be used in preparing the suspension. When the fineparticles are added to the sedimentation column, there is an immediate drop in the transmission of the light beam to the level $\mathbf{I}_C$ as shown. At time 0, indicated on the chart, the stirring of the suspension ceases and the strength of the transmitted beam remains the same until the largest fineparticle present in suspension moves past the monitoring zone, time $\mathbf{T}_L$ as indicated in the diagram. If the monitoring is being carried out at a depth **H** below the top of the fluid in the sedimentation column, then the Stokes diameter of the largest fineparticle present equals a constant **K** time $\sqrt{H/T}$. The constant **K** can be calculated from the known parameters of the suspensions such as the viscosity, the density of the suspending fluid, and the density of the fineparticles. In an ideal situation the transmission of the suspension drops to the initial value $\mathbf{I}_1$ at $\mathbf{T}_S$ after the smallest fineparticle has moved through the monitoring zone. Sometimes during the adding of the powder at the sedimentation column in this type of simple instrument the exterior of the sedimentation tank is smeared, in which case there is an error introduced into the transmission of the empty cell with a corresponding failure of the signal to return to the original value of $\mathbf{I}_1$ at time $\mathbf{T}_S$.

The first stage in the transformation of the transmission-time data is to construct a graph of the log of the transmittance against Stokes diameter using $\log_{10}(\mathbf{I}_{TH}/\mathbf{I}_I) = \boldsymbol{\gamma}$. Consider an element of the $\boldsymbol{\gamma}$/**Stokes** diameter curve bounded by the diameters **d'** and

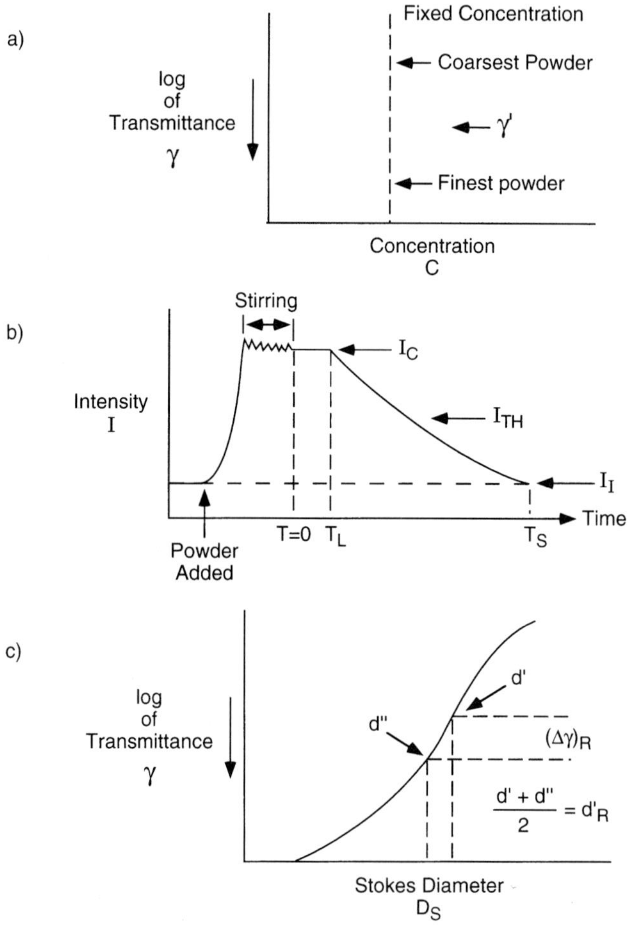

Figure 4.6. Steps taken to determine the Stokes diameter distribution from photosedimentation data. a) Calibration curve for turbidometric quality control. b) Opacity-time curve for a settling suspension. c) log of Transmittance-Stokes diameter curve.

d", it can now be assumed that the drop $(\boldsymbol{\Delta\gamma})_R$ is caused by fineparticles with Stokes diameters in the range **d' → d"**, which can be considered to have an average diameter of $\mathbf{d}_R$. To transform $(\boldsymbol{\Delta\gamma})_R$ into an estimate of the concentration of fineparticles having Stokes diameters of size $\mathbf{d}_R$, one must either know or make some assumptions about its dependence on $\mathbf{d}_R$. Assume that calibration studies using fineparticles of size range **d' → d"** have established that $\boldsymbol{\mu}_R$ and $\mathbf{d}_R$ are known for a range of values of $\mathbf{d}_R$. The weight of a powder of average Stokes diameter $\mathbf{d}_R$ is given by $(\boldsymbol{\Delta\gamma})_{R/}\boldsymbol{\mu}_R$ and the total weight of the powder is

$$W_T = \sum_{d_S}^{d_L} \frac{(\Delta\gamma)_R}{\mu_R} \tag{4.15}$$

Using these relationships, the Stokes diameter distribution can be deduced. For many control procedures, it is quite adequate to assume $\mu_R = 1$ because the technologist is usually looking for differences between powders, not their absolute fineness.

The most straightforward way of evaluating the value of μ_R is to prepare a set of small-diameter-range powders of various average size and then μ_P is measured by plotting a graph of γ against concentration for each powder. Unfortunately, it is not easy to fractionate powders accurately (see discussion of fractionation efficiency curves for elutriators in Chapter 5). In the absence of closely sized fraction of powder it is common practice to calibrate photosedimentation data by using a powder whose size distribution function has been established by another characterization procedure [24, 25].

In the development of photosedimentometers it was soon discovered that because the Stokes diameter of fineparticles was related to the square root of the reciprocal of the settling time required to reach a fixed monitoring zone at a depth H below the surface of the suspension, the use of a monitoring beam at a fixed depth required very long sedimentation times for the characterization of a wide-range powder. In 1958 Harner and Musgrave of the Eagle Pincher Company described a photosedimentometer in which the sedimentation tank was mounted on a platform and moved up and down through the monitoring beam [26]. Although it was mechanically easier to move the tank rather than the somewhat cumbersome optical system used in these early photosedimentometers, there was always a danger of jolting the sedimentation tank so that fluid spilled over the side of the tank, resulting in a smear down the side of the tank. The sensitivity of the early photosedimentometers was limited by the problems of keeping the light beam at constant power level throughout the characterization of the fineparticles. Any attempt to improve the thermal insulation of the system with accompanying longer sedimentation times usually engendered problems of drift in the power levels of the light and in the response of the selenium barrier layer photocells used in the early instruments. Usually the only choice open to the scientists was to check the power level of the beam at the beginning and the end of the experiment and to replace the bulb or try again if the power level had shifted. Large constant voltage transformers and/or power supplies were used in some of the early instruments in attempts to stabilize the power level of the lamp. For the same reason, the lamp had to be switched on for some time before the commencement of a characterization study to allow the lamp and its housing to come into thermal equilibrium with their surroundings. When using an incandescent lamp source in a photosedimentometer system, it is advisable to place an infrared filter in the path of the light beam to remove radiation that could upset the thermal equilibrium of the sedimentation column.

Heywood developed a system with optics carried on a board capable of swinging in an arc. Several sedimentation tanks were then placed in an arc and monitored with the same optical system. Rose introduced the use of an outer tank with water as a thermal insulation device. Three commercial instruments evolved out of the work of Rose. One of them is known as the Evans Electroselenium (EEL) photosedimentometer (the acronym is derived from the initial letters of the company making the equipment). In the EEL photosedimentometer the optical system was placed on a small trolley so that it can be moved along a line of sedimentation tanks for sedimentation to be monitored concurrently in a series of tanks. The second commercial system based on the work of Rose was the Boundbrook photosedimentometer developed by Morgan [27, 28]. The name of this equipment comes from the fact that Morgan was associated with Boundbrook Bearings Company. The Boundbrook photosedimentometer was the first to use matched photoelectric cells in an attempt to cope with the problem of optical drift in the equipment. Light from an incandescent electric bulb was split into two beams. One went through the sedimentation tank and the other through a dummy glass tank filled with clear sedimentation fluid to match the opacity of the two beams so that a balanced system could be employed in the measurement of the changes in transmission of the sedimenting suspension. The Boundbrook photosedimentometer used mechanical scanning of the sedimentation tank through the light beam. The Boundbrook photosedimentometer was further developed by the Goring Kerr Company [29]. This equipment proved to be useful in the quality control of metal powders. Another commercial photosedimentometer is available from the Micromeritics Corporation [30].

Once experimental studies started to indicate that the calibration of photosedimentometers could be facilitated by using large solid angles at the photoelectric detector with respect to the sedimentation tank, the basic design of photosedimentometers started to change [25]. In the Wide-Angle Scanning Photosedimentometer (WASP) described by Allen, the photocell used to monitor the transmission of light through the sedimentation column is placed as close to the column as is convenient from mechanical considerations [2]. In this equipment a reference photocell is used to monitor the light beam and correct for any optical drift that might occur over the duration of the characterization study. Again, mechanical movement of the sedimentation column is used to accelerate the collection of experimental data [31].

The availability of fiber optics and cheap solid-state integrated circuits made it possible to start considering systems having any desired optical configuration in which the sedimentation column is scanned optically over great heights without any moving parts [32]. Davies and co-workers described an instrument in which fiber optics were used to take the light beam to the sedimentation tank. An arc of fiber optics was placed around the periphery of the horizontal plane through the sedimentation tank to pick up wide-angle scattered light [33].

Naylor and Kaye have used a fiber optic-based system to make measurements on a sedimentation column of up to one meter in height. This procedures facilitates rapid collection of information on the largest and smallest fineparticles present in an industrial suspension. Light is piped to the outside of the sedimentation column in a fiber optic bundle containing many hundreds of small hair like fibers. En route to the sedimentation tank a small subset of the fibers is twisted out of the main fiber and taken to an interface with a solid state detector device. Another fiber optic bundle is placed on the opposite side of the sedimentation column and used to monitor transmitted light. This second fiber optic bundle is also interfaced with the solid-state detector device. This solid state device is able to indicate in real time the ratio of light energy going to the tank and light energy leaving the tank. An additional integrated circuit can be used to give directly the logarithm of the transmittance of the output. Alternatively, the output can be connected directly to a minicomputer. The use of the solid-state ratio circuit means that all problems related to unstable light sources have been eliminated. This enables the use of low-cost incandescent light sources when designing photosedimentation equipment. The equipment can be plugged into any available power supply. Light beams leaving a fiber optic bundle are slightly divergent; this has to be taken into account in assessing the performance of the instrument. Fiber optic bundles can be placed at any particular height and the whole of the sedimentometer tank plus fiber optic assembly potted in a matrix of expanded plastic foam, thus achieving mechanical robustness and good thermal insulation. The whole assembly is so robust that dropping the equipment has no effect on its performance. Measurements can be rapidly taken at any height in the tank.

To avoid the problems of interpreting the data on concentration changes monitored by means of light beams, several workers turned to the use of X-rays to monitor the changes occurring within a settling suspension [34, 35, 36].

4.3 Sedimentation Characterization Based on Cumulative Monitoring of Sediments from an Initially Homogeneous Suspension

In the sedimentation characterization based on cumulative monitoring of sediments from an initially homogeneous suspension a record of the weight of sediment $\mathbf{W}_t$ at time **t** is used to deduce the Stokes diameter distribution of the suspended fineparticles. At time **t**, the sediment contains all fineparticles greater than $\mathbf{d}_{Ht}$ (where $\mathbf{d}_{Ht}$ is the Stokes diameter calculated from **H**, the height of the settling chamber, and **t** and the appropriate values of the other parameters of the suspension) and a portion of the fineparticles less than $\mathbf{d}_{Ht}$. If $(\mathbf{dW/dT})_t$ is the rate at which fineparticles are arriving at the base of the column at time **t**, then the amount of fineparticles already present in the slurry smaller than $\mathbf{d}_{Ht}$ is given by the relationship:

$$\left(\frac{dW}{dt}\right) \cdot t \tag{4.16}$$

Therefore, the weight of fineparticles greater than $\mathbf{d}_{Ht}$ is given by

$$W_{Ht} = W_t - \left(\frac{dW}{dt}\right) \cdot t \tag{4.17}$$

If now, $\mathbf{W}_t$ is the total weight of fineparticles in suspension, then

$$R_{Ht} = \left(\frac{W_t}{W_T}\right) \times 100 - \left(\frac{dW}{dt}\right)_t \cdot \frac{t}{W_t} \cdot 100 \tag{4.18}$$

where $\mathbf{R}_{Ht}$ is the weight percentage of fineparticles greater than $\mathbf{d}_{Ht}$.

In the 1960's and 1970's, this group of methods were intensively studied and several commercially available instruments were developed but they are no longer in widespread use. Interested readers can find a detailed survey of the method in references 1 and 2.

4.4 Line Start Methods of Sedimentation Fineparticle Size Characterization

In this group of methods a thin layer of a suspension of the fineparticles to be characterized is floated onto a column of fluid through which the fineparticles settle. The settling fineparticles can be monitored by passing a light beam or x-ray beam through the column of initially clear fluid and the amount of the fineparticles of all the base of the column. In the days before the availability of huge power, low cost, data processing equipment the big attraction of this suspension configuration, particularly with centrifugal sizing equipment, was the simplicity of the data processing required with the line start procedures since each size of sedimenting fineparticle arrived at the monitoring point in sequence. Because of this attractive feature of the line start method to question of the stability of the suspension layer floated onto the top of the column, which was the source of considerable controversy, received intense study. Basically, one has a choice. The first choice facing investigators was that one could work with low solids concentration in the layer of suspension so that its density was not sufficient to cause gravity driven streamers of suspension to travel down through the column of fluid to create pseudo two particles apparently having large stokes diameter than the real fineparticles in the suspension. Data on gravity driven streaming errors in line start sedimentation techniques for various layer solids concentration in water were reported by Kaye and Davies [11]. They used a glass powder suspension in water floated onto a column of water. The fineparticle sedimentation dynamics were monitored and recorded on a balance pan suspended at the bottom of the clear column of fluid. Their data is summarized in Figure 4.7(a)

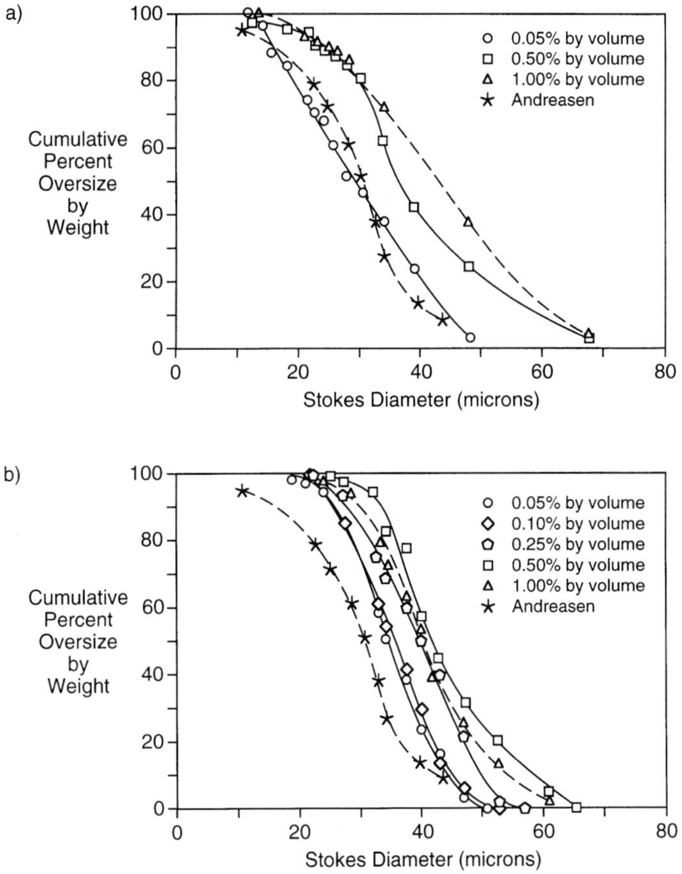

Figure 4.7. Stokes diameter distribution data obtained for line start sedimentation of a glass powder performed at various concentrations. a) Data for a water suspension floated onto a water column. b) Data for a water suspension placed on a brine column.

It can be seen that the measured size distribution obtained using 0.05 % solids, agrees tolerably well with the Andreasen data. Thus the median diameter for the two sets of data are 28 and 29 μm, respectively. However, the two sets of data for suspension concentrations of 0.5 and 1.00 percent by volume solids demonstrate a marked apparent coarsening in the measured Stokes diameter distribution. Thus the shift in the median diameter for the three sets of data is 28, 37, 45 μm. There is also a large increase in the measured number of apparent large fineparticles which is obviously an artifact of the measurement procedure. It represents the effect of suspension streamers moving rapidly down the clear fluid resulting in the registering of pseudo-fineparticles.

Various investigators claimed that suspension streamers could be prevented by using a lower density fluid in the suspension layer as compared to a miscible liquid of higher density in the main column of fluid. Such a strategy will often suppress visible streams but the clusters leave the layer of suspensions can still give inaccurate results. This is illustrated by another set of data reported by Kaye and Davies summarized in Figure 4.7(b). In this set of experiments Kaye and Davies used distilled water plus a dispersing agent, to prepare the suspension which was floated onto a brine solution with a density of 1.18 g cm^{-3}. From the data of Figure 4.7, it can be seen that the Stokes diameter distribution measured at the lowest suspension concentration of 0.05 % is coarser than for the simple line start procedure using water in the sedimentation column shown in Figure 4.7(a). The largest effective size measured using a different liquid to prepare the layer of suspension is lower than that for the all-water system. This is consistent with the hypothesis that the use of a different fluid in the line start layer suppresses gross streaming instability. It can be seen that the measured distributions of Figure 4.7(b) are generally deficient in fines compared to those in Figure 4.7(a). This is indicative of the fact that the streams are composed of dynamic clusters of fineparticles and that streams, even if not particularly visible in a fluid medium different from that of the fluid through which they move probably retain trapped smaller fineparticles that circulate with the larger members of the cluster. As will be discussed in greater detail later in this chapter, line start methods using lower density fluids to create the suspension layer are widely used in disc centrifugal methods and some investigators do not always check that accelerated sedimentation, because of dynamic clustering streamer formation, is not occurring in their system.

One of the earliest line start sedimentation methods was devised by Werner [37]. His system is shown in Figure 4.8(a). An external water jacket (not shown) around the sedimentation column is used for thermal protection of the equipment. The sedimentation column has a tapered base, the final length of which is a narrow-bore tube. The suspension of fineparticles to be characterized forms a narrow layer on top of the column. The amount of powder arriving at the bottom of the column at different periods of time is monitored by measuring the height of the sediment in the narrow-bore tube. It is assumed that the height of the sediment is proportional to the weight

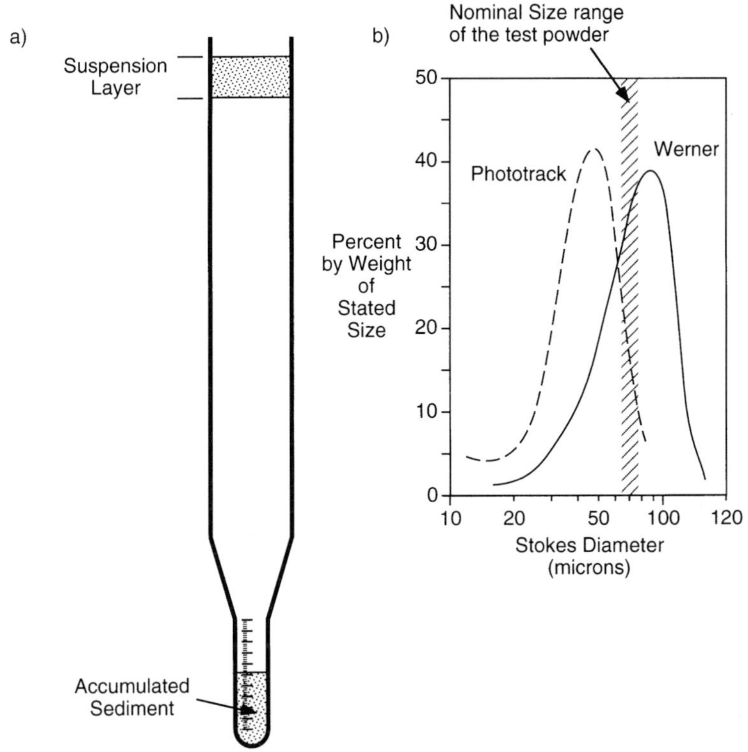

Figure 4.8. Appearance of the Werner sedimentation tube along with data comparing the performance of the Werner tube to the Phototrack equipment.

of powder that has sedimented at any given time. There are several problems associated with this equipment. First, some of the fineparticles sediment onto the shoulders of the tube. When sufficient quantities have accumulated, they slide down into the base of the tube, and the exact time of their arrival in the column is not related to their size. Sludge dumping from the shoulders creates convection currents, thus delaying the arrival of some of the fineparticles caught in the upward-moving portion of the convection currents.

The justification for relating the height of the column of sediment to the amount of the fineparticles arriving at the base is that, along the length of the sediment column, fineparticles in contact with each other are approximately the same size, and it has been shown that, in general, monosized fineparticles randomly packed to the same porosity irrespective of their absolute size for a wide range of sizes. The accumulation of the sediment in the base of the column can consolidate the lower levels of the sediment, thus causing error in the estimate of powder quantities.

If the fluid used to prepare the suspension is the same as that in the column, the layer of suspension will have a higher density than the column fluid. In this situation, as discussed earlier in this section, the assembled line start system in the column will be essentially unstable and tendrils of suspension will tend to "bleed" down the walls of the tube. Allen comments on the Werner instrument,

> *"Some of the suspensions settles en masse in the form of pockets of fineparticles which fall rapidly through the clear liquid leaving a trail of fineparticles" [2].*

The distortion with the measured size distribution caused by the instability of the line start system in the Werner tube is illustrated by the data of Figure 4.8(b) reported by Stairman. He prepared a fraction of sand using sieves. The powder passed through a sieve with apertures of size 76 μm and retained by 66 μm. The size distribution by sedimentation techniques first sized with an instrument known as a Phototrack and then by the Werner tube technique. The Phototrack starts with a very dilute homogeneous suspension and records the settling velocities of the powder grains by recording their path using a light beam and photographic film so that the measured size distribution is not affected by any fineparticle-fineparticle interaction in the sedimenting suspension. As can be seen from the data of Figure 4.8(b), the distribution obtained with the Werner tube is much coarser than the nominal size range of the powder and the Phototrack determined data.

Although the Werner tube is not widely used in modern fineparticle science, the Mine Safety Association (MSA) particle size analyzer developed by Whitby and coworkers used a miniature version of the Werner sedimentation tube that can be used in a centrifuge tube [38, 39]. The MSA equipment was used to characterize fineparticles by a combination of gravity and centrifugal sedimentation, and problems similar to those encountered when using the Werner sedimentation tube occurred with the MSA equipment. (The company discontinued making the equipment in the 1970's but many health and safety publications dating from the 1950's contain data on hazardous dust generated using the equipment.)

A sedimentation technique which used a line start procedure to characterize metal powders, called the Micromerograph was widely used between the late 1950's and the late 1980's. It is no longer in use but many scientific studies of powder metals and ceramic powder technologies used data generated by the procedure [40]. The name derives from the fact that the term "Micromeritics", literally, "small measurements", an earlier term for powder technology. The name is still used in the name of the company which makes particle size determination equipment [30]. The basic system of the Micromerograph is shown in Figure 4.9. The powder to be characterized is dispersed at the top of the sedimentation column, usually containing dry nitrogen, by blowing a jet of high-velocity gas through powder placed in the sample holder. Theoretically, the cloud of fineparticles formed in this manner constitutes a thin layer of

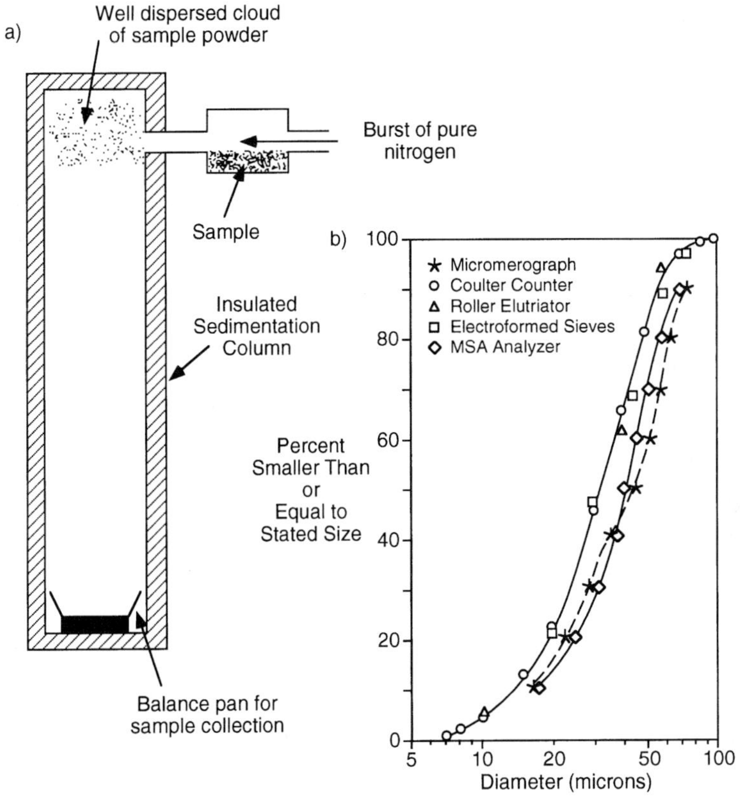

Figure 4.9. The Micromerograph system uses a short burst of pure nitrogen gas to create a well dispersed cloud of the sample powder which forms what is essentially a line start sedimentation system. a) Appearance of the Micromerograph system. b) Comparison of several system with data for the Micromerograph.

suspension at the top of the column. The airflow through the sample holder is designed in such a way that the powder is deagglomerated by the aerodynamic shear forces in the narrow feed channel. The fineparticles are collected on the balance pan of a sensitive electronic weighing device placed at the base of the column [40]

Studies of the configuration of the injected cloud using high-speed photography showed that the cloud of fineparticles moved down one seventh of the column before a stable cloud was obtained [41]. This failure of the initial cloud to remain in a narrow region at the top of the column would cause the real height of fall to be less than the column height. This factor would cause the measured size distribution to appear coarser than it should be.

A source of difficulties that arise in interpreting the experimental data obtained from a Micromerograph is that the total weight eventually recorded by the balance does not always equal the weight of powder dispensed into the column. It is usually assumed that the missing powder is attached to the walls of the sedimentation column by electrostatic force. Although the settling column is made of aluminum, aluminum surfaces are usually covered by a thin layer of oxide so that in essence the surface of the settling column has insulating properties with respect to the settling dust fineparticles. Therefore, electrostatic charging could cause them to adhere to the sides of the sedimentation column. It has been suggested that the use of anti-static agents improved the accuracy attainable with the Micromerograph [42, 43].

If the difference between the powder that settles onto the balance pan and the initial sample weight is small, it was assumed that the powder lost was typical of the whole powder. However, if the weight discrepancy is greater that 5 %, any assumptions made concerning the size distribution of the missing powder had to be checked at least once by an examination of feed powder and that lost to the walls of the equipment by microscopic examination. If it was assumed that the powder lost is representative of the entire powder, then the total weight of powder retained ultimately on the balance pan was used to calculate the percentage weight on the pan at any specific time representing the portion of powder greater than a given Stokes diameter.

Sometimes the weight on the pan changed exponentially, and experimenters were tempted to take a short cut and extrapolate to the ultimate weight retained at infinite time by graphical techniques. If such an extrapolation was coupled with assumptions concerning the missing powder, the absolute accuracy of the ultimately derived Stokes diameter distribution of the powder was questionable.

An operational disadvantage of the Micromerograph was that the sedimentation column was usually cleaned out at the end of a characterization study by a blast of gas. If the powder had any toxic attributes, it was very difficult to contain the emergent cloud.

Ullrich [44] compared the performance of the Micromerograph with other methods of characterizing fineparticles. He used a spherical aluminum powder so that the data from the various characterization procedures using different physical principles could be correlated. His data, summarized in Figure 4.9(b) showed that data from the Coulter counter (described in Chapter 6) the Roller elutriator (described in Chapter 5), and the electroformed sieves correlate very well, considering the very different physical parameters measured in these three methods of characterization. The data obtained with the MSA analyzer and the Micromerograph differed markedly from the other three sets of data. If there had been no systematic error involved in the use of the line start procedure, the MSA data and the Micromerograph data should have been directly comparable to the elutriator-generated data. In fact they exhibit a shift of the order of 30 % on the measured median diameter compared to that from the Roller elutriator. Moreover, the Micromerograph data compare directly with the MSA equipment because they both suffered from the instability of the line start suspension system used in the experimental study.

4.5 Sedimentation Studies of Fineparticles Moving in a Centrifugal Force Field

It can be shown that

$$V_c = \frac{\omega^2 x}{g} \cdot V_g \tag{4.19}$$

where

V_g = the terminal velocity under gravitational forces in the same fluid
V_c = velocity of fineparticle along radius of centrifuge
x = distance of fineparticle from axis of revolution
D_s = Stokes diameter of fineparticle
ω = angular velocity of centrifuge (radians per second)

A laboratory centrifuge in which test tubes are hung at the end of radial arms able to spin about a central axis is termed a tube centrifuge. When the centrifuge is switched on, these tubes assume a horizontal position as they rotate rapidly. The earliest attempts to characterize the Stokes diameters of fineparticles by centrifugal sedimentation used tube centrifuges. In a characterization study, usually the centrifuge has to be taken up to speed, run for a set time at top speed, and then stopped to enable determination of the movement of fineparticles in the tube. The starting and stopping of the centrifuge results in a varied centrifugal acceleration experienced by the fineparticles during the time that the centrifuge is operating. Although one can calculate the force at any given time at any given position in the centrifuge and integrate the overall effect on the trajectory of the fineparticle, this is not a sufficient correction for the fact that during the acceleration and deceleration the fluid in the tube is made to spin by the force exerted in the centrifuge. This fluid spinning can be demonstrated by viewing with a stroboscopic light, an open centrifuge equipped with transparent tubes. With the stroboscope adjusted to the speed of the centrifuge, one can note the time at which the fineparticles are all completely sedimented onto the base of the tube. If the centrifuge is switched off at this time, the swirling currents caused by the deceleration of the centrifuge will resuspend fineparticles lodged on the bottom of the tube.

The second disadvantage of a tube centrifuge is that the fineparticles follow radial trajectories. Therefore, sedimenting fineparticles impinge on the side walls of the centrifuge tube and either adhere to the sides or initiate currents down the sides and up the middle of the tube as accumulated fineparticles on the walls slip down into the base of the centrifuge tube.

Tube centrifuge characterization systems have been used for quality control of fineparticle systems; however, any variation in the experimental technique usually leads to shifts in the measured Stokes diameter distributions.

To avoid the problems associated with tube centrifuges, many investigators have used shallow bowl, or disk centrifuges, in which the fluid supporting the fineparticles is centrifuged in a hollow disk rotating about its major axis. (See discussion of these instruments in Reference 1.) Problems associated with radial wall impingement are eliminated by such centrifuge design. To avoid the problems associated with the stopping and starting of the centrifuge, suspension to be characterized can be injected into the centrifuge already running at a steady speed. The many experimental difficulties associated with such injection procedures are a major area of controversy in the methodology of centrifugal sedimentation procedures.

The dependence of the instantaneous velocity of a fineparticle sedimenting under centrifugal force on its distance from the center of rotation makes the interpretation of the dynamics of an initially homogeneous suspension complex. To simplify data interpretation in centrifugal sedimentation studies, long-arm centrifuges have been developed so that the total movement of the fineparticles from $\mathbf{x}_1$ to $\mathbf{x}_2$ the initial and final position of the fineparticle is such that the difference between $\boldsymbol{\omega}^2\mathbf{x}_1$ and $\boldsymbol{\omega}^2\mathbf{x}_2$ can be neglected. Long-arm centrifuges have not gained wide acceptance, probably because of the mechanical problems of building stable centrifuges of large dimensions within acceptable cost limitations [45].

An alternate strategy which can be used to simplify the data process from a centrifugal experiment is to use a line start procedure. In the line start procedure, since all the fineparticles start at essentially the same distance from the center of rotation of the centrifuge, the location of the fineparticles at any time **t** after the commencement of centrifugation can be calculated by using equation 4.19. Let all the fineparticles start at a distance $\mathbf{H}_1$. By integrating equation 4.19 over the distance $\mathbf{H}_1 \rightarrow \mathbf{H}_2$, it can be shown that all fineparticles greater than

$$D_{Ht} = \sqrt{\frac{18\eta \cdot \ln(H_2 / H_1)}{(\rho_p - \rho_1)\omega^2 t}} \tag{4.20}$$

will have reached the position $\mathbf{H}_2$ at time **t**.

When H_2 represents the base of the centrifuge, then D_{ht} is the Stokes diameter of the fineparticles that will just have reached the base by time **t**. Let W_t be the weight deposited on the base at time **t**; then

$$R_{H_2t} = \frac{W_t}{W_T} \times 100 \qquad (4.21)$$

where W_T is the total weight of powder in suspension and R_{H2t} is the percentage of fineparticles greater than D_{ht}. By studying the rate of accumulation of sediment at the base of the centrifuge, equations 4.20 and 4.21 can be used to calculate the Stokes diameter distribution of the sedimenting fineparticles.

The first line start sedimentation procedure based on centrifugal sedimentation was described by Marshall [46], who was aware of the problems involved in floating a layer of suspension onto a column of clear fluid because of reported difficulties with the Werner sedimentation tube discussed in the previous section. He attempted to achieve a stable suspension system by using a suspension of lower density than the liquid column through which the fineparticles were to be sedimented. He used a sucrose solution for the column fluid and a water-based suspension of fineparticles. This strategy as discussed earlier only eliminates gross inversion of the layer of suspension.

Many systems based on a line start suspension system, with incremental monitoring of the sedimenting fineparticles by use of a light beam, have been described. All the problems associated with the extinction coefficient, the solid angle subtended at the photoelectric detector, and the wavelength of the scattered radiation discussed in connection with gravity photosedimentometers are encountered in centrifugal photosedimentation techniques. Again, as in the case of gravity photosedimentometers, there are two approaches to the problem of interpreting the optical signals in terms of fineparticle concentrations. One relies on empirical calibration of the transmission signal in terms of true concentrations of fineparticles, and the other attempts to calculate the scattering power of the fineparticles from the light-scattering theory. In the case of centrifugal sedimentation, many of the fineparticles are of the order of the wavelength of light or less and the theoretical calculations become very difficult. In particular, as one considers the fineparticles down to sizes smaller than 0.1 mm Stokes diameter, the light scattering power of the fineparticles is very small. This factor tends to create a lower size limit below which the characterization procedure is not useful.

The basic system of a centrifugal disk photosedimentometer is illustrated in Figure 4.10(a). Movement of fineparticles in the spinning disk is monitored by the light beam as shown. In a line start procedure the disk is run up to speed partially filled with clear fluid termed the spin fluid. Under centrifugal forces the free surface of the fluid becomes a cylindrical surface coaxial with the axis of the rotation. The suspension to be characterized is injected into the centrifuge through the axial entry port shown in Figure

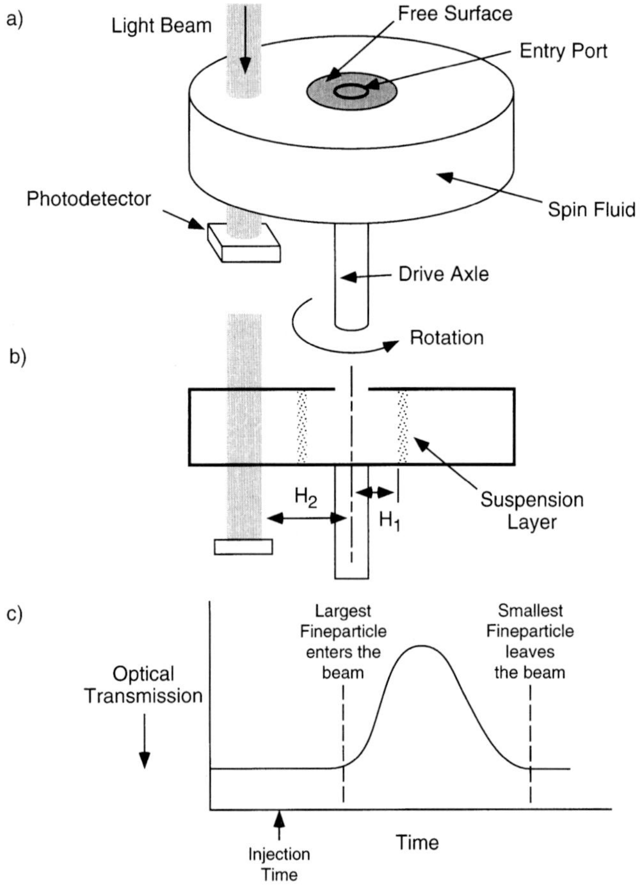

Figure 4.10. A basic photocentridisc, as used to perform line start sedimentation studies, has a thin layer of suspension added onto the free surface of the fluid in the spinning disc. a) Appearance of the photocentridisc. b) Side view of the equipment showing the suspension layer. c) Typical signal output from a photocentridisc.

4.10(a). The cylindrical free surface of the spin fluid is moving with high energy so that the injected suspension seldom penetrates the surface before it takes up the speed of the centrifuge. In this short time period it spreads out into a thin-layer configuration, as illustrated in Figure 4.10(b). Because the volume of any particular monitoring zone is larger than a comparable width of suspension at the line start position, the dilution of the fineparticles as they move out from the center has to be accounted for in the transformation of the experimental data if more than one interrogation zone is used during a characterization study. Also, the effect of the width of the beam of light used to monitor the movement of the fineparticles has to be taken into account [47, 48, 49].

In general, it can be stated that most of the experimental work reported in the literature on the performance of centrifugal disk photosedimentometers using white light monitors and a line start strategy has demonstrated that for such equipment, it is often reasonable to take the log of the transmittance of a finite-width monitoring beam as being directly proportional to the weight of the fineparticles in the beam.

Line start-incremental monitoring system centrifugal disk photosedimentometers yield direct information on the largest and smallest fineparticle present in system. They can be operated in the fingerprint mode; that is, the pen recorder traces for good and bad powders in a quality control situation can be established and compared directly to the traces generated by powder samples taken from production run.

The first disc centrifuge using a white light beam to monitor the movement of the fineparticles was described by Kaye and co-workers. In their prototype equipment a hollow Lucite disk was mounted inside a horizontal flywheel mounted on the axis of an ordinary laboratory centrifuge. The equipment was placed in a large wooden enclosure, both as a safety precaution and also to increase the thermal stability of the immediate environment of the centridisk. Suspension could be injected through an axial entry port in the top surface of the disk [49].

In the original photocentridisc design studies undertaken by Kaye and co-workers, some consideration was given to the possibility of using a servomechanism to achieve stability in the rotation of the disk. However, it was felt that the approach of adding mass to the system in the form of a flywheel not only increased the stability of the system, but avoided problems that could arise from the operation of the servomechanism, which creates transient speed changes as the spinning disk responds to external controls. In all of the instruments with which Kaye has been directly associated, this approach of using high-mass systems to achieve stability in the equipment has been a basic feature of the instrument. It is possible that some of the discrepancies between the performances of the instruments of high mass versus those designed by other workers using feedback control to achieve stability may be due partly to this different approach to the problem of operating at constant speed. Moreover, by operating with a high-mass centrifuge system, any disturbances caused by the injection of small amounts of suspension are minimized. A massive flywheel also acts as a thermal ballast to improve the thermal stability of the equipment. The movement of fineparticles in the Lucite disk was monitored by a beam of white light passing through slots cut in the base of the flywheel.

A more sophisticated photocentridisc based on the principles used in the original design evolved by Kaye and co-workers was constructed at the Atomic Weapons Research Establishment, Great Britain. It is also known as the AWRE photocentridisk, or simply as the AWRE centrifuge [50]. This instrument has been used extensively by

Burt [48], who has published excellent experimental data on the performance of the instrument. The centrifugal chamber used in the AWRE centrifuge is a composite disk made of glass and metal. The transparent portion of the tank for the light beam consists of arcs cut in the surfaces of the tank and sealed with optical flat glass strips. Entry to the tank is through an axial entry port. The lower axis of the centrifuge has a flywheel to enhance the stability of the rotating system and to minimize the effect of added mass when suspension is injected into the tank. The light beam used to monitor the movement of fineparticles is interrupted by a mechanical chopper that alternately diverts the beam of white light through the sedimentation tank or through an optical wedge filter.

The opacity of the wedge filter is graded so that the linear angular rotation of the wedge corresponds to linear increase in the log of the transmittance of the opacity of a suspension being monitored. If the opacity of the suspension in the disk centrifuge differs from that of the optical wedge, an alternating signal is generated at the monitoring photocell, thus initiating rotation of the optical wedge until its opacity matches that of the tank. By using this type of servomechanism, all problems due to instability

Table 4.2. Comparative data for a titanium dioxide powder obtained from the AWRE centrifuge using various suspension concentrations and experimental methods.

	Suspension Concentration	Sample Volume	Equivalent Stokes Diameter μm at Percent by Weight Undersize of				
Method	**Percent by Volume**	**ml**	**95**	**84**	**50**	**16**	**5**
Three Layer	0.05	1	1.05	0.94	0.78	0.60	0.45
	0.05	2	1.02	0.93	0.77	0.56	0.38
	0.05	3	1.08	0.95	0.78	0.57	0.37
	0.025	4	0.99	0.90	0.75	0.58	0.42
	0.025	6	1.03	0.93	0.76	0.55	0.36
	0.025	8	1.06	0.94	0.74	0.54	0.38
	0.01	2	1.02	0.92	0.75	0.53	0.37
	0.02	2	0.97	0.87	0.74	0.56	0.38
	0.04	2	0.98	0.93	0.79	0.61	0.49
	0.08	2	1.03	0.91	0.74	0.50	0.30
Two Layer							
Run 1	0.0005	20	0.97	0.89	0.73	0.50	0.32
Run 2	0.0005	20	0.98	0.91	0.74	0.52	0.34
Run 3	0.0005	20	1.01	0.92	0.74	0.52	0.35

The injected suspension is warmed a few degrees so that its density is lower than the spin fluid which is distilled water containing 0.1 % dispersing agent. 20 ml of suspension is injected resulting in a 6 mm line start layer. The centrifuge runs at 1500 rpm.

The three layer system uses spin fluid consisting of distilled water with 0.1 % dispersing agent, a middle layer consisting of 2 ml of a 50 : 50 mixture of ethanol and distilled water with 0.1 % dispersing agent, and the injected suspension of fineparticles in an 80 : 20 mixture of ethanol and distilled water with 0.1 % dispersing agent.

of the light source and potential drift in the performance of the photoelectric devices are avoided.

The use of the optical wedge filter has the further advantage that since its angular motion is directly proportional to the log of the transmittance movement, the wedge can be used to drive a pen recorder so that linear motion of the pen is proportional to the weight of fineparticles in the beam.

In Tables 4.2 and 4.3 some of the data obtained by Burt using the line start procedure with AWRE photocentridisk are summarized. In all experimental work Burt used volume concentrations in line start suspension below that required to achieve stable line start systems, and this concentration limit was established by a series of experiments at various concentrations. In Figure 4.11(a) data obtained by Burt with the AWRE centrifuge for a titanium dioxide powder is compared with data obtained by V. T. Crowl of the Research Association of British Paint Colour and Varnish Manufacturers by direct measurement of individual fineparticles imaged with the electron microscope.

The data in Figure 4.11(b) show that the use of white light centrifuge disk photosedimentometer system for a system of spherical particles with the data transformation used in this system agrees very closely with the size distribution function of the spheres as measured by microscopic examination of fineparticles. The techniques

Table 4.3. Comparative data for an explosive powder (HMX) obtained from the AWRE centrifuge using various suspension concentrations and experimental methods.

	Suspension Concentration	Sample Volume	Run	Equivalent Stokes Diameter mm at Percent by Weight Undersize of				
Method	Percent by Volume	ml		95	84	50	16	5
Three Layer	1.0	2	1	23.5	18.0	7.7	3.3	2.3
			2	20.5	14.0	6.0	3.0	2.2
			3	23.0	14.0	5.3	3.0	2.2
			4	22.5	15.5	6.1	3.1	2.3
Two Layer	0.05	20	1	27.0	18.5	7.2	3.1	2.3
			2	26.5	18.0	7.4	3.1	2.3
			3	26.0	17.5	7.2	3.1	2.2
			4	27.0	18.5	7.2	3.1	2.2

Microscopic examination of the powder indicated that fineparticle range up to about 35 mm. The two layer data showed the maximum diameter to be 32–34 μm, while the three layer method gave a maximum diameter of 26–32 μm.

The centrifuge was run at 750 rpm with a spin fluid consisting of a 52 : 48 glycerol and distilled water mixture with 0.01 % dispersing agent.

The three layer system used a middle layer consisting of 2 ml of a 30 : 70 glycerol and distilled water mixture with 0.01 % dispersing agent, and an injected suspension of fineparticles in a 20 : 80 glycerol and distilled water mixture with 0.01 % dispersing agent.

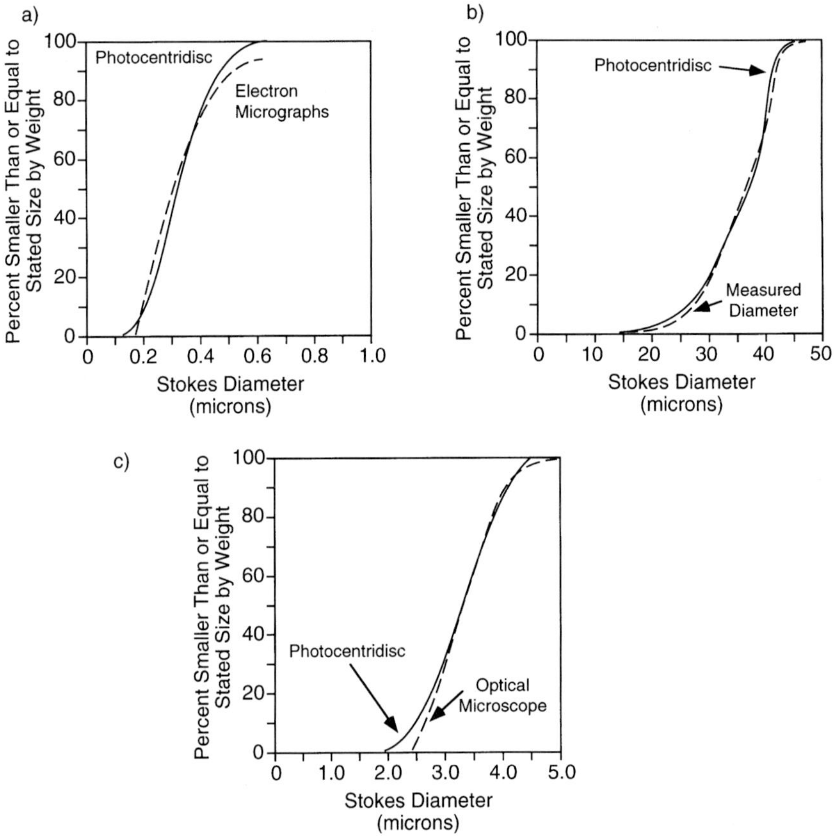

Figure 4.11. Comparison of photocentridisc Stokes diameter distributions to data obtained by other methods for the same powder. a) Data for a titanium dioxide powder. b) Data for spherical glass beads. c) Data for polyvinyltoluene latex spheres.

used by Burt to obtain the data in Figure 4.11 were as follows. The centrifugal study was carried out by sedimenting the spheres through a 46 : 54 glycerol-water mixture at 750 rpm; the microscopic evaluation was carried out on photomicrographs of the powder taken with a Zeiss photomicroscope, and one thousand fineparticles were measured and the number frequency converted to a weight frequency. The data in Figure 4.11(b) demonstrates the rather interesting fact that the centrifugal system can be used for relatively coarse particles by using a viscous fluid in the centrifuge. Although the fineparticles of this size can be quite adequately measured in gravimetric procedures, the major advantage of using a system such as that in the AWRE centrifugal disk photosedimentometer is that once the fineparticles have passed the monitoring zone, there is no need to stop the centrifuge to clean it. One can immediately inject the next sample of suspension to be characterized. In fact, by using a line start procedure that uses the same fluid in the suspension as in the body of the centrifuge, one can carry out many tests in sequence, and the equipment can easily be automated to study a long sequence of samples overnight without monitoring. (Note, however, that each time a sample suspension is injected into the centrifuge the height of fall for the next suspension is increased by the width of the line start for the previous sample. This adjustment can be readily incorporated into automated data transformation systems.) In Figure 4.11(c) the Stokes diameter distribution of close-range spherical polyvinyl toluene latex fineparticle suspension as measured by the AWRE centrifuge is compared to the distribution evaluated by the optical microscope examination.

In early 1960's the Coulter Electronics Company considered the possibility of developing the AWRE photocentridisk on a commercial basis. They produced several prototypes that were described as Coulter centrifuge or as Photofuge®. These instruments retained the use of the servo controlled optical wedge filter, and several; investigations describing data obtained with photofuges were published [50, 51].

Scarlett and co-workers reported difficulty in obtaining stable line start systems with the Photofuge® and introduced a techniques that they termed the three-layer system[52]. The name is somewhat misleading in that the spin fluid constitutes one of the layers and only two layers are floated onto the vortex of the main fluid. In this techniques a low-density fluid is used to prepare the suspension of the fineparticles to be characterized, and a middle layer of fluid of intermediate density if interposed between the spin fluid and the injected suspension. Kaye and co-workers, working in the United States with much more massive centrifuges, did not find it necessary to utilize this system since they were always able to achieve stable line start systems, using fluid identical with the spin fluid to prepare the test suspension at sufficiently low concentrations to avoid fluid suspension breakdown [53, 54]. However, many European workers have reported that the three-layer system is useful with their centrifuges. Burt has compared data obtained with the line start system and the "three-layer system" using the AWRE centrifuge. These data, summarized in Tables 4.2 and 4.3, illustrate the relative merits of the two experimental procedures with that instrument and can probably serve as a guide for probable performance if one opts for the three-linc strategy. (It should be realized that sequential analysis of a series of samples is not possible once several fluids are used in the experiments.) Burt reports that for his later work, he prefers to use the three-line start because of the smaller injected volume of suspension; however, this conclusion may be specific to his instrument and technique.

It is a little difficult to explain the success of the "three-layer" strategy. I suspect that the middle fluid plus the injected layer forms a broader initial line start by turbulent mixing than a density gradient restricted to within a small depth of the surface of the vortex of the spin fluid. Within this turbulent mixed zone the operative concentration of solids may be considerably less than that in the injected suspension. The writer's own opinion is that it is theoretically and practically better to seek out an experimental strategy and safe concentration level for the simple line start procedure than to bother with complex mixtures of fluids. A stable line start system ensures that fluid-fineparticles interaction effects are eliminated.

It is surprising that some workers working with low-mass centrifuges equipped with feedback-controlled compensated speed regulators have been unable to achieve stable simple line start systems even with very low solids concentration suspensions [55, 56]. One can only suggest either that some transient speed instability in the low-mass centrifuge caused by the injection of the test suspension or that some unspecified aspect of the injection procedure is causing problems.

The Joyce-Loebel disk centrifuge is based on the work of Atherton and Cooper [57]. It is sometimes referred to as the ICI disk centrifuge or the ICI/Joyce-Loebel disk centrifuge because the original work was carried out at the paint laboratories of Imperial Chemical Industries. This equipment consists of a transparent disk with an axial entry port mounted on a servo controlled motor running at constant speed. The equipment was originally designed to work with physical extraction of the suspension; however, as early as 1969, Peyrade [50] fitted optics to this instrument and used it in conjunction with the line start procedure to size latex fineparticles. When modified in this way, the equipment is essentially a low-mass instrument similar in operational principles to the photocentridisk described by Groves et al [47]. Later versions of the Joyce-Loebel equipment have optical incremental monitoring as an optional extra. Jones [59] introduced a version of the "three-layer system" that he termed the *buffered layer-start*. In this procedure a transient acceleration is given to the centrifuge after the middle layer has been floated onto the vortex of the spin fluid. It is claimed that this gives a density gradient in the top portion of the vortex of the spin fluid, thus helping to stabilize the injected sample of suspension. This claim would seem to support the "suggestion" made earlier in this section that the "three-layer system" in general is successful because of some turbulent dilution of the injected suspension that is then confined to the top zone of the spin fluid vortex by a density gradient. In recent years, Provder and co-workers have developed sophisticated versions of the Joyce-Loebel centrifuge and published several papers on the characterization of latex suspensions and paint pigments using their equipment [60, 61]. The Provder and co-workers equipment is available commercially from Brookhaven Instruments Corporation [62].

In 1964 Kaye and co-workers built and operated a very massive photo-centridisc [53, 54]. One feature of this instrument is that the entry port can be sealed with a rubber stopper while the centrifuge is running, thus improving the thermal stability of the system by preventing evaporation from free surface of the spin fluid. The overall mass of the system was such that after the equipment was cleaned and loaded with clean spin fluid, it required 24 hours of running to reach an adequate level of mechanical and thermal stability. However, this was compensated for by the fact that very stable systems were obtained with small-volume injections. As many as 50 samples could be characterized in sequence before stopping and cleaning the centrifuge. The excellent resolution achievable with this instrument is illustrated by the pen recorder trace reproduced in Figure 4.12. These data are from a sedimentation study of a nominally monosized latex of 2.61 μm diameter, as determined from electron microscopic studies. At the time that the original data were reported, they were somewhat controversial in that they showed the presence of doublets in the supposedly monosized suspension; however, the presence of these doublets in many lattices has now been amply demonstrated by other characterization procedures.

From the breadth of the primary peak obtained in the line-start characterization study, it can be seen that the measured range of Stokes diameters in the primary fineparticle population of the latex is 2.42–2.92 μm. The two subsidiary peaks are obviously doublets. The centrifuge has been able to resolve the orientation with which the doublets are falling with respect to the radial direction of the centrifugal force field in the centrifuge. The two subsidiary peaks in Figure 4.12 occur at 0.76T and 0.56T,

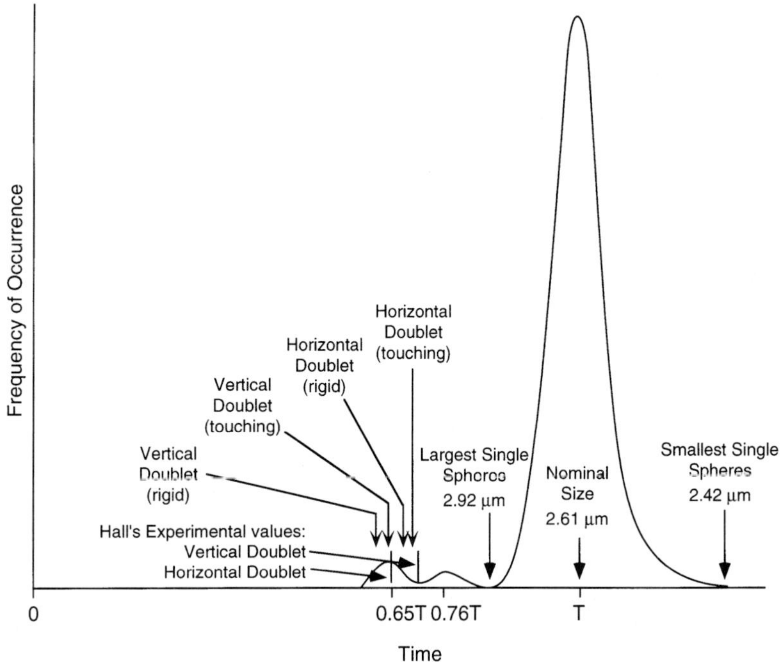

Figure 4.12. Data output from the IITRI photocentridisc showing the high resolution obtainable for a nominally 2.61 μm latex spheres. Also shown are theoretical falling speeds for touching and rigid doublets in horizontal and vertical orientations with respect to the direction of fall.

where T is the sedimentation time of the prime peak into the inspection zone. Hall has shown experimentally that two equal-sized spheres forming a doublet, aligned in the horizontal plane with respect to the direction of motion, had a falling speed just greater than 0.7 that of a single sphere, whereas a doublet falling with its orientation along the line of fall had a falling speed of 0.65 of the single sphere [63]. When it is appreciated that random agglomeration of the single spheres of the primary population into doublets could generate doublets ranging in structure from two spheres of 2.4 μm diameter, up to a doublet of two spheres of 2.9 μm in diameter, with many mixed intermediate sizes possible, the spread of velocities for doublets that have horizontal alignment as compared to Hall's experimental value for an ideal doublet is reasonable. The varied size of the components in the doublet would be less important when the doublet is aligned with the direction of flow; hence, it is not surprising that experimental values of falling speeds for the vertical alignment agree well with Hall's data. If two spheres had fused together to form a single larger sphere, such spheres would have traveled into the monitoring beam at the point shown by a slightly enlarged single sphere on the axis, as shown in Figure 4.12. Theoretical values for rigid and free doublets as calculated by Happel are shown in the diagram for comparison [64]. Typical sets of data showing the effect of a failure to use very low volumes in the injected layer when working with the instrument are shown in Figure 4.12 [65].

Groves and Yalbik [66] have used laser beams with photocentridisc equipment. They report good agreement between data obtained using this type of equipment and the other characterization procedures down to Stokes diameters of 0.02 μm.

In the LADAL x-ray centrifuge, a radioactive source is used to generate x-rays to monitor the radial migration of the sedimenting fineparticles in a disc centrifuge. Allen [67] reports that the characterization procedure is limited to powders with components of atomic number greater than 13, that is, to high atomic number elements capable of absorbing the x-ray radiation used in the monitoring strategy [2]. Allen reports that normally it is necessary to use powder concentration in the range of 0.2 to 1 % by volume to obtain measurable attenuations of the x-ray beam. The equipment can either be used with a static monitoring x-ray probe, or, alternatively, the x-ray probe can be moved along the radius of the centrifuge to decrease the characterization times.

Naylor has suggested the use of a line of fibre optic probes mounted along the radius of the photocentridisc equipment to accelerate the collection of data. Kaye has pointed out that such equipment could be used with various suspension strategies, including line start, pseudo-line start, and general homogeneous sedimentation. Naylor has pointed out that, in the latter situation, sequential scans along the radius of the centrifuge, at times separated by a minute, could be used to remove the ambiguity in the general solution of the homogeneous sedimentation equations (see discussion in reference 68).

For a discussion of the many different designs of centrifugal sedimentometers, before diffractometers and Doppler based methods made many of them of purely historical interest, see reference 1 and 2.

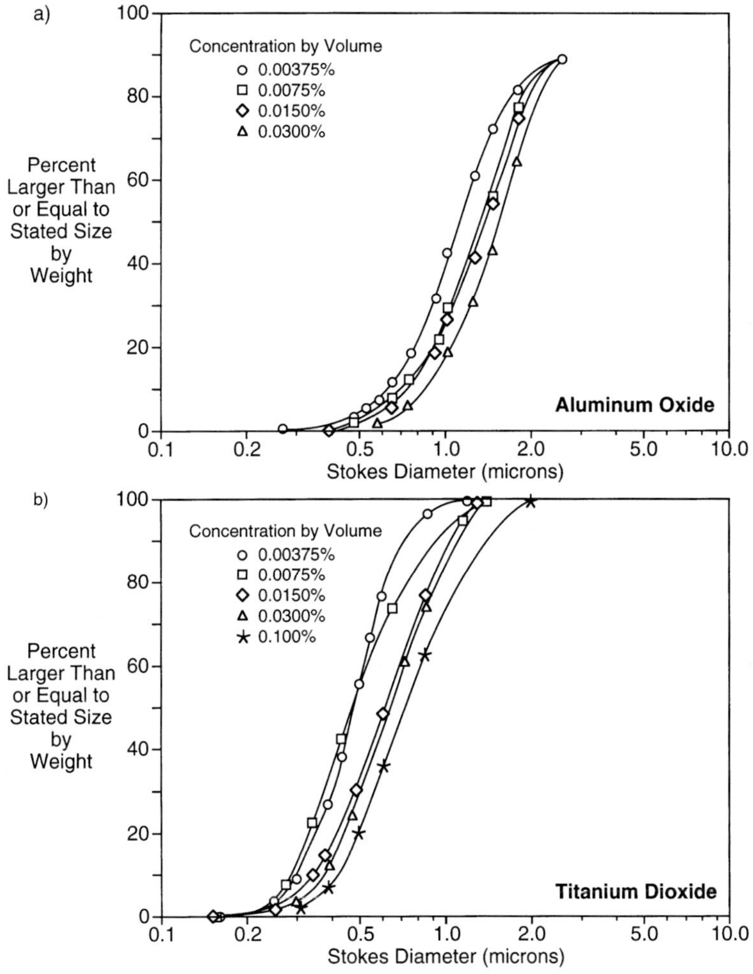

Figure 4.13. Data on the effect of suspension concentration for line start procedures in the IITRI photocentridisc.

References

[1] See for example chapters 5 and 6, B. H. Kaye, *Direct Characterization of Fineparticles,* John Wiley & Sons, New York, 1982.

[2] T. Allen, *Particle Size Analysis,* Chapman and Hall, 4th Edition, 1994.

[3] B. H. Kaye, *Chaos and Complexity – Discovering the Surprising Patterns of Science and Technology,* VCH Publishers, Weinheim Germany, 1993.

[4] G. Parfitt, *Dispersing of Powders in Liquids,* Applied Science Publishers, 1973.

[5] Extensive information on dispersing specific powders is to be found in Chapter 8 of reference 2.

[6] Information on the ISO Standards for particle size analysis, available from the International Organization for Standardization Secretariat, Building Division of DIN, Burggrafenstrasse 4-10.D-1000, Berlin 30.

[7] R. P. Boardman, *The Terminal Velocity of a Particle Falling Through a Viscous Fluid in the Presence of Other Particles,* M.Sc. Thesis, London University, England, 1961.

[8] B. H. Kaye and R. P. Boardman, "Cluster Formation in Dilute Suspension," *Proceedings of the Symposium on Interaction between Fluids and Particles*; published by Institute of Chemical Engineering, London, 1962. pp. 17–22.

[9] The interaction of fineparticles in a suspension has been reviewed in detail in Reference 1

[10] B. H. Kaye and C. R. G. Treasure, "The Thermal Stability of Sedimenting Suspensions Used to Measure Particle Size Distributions", *Materials Research and Standards,* 5:11 (1965).

[11] R. Davies and B. H. Kaye, "The Effect of Cluster Instability in Suspensions and Particle Size Analysis Measurements with the Cahn Balance" Particle Size Analysis Conference 1970, Bradford University, British Society for Analytical Chemistry, London.

[12] H. Heywood, *Symposium on Particle Size Analysis,* Institute for Chemical Engineers, Great Britain (1947) 114.

[13] Andreasen pipette equipment is available from Gallenkamp, Portrait Lane, Stockton-on Tess, Durham County, England.

[14] See catalogue of Gilson, Worthington, Ohio.

[15] A. H. M. Andreasen and J. J. V. Lundberg (1930) Ber. Deut. *Keram Ges* 111:5, 312–323.

[16] K. Leschonski, "Vergleichende Untersuchugen der Sedimentations Analyse," *Staub,* 22:11 (1962) 475–486.

[17] B. H. Kaye and R. Davies, "A Continuous Stem Pipette Equipment for Evaluating Stokes' Diameter of a Fineparticle System" *Proceedings Powder Bulk and Solids,* Rosemont, Illinois May 1978, pp. 351–352.

[18] See for example, British Standard 1377, 1948, Methods of Test for Soil Classification and Compaction (this standard was revised in 1961).

[19] ASTM Procedure of Soil Testing, Part IV, 1961, p. 1272.

[20] B. H. Kaye and G. W. James, "Investigations into the Diver Method of Particle Size Analysis," *Br. J. Appl. Phys.*, 13 (August 1962) 415–419.

[21] S. Berg, "Determination of Particle Size Distribution by Examining Gravitational and Centrifugal Sedimentation according to the Pipette Method and with Divers," ASTM, Special Technical Publication No. 234, 1958, pp. 143–171.
[22] For a general discussion of all the various theories of light scattering, see the book by H. C. Van de Hulst, *Light Scattering by Small Particles,* Wiley, New York, 1957.
[23] H. E. Rose, *The Measurement of Particle Size in Very Fine Powders,* Constable and Company, London 1953.
[24] K. Leschonski, K. L. Metzger, and U. Schindler, "A New On-Line Particle Size Analyzer," Salford Conference on Particle Size Analysis, September 1977, Proceedings, M. J. Groves, Ed., Heyden and Son, London, 1978.
[25] B. H. Kaye and T. Allen, "Optical Scattering Cross-section of Small Particles and the Design of Photosedimentometers," *Analyst,* 90 (1965) 147.
[26] H. R. Harner and J. R. Musgrave, "Particle Size Measurements," ASTM Special Technical Publication No. 234 (1958) 172–179.
[27] V. T. Morgan, Iron & Steel Institute Special Report No. 58, (1954) 81
[28] V. T. Morgan, Contribution to the symposium discussion in the Physics of Particle Size Analysis Supplement No. 3 *Br. J. Appl. Phys.* (1954) S207, S208.
[29] Goring Kerr Ltd. Station Road, Gerard's Cross, Buckinghamshire, England.
[30] Commercial information on this photosedimentometer, Sedigraph 5500L, available from Micrometrics Corp. 5680 Goshen Springs Rd. Norcross, Georgia 30093.
[31] Commercial equipment using wide-angle optics available from LADAL Scientific Equipment Ltd. Warlings, Warley Edge, Warley, Halifax, Yorkshire England.
[32] B. H. Kaye, "Automatic Decision Taking in Fineparticle Science," *Powder Technology* 8(1973) 293–306.
[33] R. Davies, private communication.
[34] J. P. Olivier, G. K. Hickin, and C. Orr Jr., "Rapid Automatic Particle Size Analysis in the Sub Sieve Range," *Powder Technology* 4 (1970/71) 257–263.
[35] Commercial information on the Sedigraph 5000 is available from Micrometrics see ref. 30.
[36] Commercial information on LADAL s-ray sedimentometer available from ref 31.
[37] D. Werner, "A Simple Method of Obtaining the Size Distribution of Particles in Soils and Precipitates," 1925 Transactions of the Faraday Society vol. 21 pp. 381
[38] K. T. Whitby, A. B. Algren, and J. C. Annis, 1958 Special Publication of ASTM, No. 234, pp. 117.
[39] Mines Safety Appliance Company, 201 Braddick Avenue, Pittsburgh 8, Pennsylvania.
[40] Technical Literature on Micromerograph, from Franklin Electronics, Inc. East Fourth Street, Bridgeport, Mont. W., Pennsylvania.
[41] R. S. Eadie and R. E. Payne, "New Instrument for Analyzing Particle Size Distribution", *Chem Eng.* 1 (1956), 306.
[42] S. M. Kaye, D. E. Middlebrooks, and G. Weingarten, "Evaluation of the Sharples Micromerograph for Particle Size Distribution Analysis", Technical Report, FRL-TR-54, Picatinny Arsenal, Dover, New Jersey, Feb 1962.

[43] G. L. Duchesne, Cardre Technical Note 1622, Canadian Armament Research and Development Establishment, Valcartier, Quebec, Dec. 1964.
[44] W. J. Ullrich "Rapid Particle Size Analysis of Metal Powders with an Electronic Device", reprint No. 6 International Powder Metallurgy Conference, New York, June 14–17, 1965.
[45] For an introduction to the mathematics of centrifugal sedimentation, see H. J. Kamak, *Anal. Chem.* 23 (1951), 844.
[46] C. E. Marshall, Proc. Roy. Soc., Section A, 126 (1930), 427.
[47] M. I. Groves, B. H. Kaye. and B. Scarlett. "Size Analysis of Sub-sieve Powders using a Centrifugal Photo Sedimentometer," *Br. Chem. Eng.* 9 (1964), 742–744.
[48] M. W. G. Burt, "The Accuracy and Precision of the Centrifugal Disc Photo sedimentometer Method of Particle Size Analysis," *Powder Technology.*, 1 (1967), 103–115.
[49] B. H. Kaye, Physical Problems of Particle Size Analysis, Ph. D. Thesis. University of London, 1962.
[50] Coulter Electronics, 590 20th Street, Hialeah, Florida 33010.
[51] B. R. Statham, *Proc. Soc. Anal. Chem.*, 9 (1972), 40–43.
[52] B. Scarlett, M. Rippon, and P. J. Lloyd, Particle Size Analysis, proceedings of a conference held at Loughborough University, England; published by the Society for Analytical Chemistry, London, p. 242.
[53] B. H. Kaye and M. R. Jackson, "Size Analysis of Latex Emulsions Using a Centrifugal Disc Photosedimentometer," *Powder Technology*, 1 (1967), 81–88.
[54] B. H. Kaye. M. R. Jackson, and R. Kahrun, "The Stability of Line Start Suspension Systems in the Centrifugal Disc Photosedimentometer," *Powder Technology*, 2 (1968–69), 290–300.
[55] J. Beresford, "Size Analysis of Organic Pigments Using the l.C.C.-Joyce Loebel Disc Centrifuge," *J. Oil Colour Chem. Assoc.*, 50 (1967), 594–614. For a critique of some of the data reported in this paper, see reference 15.
[56] B. H. Kaye, "Statistical Analysis of Recent Data on Concentration in Suspensions," *Powder Technology*, 3 (1969–70), 78–82.
[57] E. Atherton and A. C. Cooper. 8r. pat. 983,760 (1962). Commercial information from Joyce Loebel Ltd. Princes Way, Team Valley, Gateshead 11, Durham, England.
[58] Peyrade, J., private communication, Pechiney-Saint Gobain, Centre de Recherches d'Aubervilliers, 12–14 Rue des Gardinoux, 93 Aubervilliers, France.
[59] M. H. Jones, *Proc. Soc. Anal. Chem.*, 3 (1966), 116.
[60] M. J. Devon, T. Provder, A. Rudin, "Meaurement of Particle Size Distributions with a Disc Centrifuge; Data Analysis Considerations", Chapter 9 in *Particle Size Distribution II; Assessment and Characterization,* T. Provder (ed.), ACS Symposium Series 472,American Chemical Society, Washington D. C., 1991.
[61] M. J. Devon, E. Meyer, T. Provder, A. Rudin, B. B. Weiner, "Detector Slit Width Error in Measurement of Latex Particle Size Distributions with a Disc Centrifuge", Chapter 10 in *Particle Size Distribution II; Assessment and Characterization,* T. Provder (ed.), ACS Symposium Series 472,American Chemical Society, Washington D. C., 1991.

[62] Brookhaven Instruments Corporation, 750 Blue Point Road, Holtsville, New York, USA 11742-1896

[63] E. W. Hall, Ph. D. Thesis, 1956, University of Birmingham.

[64] J. Happel and H. Brenner, Low Reynolds Number hydrodynamics with Special Application to Particulate Media, Prentice-Hall, Englewood Cliffs, New Jersey (1965).

[65] R. Davies, private communication.

[66] M. J. Groves and H. S. Yalabik. "The Size Analysis of Dispersed Submicrometre Pigments and Lattices." *Powder Technology*, 17 (1977), 213–218.

[67] Commercial information is available from LADAL Scientific Equipment Ltd., Warlings, Warley Edge, Warley, Halifax. Yorkshire HX27RL, England.

[68] B. H. Kaye, "Monitoring Particulate Solid Properties in Automatic Control Loops, Past, Present and Future," in Proceedings of the Conference on Instream Measurements of Particulate Solids Properties, Bergen. Norway, August 1978. Information from Chris Michelsen Institute, Bergen.

5 Characterizing Powders and Mists Using Elutriation

5.1 Basic Principles of Elutriation

Elutriation is the process of separating fineparticles into fractions of different mean diameter by suspending the fineparticles in a moving fluid. Forces operating on fineparticles in an elutriator can be combinations of gravity, viscous drag, electrostatic and centrifugal forces. In the 1940–60's elutriators were used to characterize powders by using a moving fluid to fractionate the powder into two different fractions at a specific Stokes diameter. The relative magnitude of the two fractions was used to establish a point on the Stokes diameter distribution function of the fineparticle system. The complete size distribution function was evaluated by splitting the fineparticles into many different-sized fractions. (It is generally assumed in the literature of elutriation processes that all flow conditions in the elutriator are laminar.) Elutriation studies do not readily yield information on the largest and the smallest fineparticles present in a fineparticle system, but they can yield gravimetric estimates of weights greater or smaller than a specified Stokes diameter. Although elutriators are no longer used to characterize the size distribution of powder systems the need to fractionate powders into close size range fractions is an important task in modern powder technology. Also elutriation principles are important in sizing aerosol systems and submicron particles in techniques such as field flow fractionation and capillary hydrodynamic fractionation.

In a vertical elutriator the fluid containing the suspended fineparticles flows upward. For a specific fluid-velocity, fineparticles smaller than a given size move upward with the fluid whereas the larger fineparticles settle to the bottom under the influence of the gravitational force. The Stokes equation (see Chapter 4) is used to calculate the Stokes diameter of the fineparticle just balanced by the upward-moving fluid. This size is defined as the cut size of the elutriator when operating at the specified fluid flow velocity. Figure 5.1 shows the operation of a simple vertical elutriator.

Theoretical difficulties associated with the performance of a vertical elutriation are:

1. The fluid flow velocity across the cross section of the elutriating chamber is not constant. Thus in a cylindrical chamber the distribution is parabolic (see discussion of fluid flow in reference 1). Random fluctuations in fluid flow cause migration of larger fineparticles just supported by the higher flow in the center regions

to the outside of the chamber where the lower fluid flow is no longer able to sustain their buoyant weight. These fineparticles then fall and create a return flow of fluid that further disturbs the overall flow pattern of fluid. Attempts have been made to even out the flow rate across the main tube of an elutriator by placing a coarse grid across the tube [2].

2. If an elutriator is operated for long periods of time, the cut size associated with maximum flow along the center of the elutriator column dominates the fractionation achieved with the elutriator rather than the cut size calculated from the average flow.

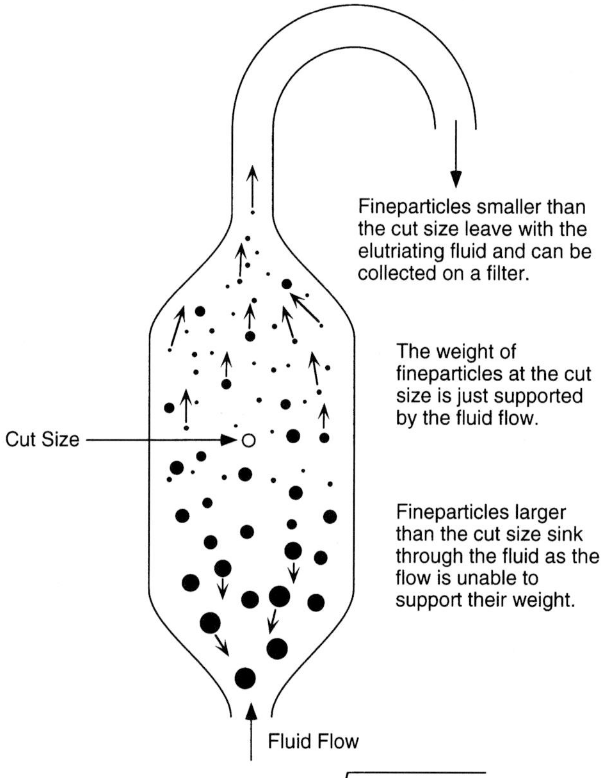

Stokes Equation: $$d = \sqrt{\frac{18\,\eta\,u}{(\rho_P - \rho_F)\,g}}$$

d : Diameter of cut size sphere η : Viscosity of the Fluid u :Velocity of the Fluid

ρ_P : Density of the Particle ρ_F : Density of the Fluid g :Acceleration due to Gravity

Figure 5.1. In a vertical elutriator upward moving fluid fractionates powders into portions greater than and less than the "cut size" determined by the Stokes Equation from the properties of the particle, the fluid, and the fluid velocity.

3. To obtain steady flow conditions, relatively long feed pipes to the elutriator are necessary. If there are any sharp bends in the feed pipework leading to or from the main tube, flow conditions are hard to predict and fractionation performance has to be determined empirically.

A practical problem associated with the operation of the elutriator is that the solids concentration of the suspension affects the performance of the elutriator; usually the higher the concentration of solids, the less predictable the fractionation efficiency. If the fluid flow rates required to fractionate the powders are very low, requiring several days to achieve the required fractionation, it is difficult to achieve thermal isolation and stability of the system. It is also difficult to generate, maintain, and monitor the low flow rates required to operate the system.

In horizontal elutriators a stream of suspension is passed over a reservoir of clear fluid. The larger fineparticles leave the suspension stream and are collected at the bottom of the settling chamber. The fractionation efficiency of this device is not good since some fineparticles of all sizes will leave the suspension stream as it moves over the clear fluid in the collection chamber. The solids concentration of the feed stream must be kept low, otherwise, bulk penetration by the feed layer into the fluid in the container will occur (see discussion of the stability of the line start suspension system in Chapter 4). It is possible to use a more dense fluid in the reservoir to improve the stability of the system but this system has not been used in size characterization studies. Fineparticles that drop through the full height of the fluid stream in the time taken for the fluid stream to move across the elutriating chamber are completely removed from the feed stream.

Cyclones, widely used in powder science and technology, are centrifugal elutriators. The operating principles of a cyclone are illustrated in Figure 5.2. The stream of suspended fineparticles, entering the top of the cyclone tangentially, is whirled downward, and escapes through an inner return flow zone. The lines of flow shown indicate general direction of flow and are not intended to show actual flow patterns. Fineparticles that are too large to follow the changes in the flow pattern impinge on the walls. The smaller fineparticles escape with the discharge stream. Fineparticles hitting the walls slide down and collect at the base. The efficiency of separation and the size of fineparticles remaining in the body of the cyclone, i. e. the cut size of the cyclone, depend on the exact flow pattern established as well as the properties of the particles in the fluid.

In practical terms the cut size fractionation of an elutriator is never a clean cut fractionation and the oversize fraction often contains some of the particles smaller than the cut size and vice-versa. In industrial situations the efficiency of an elutriative fractionation is measured by what is known as a coarse grade-efficiency curve. This curve is often referred to as Tromp's curve after its originator [3]. To be able to calcu-

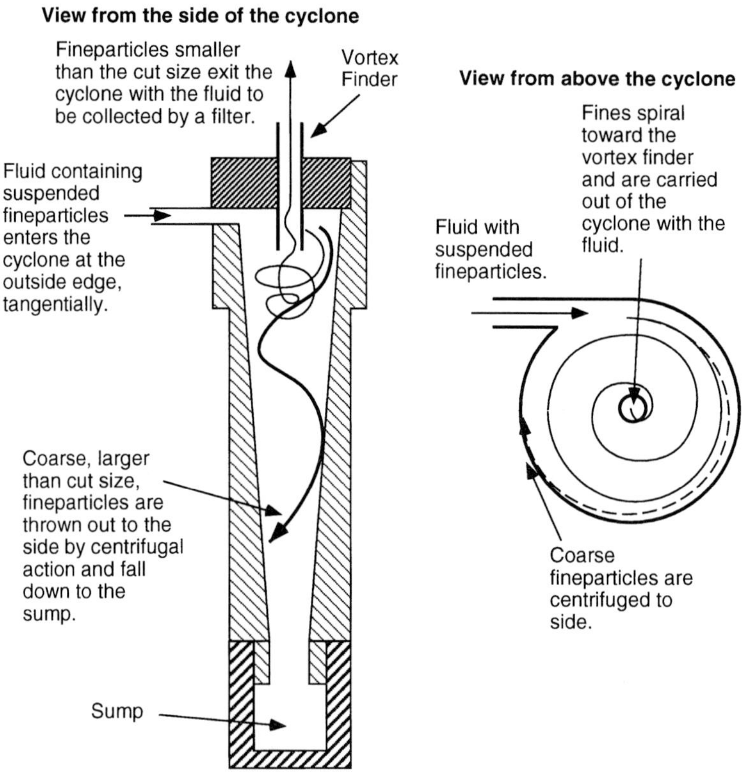

Figure 5.2. In a cyclone, rotating fluid separates fineparticles above a given cut size by centrifuging the larger than cut size fineparticles to the side so that they fall down into the sump.

late Tromp's curve one must know the size distribution of the oversize and undersize fractions created at a nominal cut size. The procedure for calculating Tromp's curve is illustrated in Figure 5.3. The coarse grade-efficiency curve is a dimensionless function that runs from the lower limit of the oversize fraction to the upper limit of the undersize fraction. The variable shown as **x25**, in Figure 5.2(c) is the Stokes diameter of the fineparticle for which the coarse grade-efficiency curve has reached 0.25. The variable **x75** is defined as the Stokes diameter for which the grade-efficiency curve has reached the value 0.75. The "index of fractionation efficiency" is defined as the ratio **x25:x75**. It would have a value 1 for an ideal elutriator. Because elutriators split the fineparticle system to be characterized into fractions and since the act of fractionating a fineparticle system is often referred to as classification (i. e., placement of the various-sized fineparticles into several classes), elutriators are sometimes referred to as "classifiers." In this book the term "classifier" is used if it is part of the usual commercial name of an instrument but normally the term "elutriator" is used for this type of instrument.

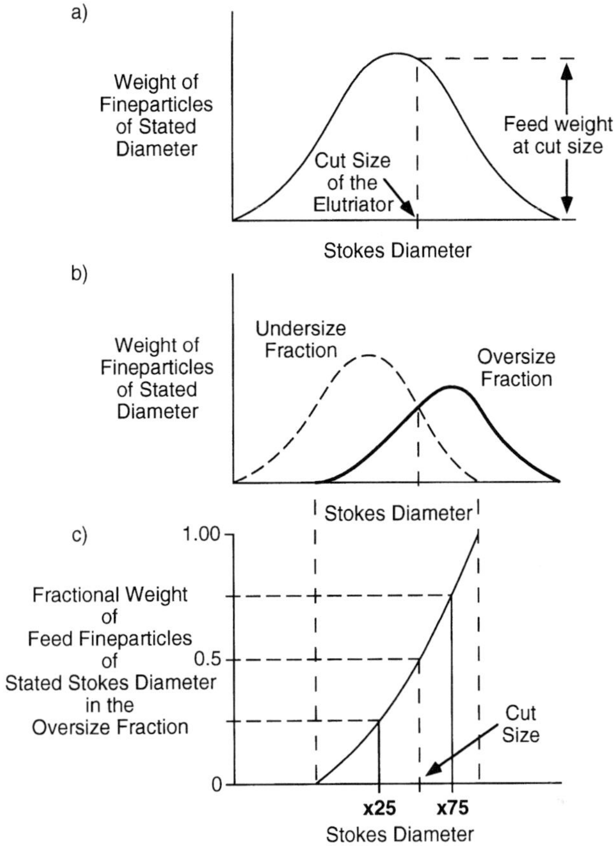

Figure 5.3. The basic steps for generating a Tromp's curve to evaluate the efficiency of an elutriator. a) Size distribution of the powder sample. b) Size distributions of the undersize and oversize from the elutriator. c) Tromp's curve for the elutriator spans the size range between the lower limit of the oversize fraction and the upper limit of the undersize fraction. The "Index of Fractionation Efficiency" is the ratio of the values **x25:x75**.

One of the earlier vertical elutriators was described by Gonell [4]. In the Gonell elutriator the powder to be fractionated is placed at the base of the elutriating column. Air from a jet impinges down into, and through, the powder fluidizing it upward through the column. A long column is used to suppress flow disturbances created by the conical entrance of the elutriating chamber. A general problem of air-operated elutriators is that the moving dry powders around with dry air generates electrostatic charges. This causes powder to cling to the side of the elutriating chamber. Rapper hammers were used in the Gonell elutriator to remove powder from the side of the chamber. Data on the performance of the Gonell elutriator has been reported by

Stairmand [5]. A commercial version of the Gonell elutriator, manufactured by the Alpine Corporation, has been described by Lauer [6]. In the Alpine Company literature it is stated that they do not provide flow accessories for fractionating at theoretical cut sizes smaller than 12 microns Stokes diameter since they find experimentally that below this size agglomeration of the powder prevents efficient fractionation. The literature also states that when dealing with powders as small as 12 microns, fractionating times for high-efficiency separation can be as high as 5 hours. Comparative data on the performance of the Gonell elutriator as compared to other methods is given in Table 5.1. Using a special version of the Gonell elutriator, Weilbacher investigated the flow conditions within the elutriation chamber [7]. He established that strong turbulence and unstable flow conditions characterize the air inlet region of the chamber of the Gonell elutriator and states that

> *"the flow conditions of the lower end of the tube govern the separation achieved with the instrument but here the flow regime is not only indeterminate, but varies with time".*

Weilbacher confirmed that the parabolic laminar velocity distribution forms near the upper end of the Gonell elutriator. He noted that the arrangement at the top of the tube for the removal of fineparticles fractionated in the chamber moved them through the boundary region of low velocities at the side of the tube, thus causing resedimentation into the boundary region and down the elutriation column thus prolonging the separation process.

Table 5.1. Comparative data on elutriator performance reported in British Standards Specification Number 3406, Part 3, 1963, "Air Elutriation Methods". (Data for a silica powder.)

	Percent in Range by Weight			
Stokes Diameter μm	**Pipette Characterization Andreasen Equipment**		**Gonell**	**Microscope**
	Lab 1	**Lab 2**		
105–75	–	–	–	3
75–53	7	8	8	5
53–37	18	14	17	13
37–26	14	10	15	17
26–19	12	15	13	18
19–13	13	13	11	14
13–9.4	8	8	6	10
9.4–6.6	4	8	6	6
6.6–4.7	7	6	3	5
4.7–3.3	4	3	4	3
3.3–2.3	3	3	2	1.5
2.3–1.7	2	1	13	2.5
1.7	8	11	13	2.5

The system shown in Figure 5.4, which uses only a short, straight fractionating column in a conical elutriating chamber, was designed to overcome the unsatisfactory features of the Gonell elutriator. The ratio of the height of the fractionating region of the chamber to the diameter of the elutriating chamber is 0.1. The base of the elutriating chamber is equipped with a high-impedance air filter. This dampens out any turbulence caused by fluctuations in the air supply to the base of the elutriating chamber. The top of the equipment can be readily separated from the base so that the material to be fractionated can be inserted on top of the filter. Air fluidizes the material on the

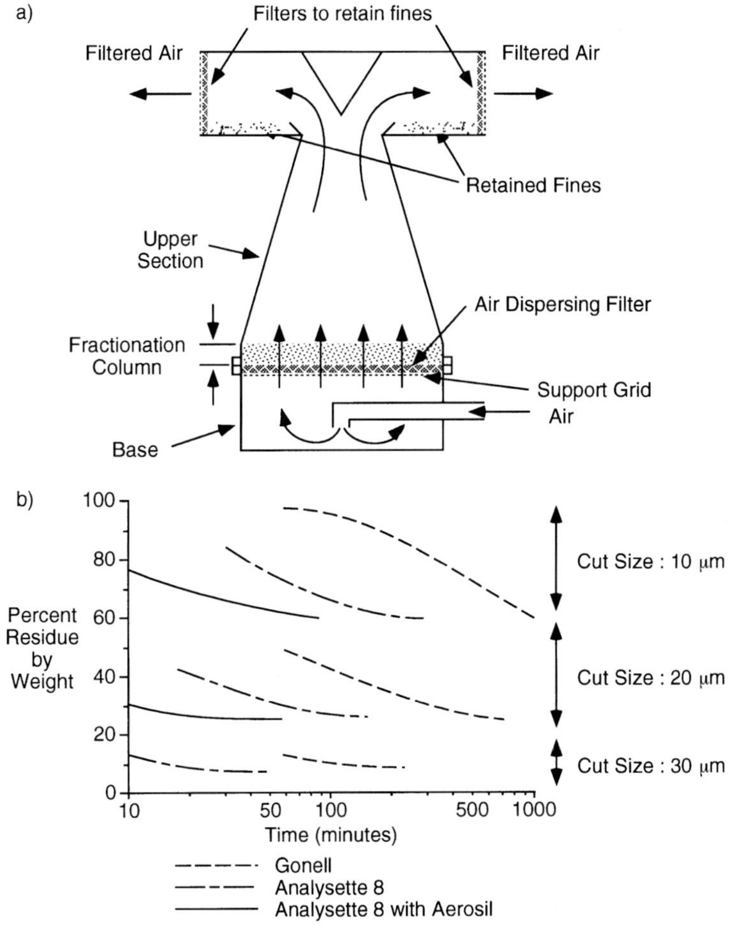

Figure 5.4. The Analysette® 8 is an air elutriator with a short fractionation column that was described by Leschonski and Rumpf. a) Side view of the structure of the Analysette 8. b) Comparative performance data for three types of elutriators.

filter, and the fines are carried up the fractionation column. After spending a short time in the straight portion, they reach the angle of the collection cone region of the elutriating chamber. Here the flow rate starts to increase so that any fineparticles that are able to move to the top of the fractionating column are then swept out of the equipment and into the collection region at the top. The fines are collected on a filter as shown. Performance data for this new elutriator, reported by Leschonski and Rumpf, are summarized in Figure 5.4 [8, 9, 10].

It is common practice in the air elutriation of fine powders to add what is known as a deagglomerating agent or a flow agent to improve the flow characteristics of the powder. A frequently used flow agent is colloidal silica. The exact mechanisms by which the improvement of flow is achieved by the addition of these very fine powders is not fully under stood [10] In their discussion of the performance of the elutriator shown in Figure 5.4, Leschonski and Rumpf state that the equipment should not be used to fractionate powders smaller than 10 microns [10].

The Roller elutriator shown in Figure 5.5(a) was first described in the late 1930's [12]. It has been widely used and is specified in both ASTM [13] and British Standards [14] procedures for characterizing fineparticles. A special feature of the Roller elutriator is the "goose neck," a flexible feed tube to the elutriating chamber that is vibrated rapidly during the operation of the elutriator. The Roller equipment is noisy because of the mechanical system used to jolt the goose neck, and the equipment is usually operated in a sound-muffling box. Fines passing through the elutriator are captured on a special filter. Data on the fractionation efficiency of the Roller elutriator are summarized in Figure 5.5(b), (c) and (d). The performance of the Roller elutriator is regarded as adequate for many industrial purposes, and it is widely used in the ceramics and metallurgical industries [15].

The Gonell and Roller elutriators are single-chamber instruments. With both instruments, different cut sizes can be achieved by changing the air velocity. To extend the range of operational cut sizes, the manufacturer of the Roller elutriator supplies several chambers of varying diameter. The Aminco classifier is an adaptation of the Roller elutriator in which eight different-sized elutriating chambers are linked together in series [16]. The Haultain Infrasizer is a cascade air elutriator [17]. This equipment was developed by Professor Haultain at the University of Toronto in the late 1930's. The special feature of this apparatus is the dispersion system at the bottom of each of the conical elutriation chambers. It utilizes a golf ball. Air entering the elutriating chamber elevates the ball slightly, creating a narrow annulus around the ball! Shearing action caused by the rotation of the rough surface of the golf ball aids the deagglomeration of fine powders [18].

One of the first liquid operated elutriators described in the scientific literature was that designed by Schone. Data on the performance of the Schone elutriator are summarized in Reference 5. The equipment is relatively efficient in the separation of the

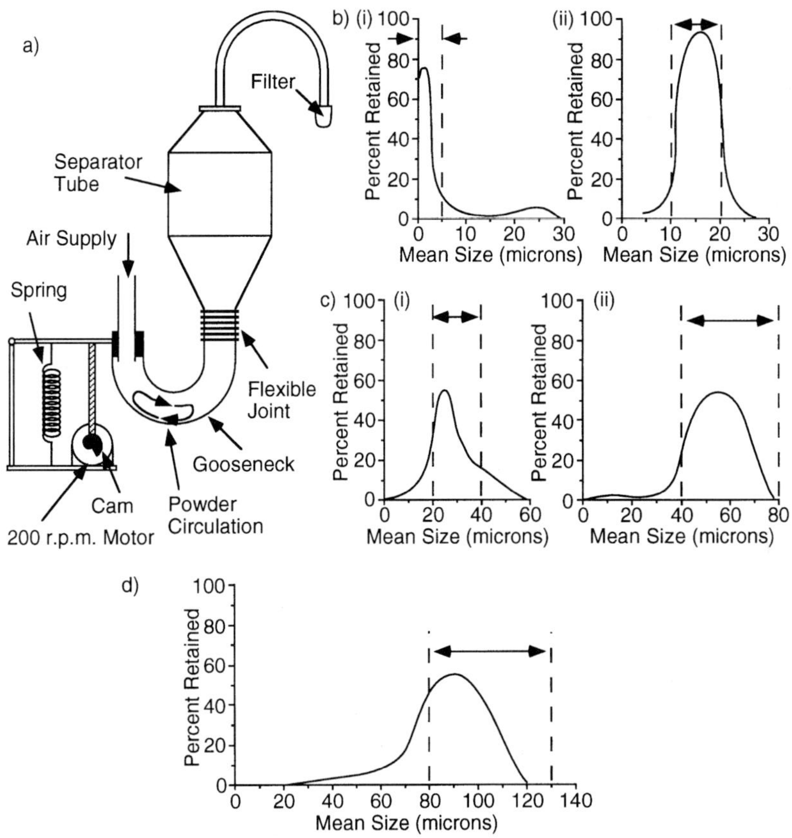

Figure 5.5. Data and performance of the Roller elutriator. (The nominal range of the fractions are indicated by the dashed vertical lines.) a) Sketch of the Roller elutriator. b) Performance data for (i) Less than 5 micron fraction (ii) 10 to 20 micron fraction. c) Performance data for (i) 20 to 40 micron fraction (ii) 40 to 80 micron fraction. 9) Performance data for greater than 80 micron fraction.

oversize fraction greater than 80 microns but the efficiency of fractionation at 10 microns is not good. The design of a liquid elutriator that can be used to make narrow-size range fractions of small fineparticles, the Blythe elutriator, is shown in Figure 5.6 [19]. A unique feature of this elutriator is that the technologist can follow the progress of the elutriation process by taking samples from the sump at the base of each elutriating column. Paddle stirrers are used to agitate the suspensions at the base of each elutriating column. Each elutriating chamber plus exit tube is a link in a siphon chain. The total flow through the siphon chain is established by feeding the sump of the first chamber with a constant-pressure apparatus and by adjusting the screw clip

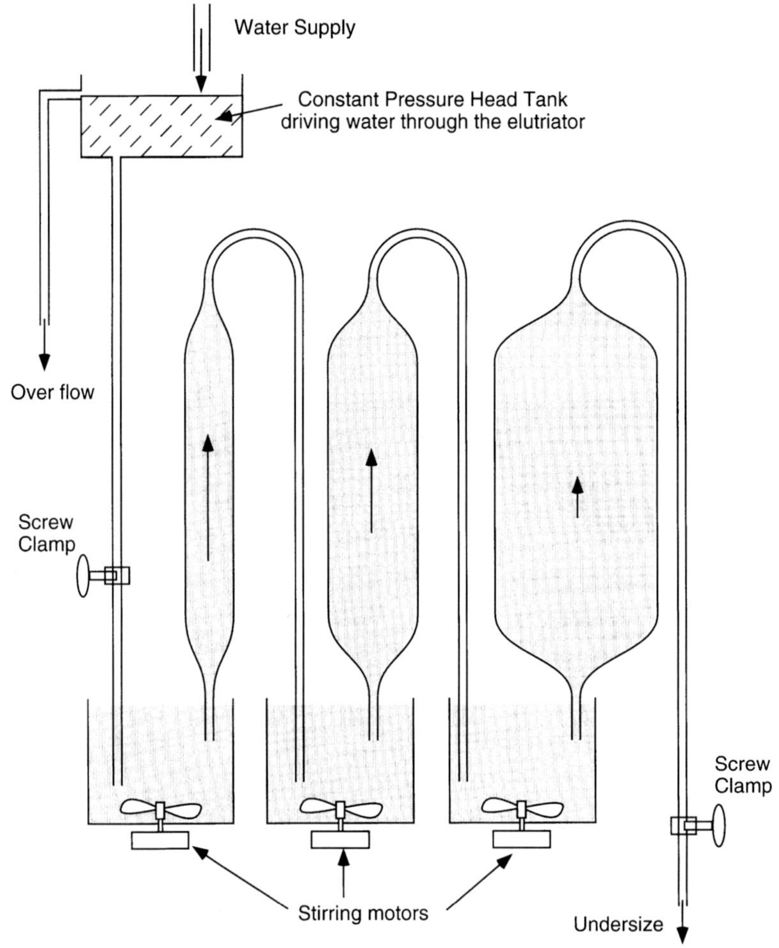

Figure 5.6. The Blyth elutriator uses a siphon action to cause the suspension to flow through the elutriation chambers.

on the final exit tube of the system. When the system is operating, there is a slight oscillation of fluid in the individual elutriating chambers superimposed on the through flow of fluid because of the use of the siphon mechanism to draw suspension through the system. Thus the siphon action tends to draw fluid out of the container immediately below an elutriating chamber. This flow of fluid into the next sump builds up a reverse pressure that attempts to alter the flow direction of the siphon. Fluid will flow in one direction until the reverse pressure generated is sufficient to cease the flow of fluid and reverse its direction. When flow is reestablished in the forward direction, fluid will flow until the redeveloped reverse pressure stops and reverses the flow once again. The elutriator system is initially filled with clear liquid and the powder to be fractionated placed in the first sump. The moving boundary of the suspension climbing the elutriator is clearly visible in the initial stages of elutriation. When this boundary reaches the top of the column, turbulent flow conditions established at the outlet from the elutriator column are often clearly visible. If the top of the elutriating chamber is not carefully designed, a density-driven convection current down the side of the elutriating chamber forms at the exit and the movement of fineparticles down the side of the elutriating chamber are clearly visible. The Blythe elutriator can be used with relatively low flow rates operating over a period of 3 days to fractionate fineparticles as small as 2–3 microns Stokes diameter. The size fractionation limit is governed by both the density of the fineparticles to be fractionated and the difficulties of providing a thermally stable environment for the equipment. It has been used on a commercial basis to fractionate in the processing of abrasive powders for polishing optical instruments.

The Hexlet sampler, used to sample aerosols, is a horizontal elutriator. It splits the aerosol fineparticles into two fractions. Its design, shown in Figure 5.7(a), is based on the work of B. M. Wright at the Pneumoconiosis Research Unit of the Medical Research Council, Cardiff, Wales [20]. The intake of the equipment is of a stack of parallel plates. Fineparticles starting at the entrance to the rectangular duct, formed by adjacent plates, sediment toward the bottom plate as the aerosol stream moves through the equipment. Fineparticles above a certain critical size are all deposited onto the plates as the air stream reaches the end of the rectangular duct. The sedimentation velocity of the fineparticles in the vertical direction is calculated by using Stokes equation. The penetration of the horizontal elutriator chambers for spheres of density 2.25 when the equipment is operating at 50 *l*/minute is illustrated in Figure 5.7(b). 2.25 is representative of the density of the type of dust encountered in coal mines. Under these operating conditions, fineparticles of Stokes diameter greater than 7 microns are all precipitated onto the plates of the elutriator. The bottom plates of the elutriator ducts are oiled to prevent re-entrainment of fineparticles into the air stream leaving the elutriator. The reason for choosing 7 microns as the cutoff in the elutriator design is that, in Great Britain, fineparticles larger than this are not considered dangerous to the lungs. There is some controversy over the magnitude of this size limit for dust capable of causing damage to the lungs, and European and North American standards opt for a slightly lower critical size (5 microns). Fineparticles leaving the elutriation stage of the Hexlet air sampler are captured by a high-efficiency tubular filter known as a Soxhlet thimble. In mine safety practice the weight of powder retained by the Soxhlet filter in a given period is described as the "burden" of dangerous dust in the working environment. The flow rate through the horizontal elutriator can be altered to achieve different cut sizes [20].

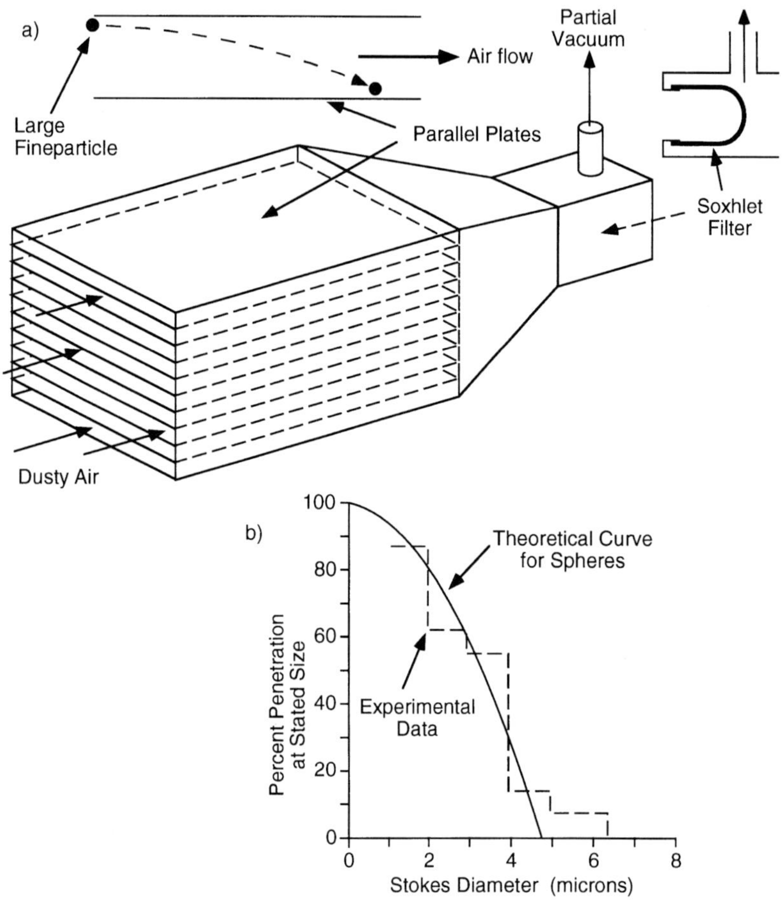

Figure 5.7. In the Hexhlet aerosol sampler, the distance between the plates is such that fineparticles larger than 7 microns will settle out of the air flow before they can reach the filter. a) Appearance and operation of the Hexhlet sampler. b) Performance data for the Hexhlet sampler with spheres of density 2.25 sampled at 50 litres per minute.

In the Timbrell aerosol spectrometer [21], shown in Figure 5.8, the aerosol to be characterized is aspirated into the center of a flow of clean air before entering the elutriation chamber and, as this stream passes through the elutriation chamber, the fineparticles sediment to the bottom according to size. The largest are deposited at the entrance to the chamber. The flow rate through the system drops as the aerosol moves along the wedge-shaped elutriation chamber to give the smaller fineparticles longer residence time to facilitate their deposition on the base of the chamber. The commercial equipment is provided with charts showing the location along the duct at

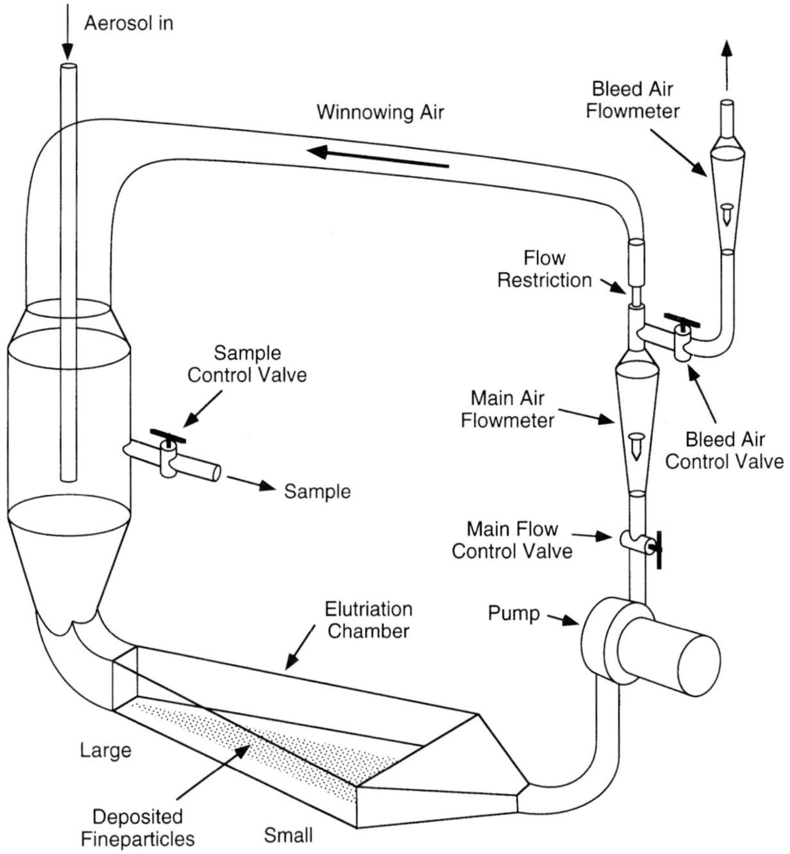

Figure 5.8. The Timbrell Aerosol Spectrometer uses a tapered elutriation chamber to lower the flow rate along the chamber so that ever smaller fineparticles can settle to the bottom.

which fineparticles of a given Stokes diameter will be deposited for a given flow rate through the system. The number of fineparticles deposited at a given location is evaluated by microscope inspection.

The Whitby aerosol analyzer shown in Figure 5.9, is an elutriator in which the significant force applied to the fineparticle in the moving viscous fluid is electrostatic in origin. It is used to fractionate very small aerosol fineparticles of Stokes diameter of 0.015–2.0 microns [22]. The effects of gravitational forces on such fineparticles can be neglected. The aerosol is initially drawn into an ionization chamber in which the ionization conditions are controlled so that the fineparticles of the aerosol are charged in such a way that the electrical mobility of the fineparticle is proportional to its size. There is some controversy regarding the claim that the ionization process achieves effective charging of the aerosol fineparticle in this way [23]. The flow of charged fineparticles out of the ionization chamber into the fractionation chamber is monitored as shown in Figure 5.9. Down the center of the cylindrical chamber is a rod that can be charged to a high potential. An auxiliary air supply system is used to create a clean air sheath around the rod, as shown at the top right-hand side of Figure 5.9. The outer limits of this sheath are indicated by dotted lines in Figure 5.9. The charged aerosol fineparticles moving through the fractionation chamber are attracted to the central rod. Depending on the electric potential of the rod, all fineparticles less than a given size are collected by the rod whereas the larger ones escape to the charge collecting filter. The magnitude of the current generated at the collecting filter is a measure of a number of fineparticles reaching the filter. An aerosol is sized by carrying out a series of experiments measuring the charge on the current-collecting filter at different potentials applied to a central rod. The equipment is calibrated using aerosols of known characteristics.

The cross-flow classifier shown in Figure 5.10, developed by various investigators, is an industrial-scale horizontal elutriator [24, 25, 26]. Fineparticles to be fractionated are entrained in a high-speed air jet, which is fired across a winnowing stream of air.

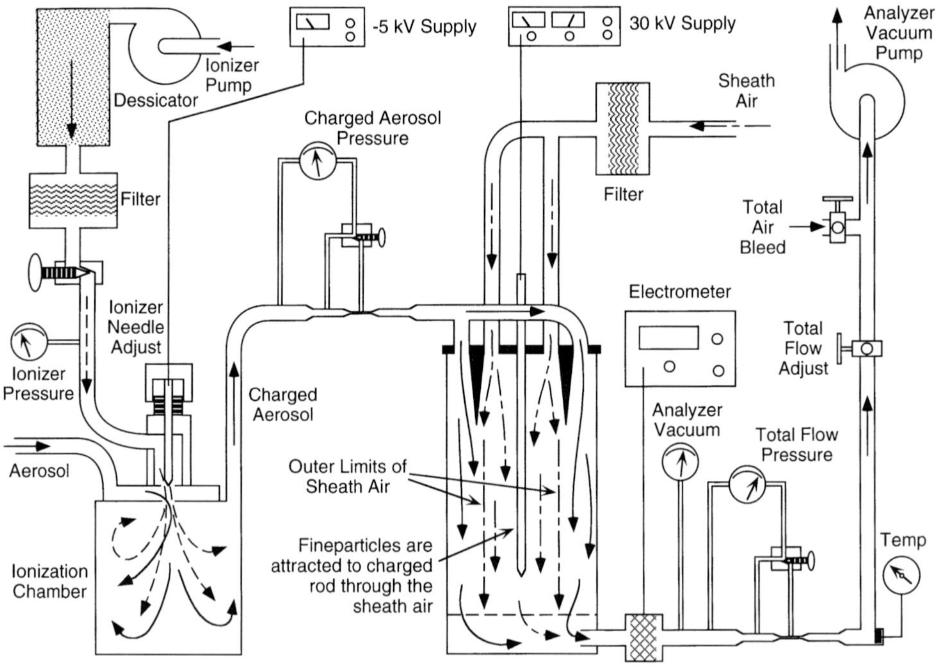

Figure 5.9. Schematic diagram of the Whitby Aerosol Analyzer.

Typical velocities used are 60 m/second for the high-speed air jet and 20 m/second for the winnowing air stream. The dispersed array of fineparticles created by the interaction of the high-speed jet and the winnowing air is split into eight fractions by using sharp-edged partitions as shown in Figure 5.10(a). The largest fineparticles have the most kinetic energy and are caught in compartment 8, with the smallest fineparticles being blown into compartment 1. Performance data are given in Figure 5.10(b). Leschonski and co-workers used a photometer to evaluate the fineparticle concentration across the dispersion fan created by the interaction of the high-speed air jet and

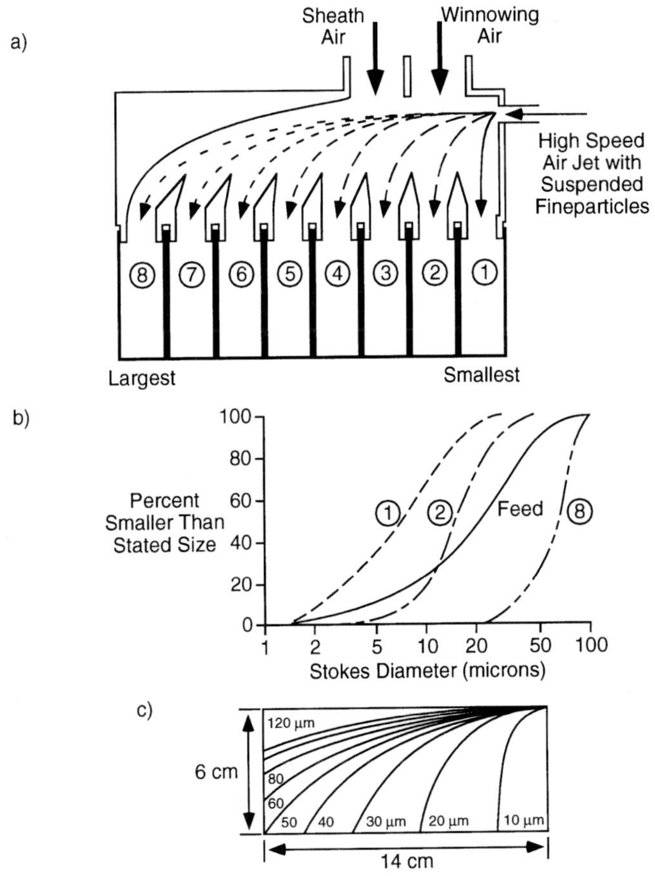

Figure 5.10. The Cross-Flow classifier uses winnowing air to separate a sample into compartments by size. The smallest fineparticles are swept into the first compartment and the largest can travel across the downward air current to the last compartment. a) Sketch of Cross-Flow classifier. b) Typical performance data. c) Fineparticle size within the dispersion fan injected at 20 m/s.

the winnowing air. In Figure 5.10(a) data generated by a typical track of the photometer across the dispersion fan is illustrated. The Lambert-Beer law is used to relate the transmission drop at any position within the dispersion fan to the concentration of fineparticles at that location. Leschonski and co-workers calibrated the photometer readings by using a powder of known size distribution as determined by the Sedigraph 5000 x-ray sedimentometer (see Chapter 4). Leschonski reports that with equipment calibrated in this way, good agreement is obtained between size characterization by photometric inspection of the dispersion fan in the cross-flow classifier and characterization by Sedigraph 5000. (Note: In essence, this is a direct empirical correlation between two experimental procedures, and the agreement between the two sets of data gives no information on the presence of any systematic errors in either procedure, only on the invariance of the cross correlation when the two procedures operate under certain conditions.) Leschonski reports that the photometer scan of the dispersion fan required only 3 seconds and that, operating at throughputs of 2–5 kg/hour, size characteristic distributions were reported every minute, with each distribution being the average of 10 sets of individual measurements [26].

Impactors, also called impingers, are samplers and size characterization equipment used to study aerosols which are essentially centrifugal elutriations. The basic concept of an impactor device used to study aerosols is illustrated in Figure 5.11. A jet of aerosol is deflected by a surface to create a quarter turn of a centrifuge. Calculations of the exact cut size of the smallest aerosol particle deposited on the deflection slide is a complex problem and commercially available Impactors results have specified operating conditions and calibration charts.

The Konimeter, manufactured by Sartorius, is a widely used jet impactor in which a fixed volume of air is drawn rapidly through a nozzle placed 0.5 mm from a collection disk [27]. The sharp change of direction imposed on the air stream by the collection plate centrifuges the fineparticles onto the collection plate. A spring-loaded piston is used to aspirate the sample through the nozzle. This enables the equipment to be used in industrial areas without the risk of generating electrical sparks which could ignite potentially explosive atmospheres. The equipment has been widely used in the sampling of dust in coal mines. In English-speaking countries the Konimeter is often known as a Conimeter.

In cascade impactors the aerosol flow is arranged so that it passes through a series of jets of decreasing orifice size. Thus in the Sciotec cascade impactor, a nest of interlocking cylinders, each of which contains a jet and a slide collector, are assembled with the widest-mouthed jet at the top [28]. Smaller and smaller fineparticles are deposited at each successive stage of the instrument. As normally operated, it has six stages of impaction. Aerosol is sampled at the rate of 12.56 ft^3 of air per minute. Calibration graphs are provided with the system so that aerodynamic diameters deposited at any given stage with any specified flow rate through the instrument can be rapidly calculated.

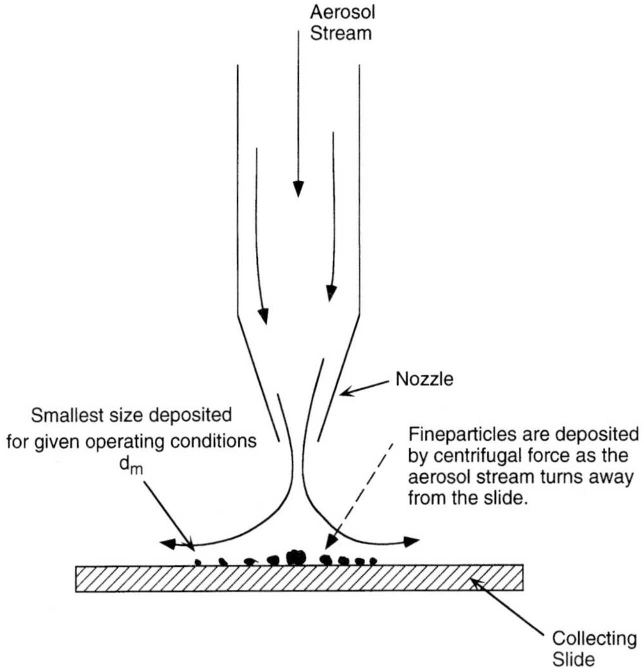

Figure 5.11. The operation of a basic jet impactor involves fineparticles being thrown out of the aerosol stream by centrifugal force as the stream is turned away from the collecting slide.

Another widely used cascade impactor, the Anderson sampler, is constructed from a set of nesting cylindrical sections [29]. The impaction jets are holes drilled in the base plates of the cylinders. There are 400 holes in the base of each element of the sampler. A shallow glass dish, known as a petri dish, is placed under the jets. These dishes are either smeared with Vaseline to help retain the deposited aerosol or can be filled with agar jelly so that, if biological aerosols are being sampled, the bacteria colony can be grown directly on the collecting medium. As the aerosol streams move down through the column the air passes through smaller and smaller jets. The flow rate through the system can be set at any required level. The structure of the May cascade impactor is illustrated in Figure 5.12(a). The microscope slide used for collecting deposited aerosol at each stage can be removed from the body of the impactor by removing the rubber cap. During the deposition operation the microscope slide is held in position by spring clips (not shown in Figure 5.12(a)). The aerosol to be characterized is drawn through four successive stages of the impactor, and the jet orifice decreases at each stage of the impactor. The equipment is sometimes used with a fifth stage containing a membrane filter. The percentage penetration of various sized

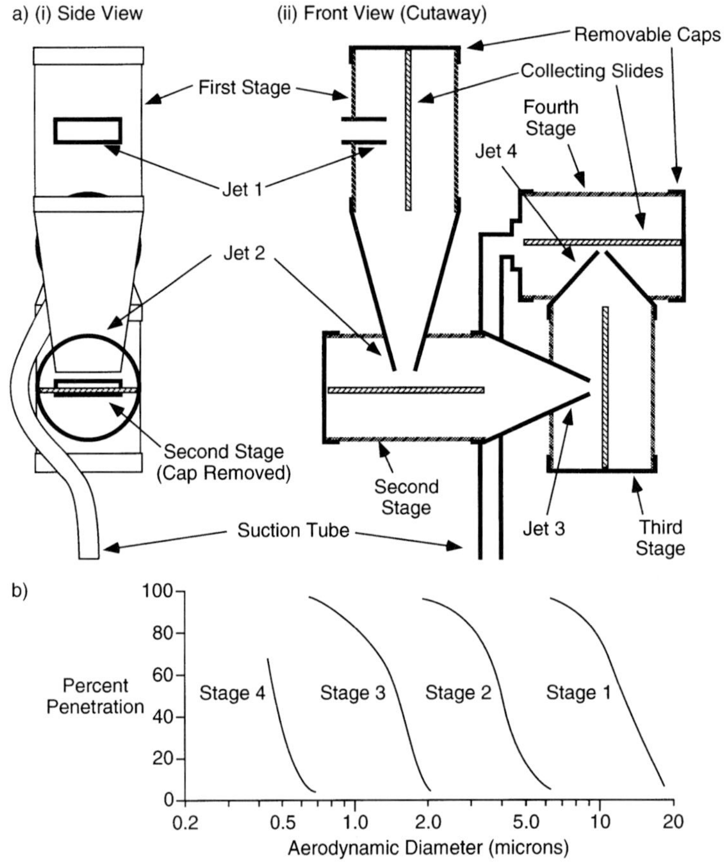

Figure 5.12. The May cascade impactor has four stages, each using a smaller jet set a smaller distance from the collecting slide in order to increase the centrifugal forces to remove smaller fineparticles from the aerosol stream. a) (i) Side view of the May cascade impactor. (ii) Front cutaway view of the impactor showing the internal structure of the stages. b) Performance of the impactor for a flow rate of 17.5 l/min.

fineparticles of the various stages, for normal operating conditions, are summarized in Figure 5.12(b). It can be seen from Figure 5.12(b) that fineparticles not normally considered a respirable hazard are deposited by the stage 1 jet [30].

In the Lundgren cascade impactor system the fineparticles are collected on revolving drums placed opposite the impactor jets [31]. Each drum has a total surface of 10 in^2. In one model, each drum rotates once every 24 hours, which is a useful feature in air-pollution studies. In another version of the equipment eight speeds of rotation of the collector drums, varying from one revolution per minute to one revolution per day, can be selected. The drums can be covered with either foil or paper for collecting the impacted fineparticles.

In the Unico cascade impactor the impacted fineparticles are collected on glass slides placed opposite to rectangular slits acting as jets. At the final stage, residual aerosol is filtered on a membrane filter. Each collection plate is attached to a geared wheel that can advance the plates simultaneously by 1/16 in. to provide a fresh collection surface whenever necessary. A chart is provided with this instrument for calculating the penetration efficiency per stage under various operating conditions for various types of aerosol [32].

In these instruments, glass slides to be used to collect droplets can be covered with magnesium oxide, deposited from a smoke, in such a manner that arriving droplets create craters in the magnesium oxide. From the dimensions of these craters, the size of the droplet can be evaluated. Another technique is to add dye to the droplets so that as they fall on a collecting filter paper they leave coloured disks, the size of which can be related to the size of the droplet. Each filter medium tends to absorb the droplet in a different way, therefore there is no unique relationship between the droplet diameter and that of the disk formed on the collection material. The spreading factor, which relates the spread-out disk diameter to that of the original droplet, has to be evaluated for any particular collection system. Another technique is to add a small amount of acid to the liquid being used to generate a test cloud of aerosol. Droplets of such an aerosol when collected on litmus paper leave circles of differing colour, the size of which can be related to the dimensions of the droplets.

Two of the major problems encountered in the use of jet impactors are referred to as bounce and re-entrainment. Bounce occurs when hard fineparticles are impacted onto the collecting surface. Because of their kinetic energy, they are able to bounce back into the air stream and be carried away. Bouncing problems can be avoided by coating the collecting surface of the impactor with grease. Strips of "two-sided sticky" tape can be placed on glass slides and used to collect solid fineparticles. The grease, however can cause problems in any subsequent study of the deposited fineparticles.

Re-entrainment occurs when aerosol flow over the collected fineparticles lifts them back into the flowing stream. When using cascade impactors, bounce and re-entrainment problems can cause sampled dust to appear finer than it is. What is known as a virtual surface has been developed to avoid problems due to bounce and re-entrainment without using grease. The basic principle of the virtual impactor is illustrated in Figure 5.13(a). Instead of a solid surface, the aerosol stream is diverted by a volume of static air in a reservoir. This static air constitutes a virtual surface, and hence the name of the device. The larger fineparticles unable to follow the aerosol stream enter the static air reservoir. Experiments carried out in the development of impactors using virtual surfaces demonstrated that a completely static reservoir of the type shown in Figure 5.13(a) suffered from turbulence at the edges of the orifice. It has been shown that

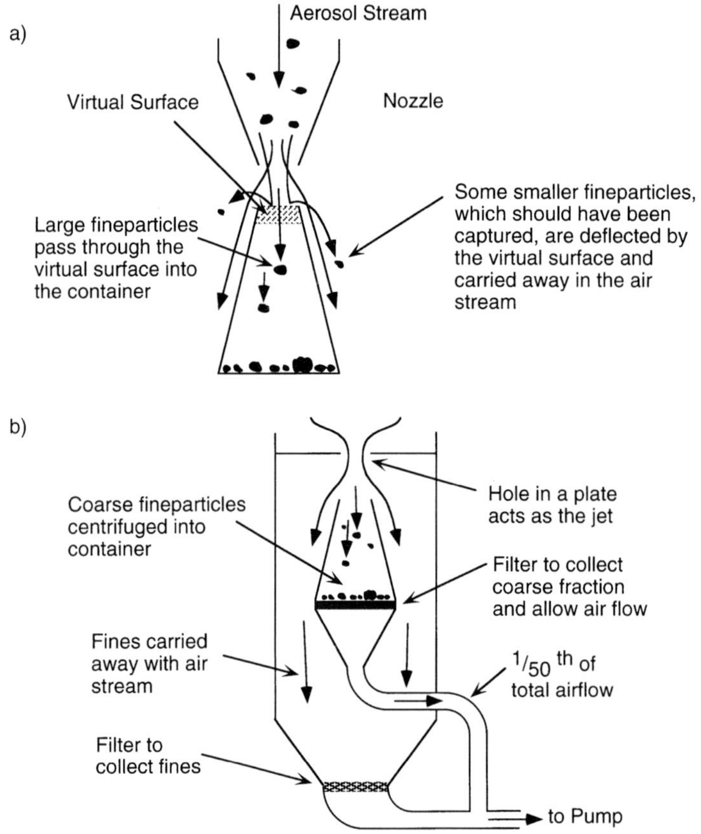

Figure 5.13. A virtual impactor uses a reservoir of air, which forms a virtual surface at the top of the reservoir, to capture fineparticles in the attempt to avoid the problems of bounce and re-entrainment. a) A simple reservoir used to create a virtual surface can resonate and cause some fineparticles to be lost. b) Using a filter at the base of the reservoir, to allow 1/50th of the total flow to pass through the reservoir, eliminates the problem of the resonating virtual surface.

better fractionation at the virtual surface is obtained if there is a very slow flow rate of gas out of the virtual impactor reservoir, as shown in Figure 5.13(b). This low flow rate of air out of the oversize zone is typically one-fiftieth of the airflow over the virtual surface [33, 34]. An aerosol sampling device utilizing the principles of virtual impaction is the dichotomous sampler illustrated in Figure 5.14.

The Centripeter® impactor uses virtual collection surfaces [35]. The construction of this instrument is illustrated in Figure 5.15 [36]. In this device the aerosol stream is forced to turn sharply at the orifice entrance. The centrifugal action created by this sharp diversion is greater than that in a normal jet impactor of comparable orifice di-

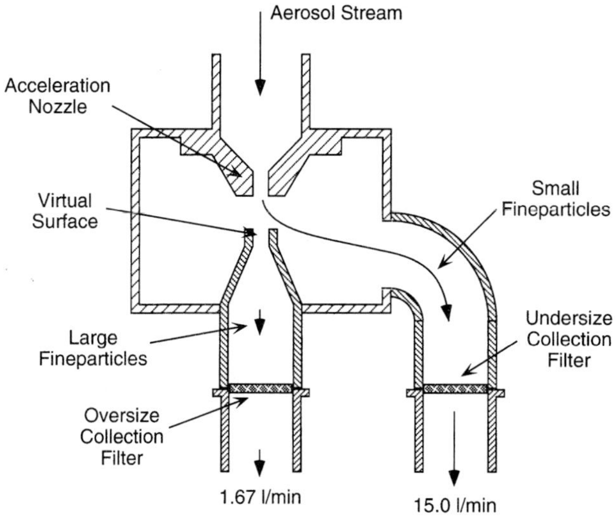

Figure 5.14. The Dichotomous sampler uses a virtual surface with a small airflow through the collection chamber, to help separate respirable and non-respirable fineparticles onto separate collection filters.

ameter. The word "centripetal" is based on two Latin words that mean "center seeking", and hence the name of this device in which the fineparticles are forced to seek the center of the airflow because of the forces created in the sharply turning air in the feed nozzle of the equipment. At each stage of the centripetal impactor the oversize fineparticles impacted into the reservoir are slowly drawn to the collection filter by a low flow rate of air down through the reservoir forming the virtual surface. The undersize fineparticles move with the air stream diverted around the collection orifice down to the next stage.

In the "3200 Particle Mass Monitor system" the collection surface is a piezoelectric crystal coated with metal on two sides connected into an oscillator circuit. A piezoelectric crystal alters its shape when it is subjected to an electrical field. Thus, if an alternating electrical field is applied to such a crystal, the crystal expands and contracts in response to this electric field; that is, it acts as an oscillator. The frequency at which a piezoelectric crystal oscillates is related to the mass of crystal. If the effective mass of the crystal changes, as a result of the deposition of fineparticles onto its surface, while it is oscillating, the frequency at which it oscillates shifts. This frequency shift can be related to the mass of fineparticles deposited on the crystal. The aerosol being studied is deposited onto the crystal by a combination of electrostatic forces and impaction forces. The change in oscillation frequency is proportional to the mass of deposited aerosol fineparticles. In this instrument a reference crystal is used to allow for any drift

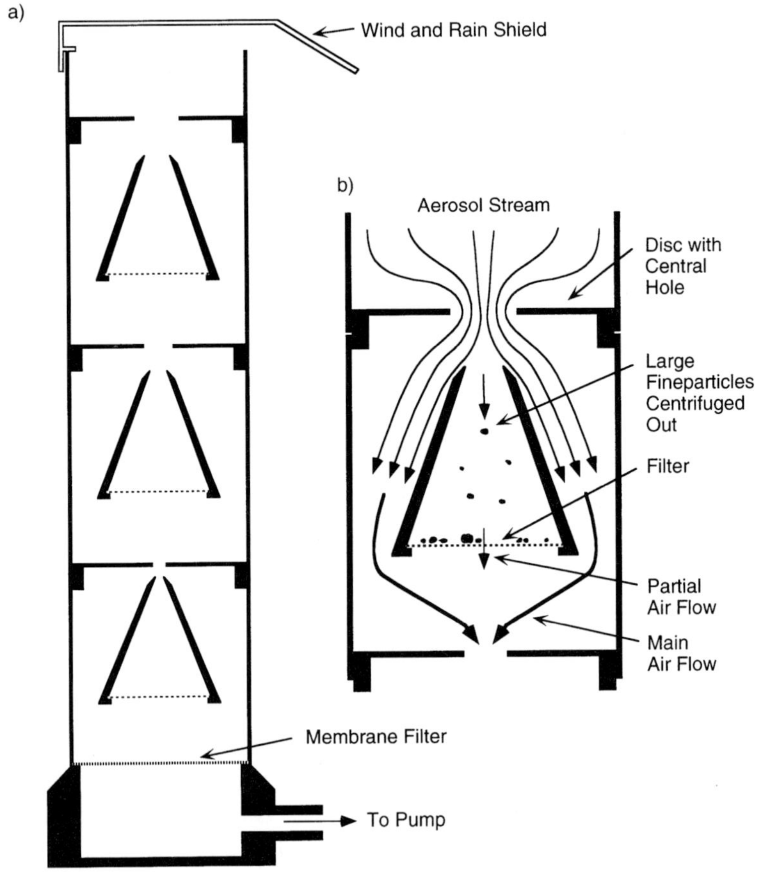

Figure 5.15. Sketch of the Centripeter cascade virtual impactor. a) Cutaway sketch of the interior of a three stage Centripeter. b) Operational details of a single stage of the Centripeter.

in the temperature of the instrument. The frequency of the deposition surface crystal is constantly compared to that of the reference crystal [37].

In Figure 5.16(a) the system used in an instrument known as TEOM being used by the Bureau of Mines and other scientists to evaluate respirable dust is shown. The term TEOM stands for Tapered Element Oscillating Microbalance. The name describes the essential element of Figure 5.16(a) shown separately in Figure 5.16(b). It is interesting to note that the TEOM monitor evolved from space research projects aimed at measuring the mass of dust grains encountered in the tails of comets. In space one cannot weigh objects because they do not have weight in the absence of a large gravitational field. The TEOM device measures the mass of a fineparticle from the change in the oscillating behaviour of the element as the dust accumulates on the filter at the

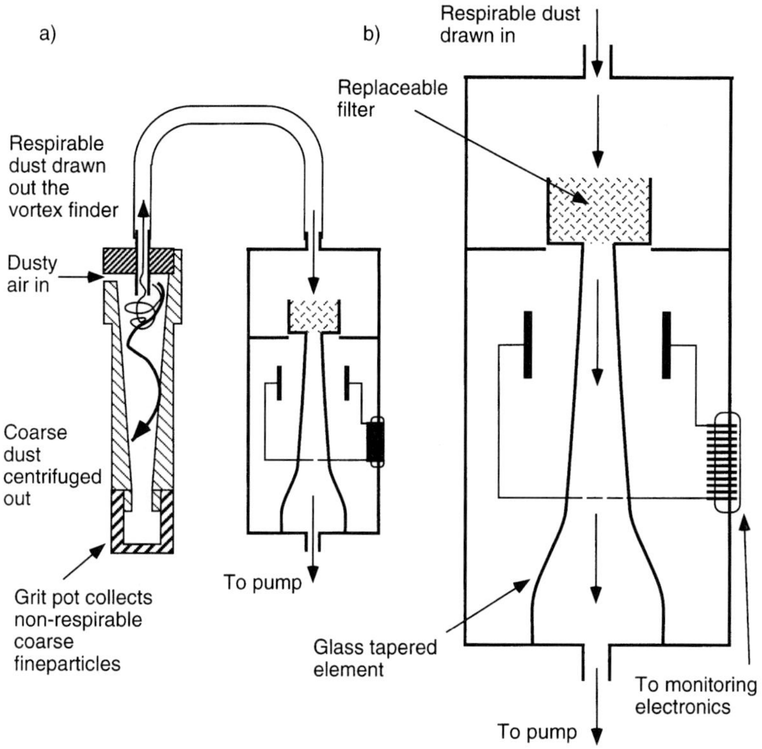

Figure 5.16. The Tapered Element Oscillating Microbalance (TEOM) can be used to monitor the amount of material being accumulated in the filter mounted on top of the glass element by measuring the change in the oscillation frequency of the element due to increasing mass. a) A system, incorporating a TEOM, which is used to monitor the respirable dust present in a working environment. b) Detail of the TEOM.

top of the tapered element. Because of the way in which it works, the orientation of the device is immaterial; it can be used upside-down or on its side depending on the available space for mounting the device. When measuring dust in the work environment the device is equipped with a prestage cyclone that removes anything other than respirable dust from the air stream. Fineparticles which have respirable diameters, that is, smaller than 5 microns aerodynamic diameter, are deposited on the filter and at the end of a shift, the miner brings the TEOM element to a central point where the deposited mass from operation during a working shift is measured. [38–41]

Several sophisticated centrifugal instruments for concurrent sampling and sizing of aerosols were developed in the period from 1950 to 1980 [42–46]. These tend to have been displaced by laser diffraction studies (see Chapter 7) and time-of-flight aerosol spectrometers (see Chapter 6). For a review of the centrifugal instruments used in aerosol studies, see reference 47.

Another type of centrifugal elutriator employing air or gas as the fluid medium is the "counterflow classifier". The basic system used in this type of elutriator is illustrated in Figure 5.17. The suspension of fineparticles to be fractionated is injected tangentially as shown. The air stream spirals into the exit hole at the center. As the fineparticles move with the air, they are subjected to centrifugal forces that tend to move them outward to the periphery of the chamber and drag forces that tend to move them inward and out through the axial exit port. The fluid drag forces are proportional to the surface of the fineparticles, whereas centrifugal forces are proportional to their mass. The competition between these two sets of forces result in a cut size such that fineparticles larger than this cut size are collected at the periphery of the fractionation chamber whereas smaller ones leave with the air stream through the axial

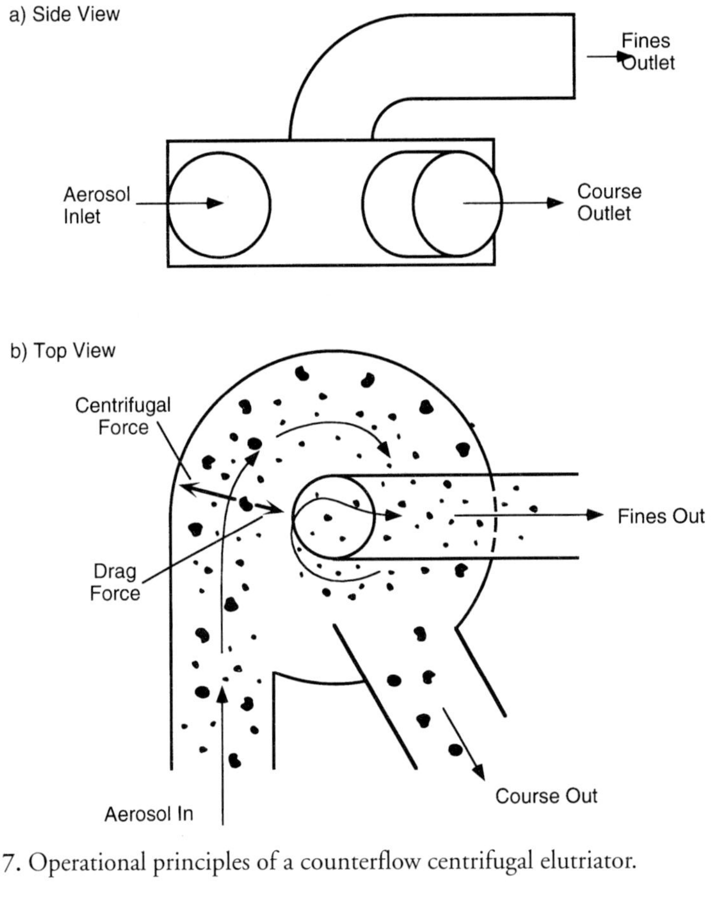

Figure 5.17. Operational principles of a counterflow centrifugal elutriator.

exit port. The cut size achieved in any particular counterflow centrifugal elutriator is usually established experimentally. Centrifugal counterflow elutriators are often referred to in the commercial literature as spiral flow classifiers.

The BAHCO classifier is a counterflow centrifugal elutriator. It is named after the Swedish company that developed it [48, 49]. The basic operating system of the BAHCO is illustrated in Figure 5.18. Except for the powder hopper, the whole of the system shown in Figure 5.18 rotates. Blades angled radially on the underside of the top plate (not shown in Figure 5.18) acts as a fan to draw air through the throttle at the base of the rotor. Entering air passes through a stack of radial disks. The narrow gap between these disks ensures that the air is spiraling as it moves into the fractionation chamber immediately above the disks. In this chamber the air-entrained powder feed

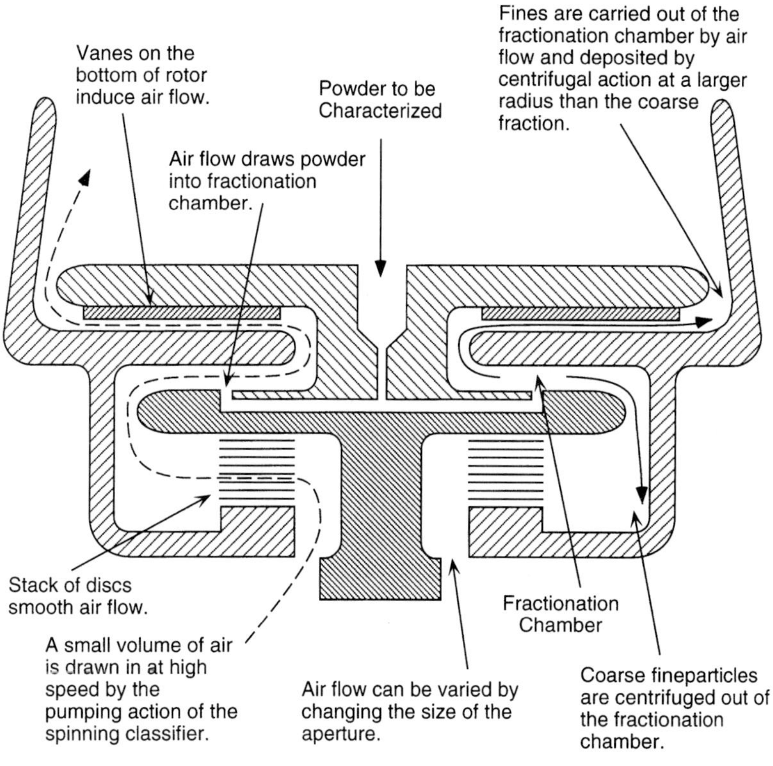

Figure 5.18. The BAHCO classifier uses centrifugal action, by rapidly spinning the entire instrument, and air flow, induced by vanes on the bottom of the rotor, to bring about separation of a powder into coarse and fine fraction. Fines are carried with the air flow out of the fractionation chamber then centrifuged at the large radius of the disc. Coarse fineparticles undergo sufficient centrifugal action to overcome the air flow and are deposited immediately to the wall of the disc.

emerges from a small orifice situated approximately half way between the periphery and the axis of the rotor, as shown. As the fineparticles acquire the motion of the spiraling air, they experience the opposing centrifugal and drag forces. The flow conditions at the point where the fineparticles enter the fractionation chamber determine the cut size of the elutriator. The airflow drops and centrifugal force increases outward from the orifice. Thus ensuring that fineparticles already moving outward will continue to do so. Similarly, fineparticles that start to move inward under the influence of the drag forces are in an accelerating airflow so that they are swept out of the fractionating chamber. At the outer edge of the top plate the airflow rate drops drastically. At the same time, the centrifugal force is at a maximum. Therefore, fines collect at the point shown in Figure 5.18. A series of fractions, either for specific studies or for purposes of characterizing a powder, is achieved by operating the instrument using various flow rates. The flow rate of air through the equipment is altered by means of collars that control the aperture at the entrance to the rotor attached to the base of the rotor.

Weilbacher and Rumpf [7] investigated factors that control the performance of counterflow centrifugal elutriators and pointed out that in an instrument such as the BAHCO, a fineparticle is only in the operating fractionation zone, determining which way it will move, for a short time. The effects of any instability in the fluid flow that caused a fineparticle to start to move in the wrong direction is, therefore, difficult to reverse. As a result of their studies, Weilbacher and Rumpf developed a new type of centrifugal counterflow elutriator. The commercial version of their equipment is known as Analysette® 9 [8]. The operation of this equipment is outlined in Figure 5.19(a). Powder to be fractionated is entrained into air fed tangentially into the fractionation chamber. The exit port on the far side of the periphery of the chamber is closed during the initial operation of the instrument. The spiraling air stream carrying fineparticles smaller than the cut size leaves the fractionation chamber at the center. The fineparticles larger than cut size move to the outer wall of the fractionation chamber and spin along the wall. When they have completed one circuit of the fractionation chamber, they are met by the incoming jet of air. This re-entrains them, and the fierce turbulence at this point tends to disperse any residual aggregates in the oversize fineparticles. It also flushes out any smaller than cut-size fineparticles that may have been attached to or trapped by the oversize fineparticles. This feed back mechanism can make oversize fineparticles circulate 2000 times per minute in the fractionating chamber. The oversize fraction is removed intermittently through the exit valve. Comparative data for the efficiency of fractionation achieved with the Analysette® 9 and the BAHCO are summarized in Figure 5.19(b). For details of other centrifugal counter flow fractionation equipment see Colon [50].

A factor limiting the efficiency of a fractionation achieved with liquid cyclones is the fact that the fineparticles spend only a short time in the high-shear region between

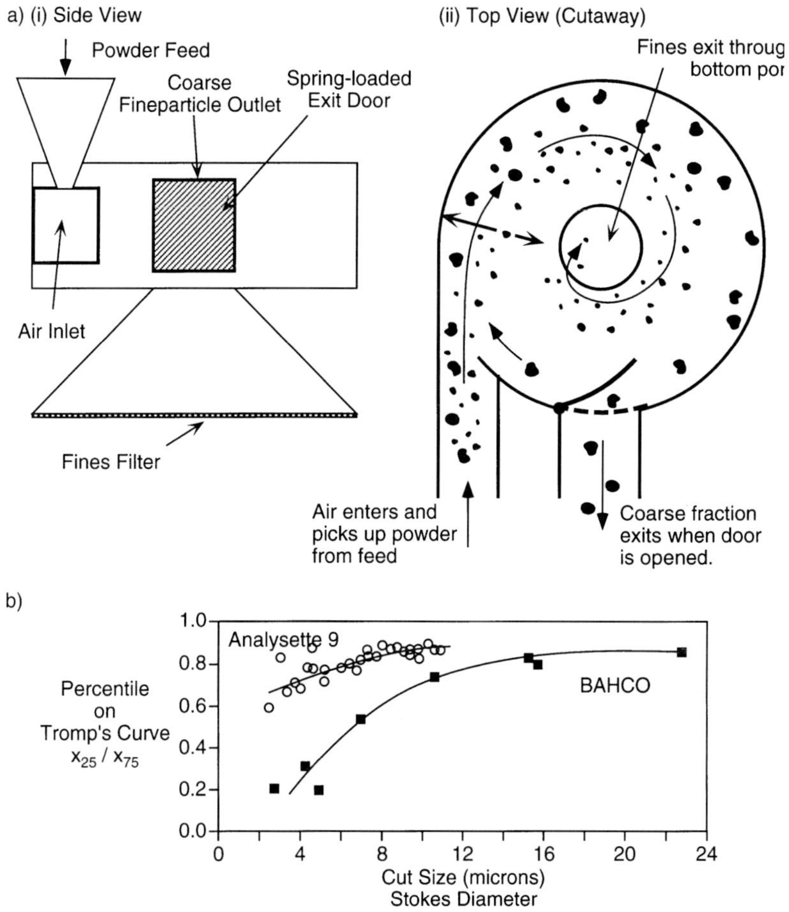

Figure 5.19. The Analysette® 9 counterflow centrifugal elutriator. a) Sketch of the Analysette® 9. (i) Exterior sketch. (ii) Top view showing the operational principles. b) Comparison of the performance of the Analysette® 9 to the BACHO classifier.

the outer layer of fluid and the inner vortex. Therefore, aggregates of fineparticles may not be dispersed and can drop down into the oversize fraction. To overcome this problem, Kelsal and McAdam inverted the cyclone as shown in Figure 5.20(a) [51]. In the inverted cyclone the oversize fraction thrown to the periphery of the cyclone migrates under the influence of gravity back into the high-shear region of the cyclone. Therefore any aggregates are given repeated opportunities to break down and participate in the fractionating action of the cyclone. The Warman cyclosizer shown in Figure 5.20(b) is based on the innovations of Kelsal and McAdam [52]. It utilizes a cascade of in-

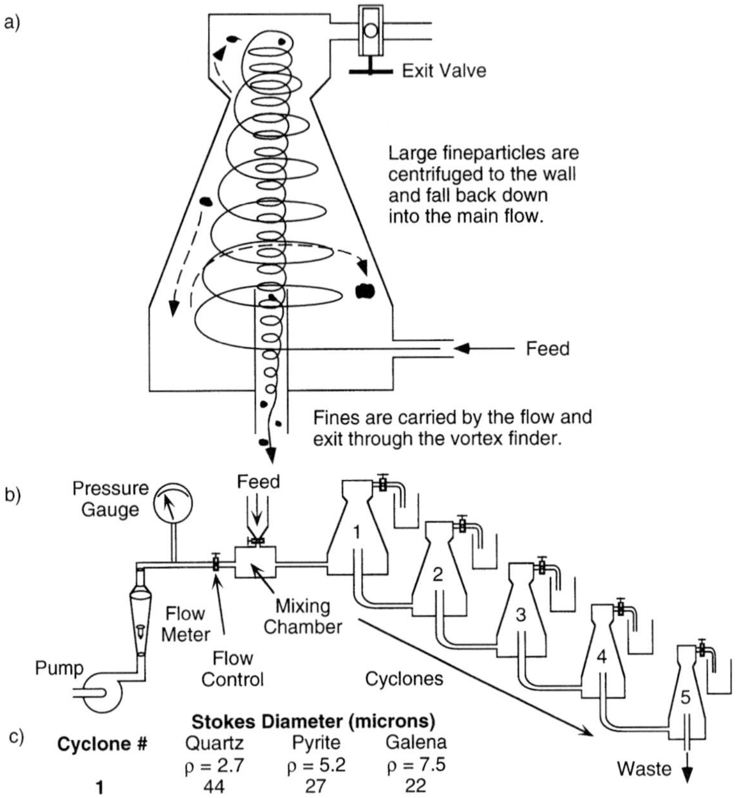

c)

Cyclone #	Stokes Diameter (microns) Quartz $\rho = 2.7$	Pyrite $\rho = 5.2$	Galena $\rho = 7.5$
1	44	27	22
2	35	22	18
3	23	14	11
4	15	9	8
5	9	6	5

Figure 5.20. The Cyclosizer uses inverted cyclones of decreasing cut size, linked in series, to fractionate a sample. a) Operation of a single inverted cyclone. Oversize fineparticles fall back into the active zone of the cyclone. b) Appearance of the Warman Cyclosizer. c) Table of the upper size retained by each cyclone, for materials of various density.

verted cyclones to fractionate a suspension of powder. To operate the equipment, a slurry of the fineparticles to be fractionated is fed into the system as shown in Figure 5.20(b). An operating period of 10 minutes is usually sufficient to achieve efficient fractionation. The instrument was developed for use in the mining industry which is concerned with relatively coarse fineparticle populations. Performance data for various mineral powders under normal operating conditions are summarized in Figure 5.20(c). The manufacturers of the instrument claim that the effective range of the instrument is 5–80 microns for material of specific gravity similar to quartz. For particles of higher specific gravity, such as Galena, the manufacturers suggest that efficient and satisfactory fractionation may be achieved down to 4 microns Stokes diameter [52, 53].

In the early 1970's H. Small developed a novel technique for fractionating fine particles which has been described as hydrodynamic chromatography (HDC) [54]. As explained at the start of this chapter when a viscous fluid flows over a surface in a tube there is a velocity gradient extending from zero at the surface up to maximum velocity in the center determined by the properties of the flow agent and geometry of the pipe. In the same way, when a viscous fluid moves down through a packing of uniform spheres the velocity of flow near the surface is zero and the flow in the center of a pore is at a maximum. The fineparticles to be fractionated in HDC are suspended in a fluid which is made to flow through the void spaces between spheres packed into a column. Because of the viscosity gradient in the pore spaces, the smaller fineparticles will spend more time traveling in the slower moving fluid near the surface of the spheres. The larger particles cannot approach as close to the surface of the packing spheres and will spend more time in the higher flow velocities occurring in the center of the pore structure. For this reason, the large particles will require less time to travel the length of the column of packed spheres than the smaller particles. In Figure 5.21(a) the fractionation of two different sized latex spheres by HDC as reported by Small are shown.

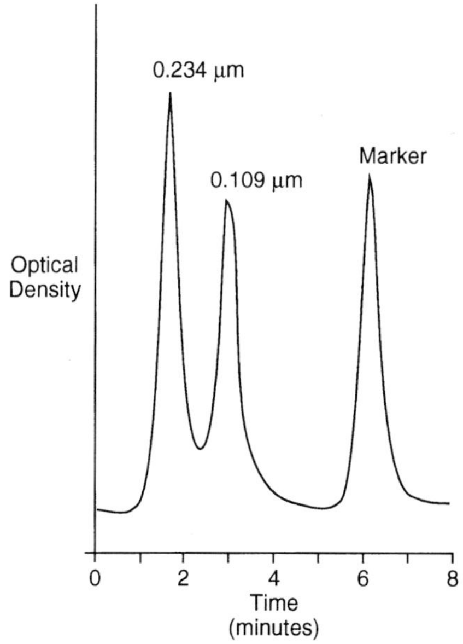

Figure 5.21. Hydrodynamic chromatography results for a mixture of two types of standard latex spheres and a marker [54].

Commenting on the development of HDC Provder states that

> *"The procedure was invented by Small in an industrial laboratory and first reported in 1976. The technology was licensed to an instrument company for commercialization. In 1986 HDC was in the process of trying to gain a foothold in the marketplace. In 1991 (when Provder was writing) HDC is not a viable commercial instrument and failed in the market place primarily because the separation mechanism, which requires fractionation and separation of particles in a packed column, had a flaw that was not fully overcome. During the fractionation process some of the particles being separated would deposit on the column packing and recovery of the injected sample was less than 100 %." [55]*

In spite of these difficulties with the development of a commercial version of the equipment, the method has been successfully used to study many colloidal sized fineparticles. Provder goes on to state that

> *"the basic concept of HDC was a good one but since 1986 a new technique for particle distribution called capillary hydrodynamic fractionation has become a commercial reality. In this technique separation and fractionation of less than one micron particles occurs in capillary tubes. Such separation avoids the problems of particle deposition on a packed bed." [55]*

In capillary hydrodynamic fractionation, particles under study are introduced under laminar flow conditions in capillary tubes. A colloidal particle suspended in this laminar flow will, by Brownian motion, move in and out of the direction of flow and the larger particles will be excluded from the slower stream lines closer to the capillary wall. Based on this exclusion effect the particles will fractionate. It is claimed that using this method, shown in Figure 5.22, several types of colloidal particles cab be fractionated in a period of eight minutes. In Table 5.2 data obtained by this method

Table 5.2. Comparison of the average fineparticle size of various samples of carbon black as determined by Capillary Hydrodynamic Chromatography (CHDC) and Dynamic Light Scattering (DLS).

Sample	CHDC mm	DLS mm
5030	0.142	0.143
5083	0.228	0.210
5249	0.205	0.198
5402	0.227	0.200
5428	0.191	0.198
6131	0.170	0.167
8178	0.171	0.198

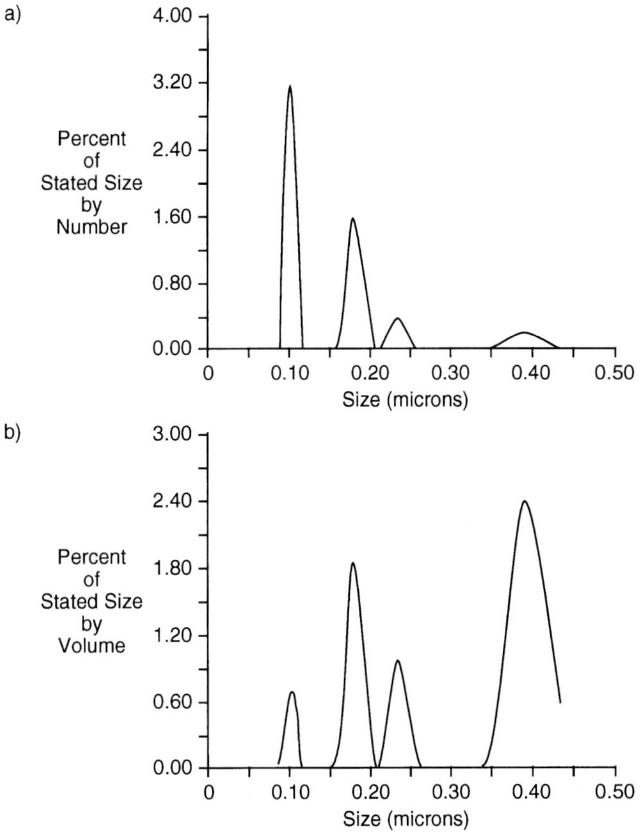

Figure 5.22. Results for Capillary Hydrodynamic Chromatography of a mixture of colloidal fineparticles fractionated over a period of eight minutes [55].

are compared to the size generated from dynamic light scattering (See Chapter 8) for carbon black pigments [55–59].

Another group of methods using elutriation technology to fractionate and characterize colloidal particles and other particles as large as 45 microns, stems from the original work of Giddings [60]. This group of methods is known as Field Flow Fractionation and is referred to as FFF or "F^3" with a prefix in order to designate the particular type of field flow fractionation being used [60]. Thus, SFFF is used to describe field flow fractionation using a gravity generated sedimentation field superimposed on the flowing fluid. In the words of Calvin Giddings who invented and pioneered the applications of field flow fractionation.

"FFF outwardly operates like chromatography separating components within a flowing stream that emerges from a column for sample detection and collection.

However it differs from chromatography in one fundamental respect – separation is induced by an external field or gradient such as an electrical or magnetic field or a thermal gradient instead of interaction with a stationary phase."

Giddings has pointed out that many different force fields can be used in FFF techniques. The technique which has received most study, sedimentation field flow fractionation, will be discussed in some detail and the interested will find details of the other methods in the references quoted [60, 61].

If one considers the parabolic flow front down a tube or a formed gap between two plates the movement of the fluid can carry the colloidal particles down the tube. In

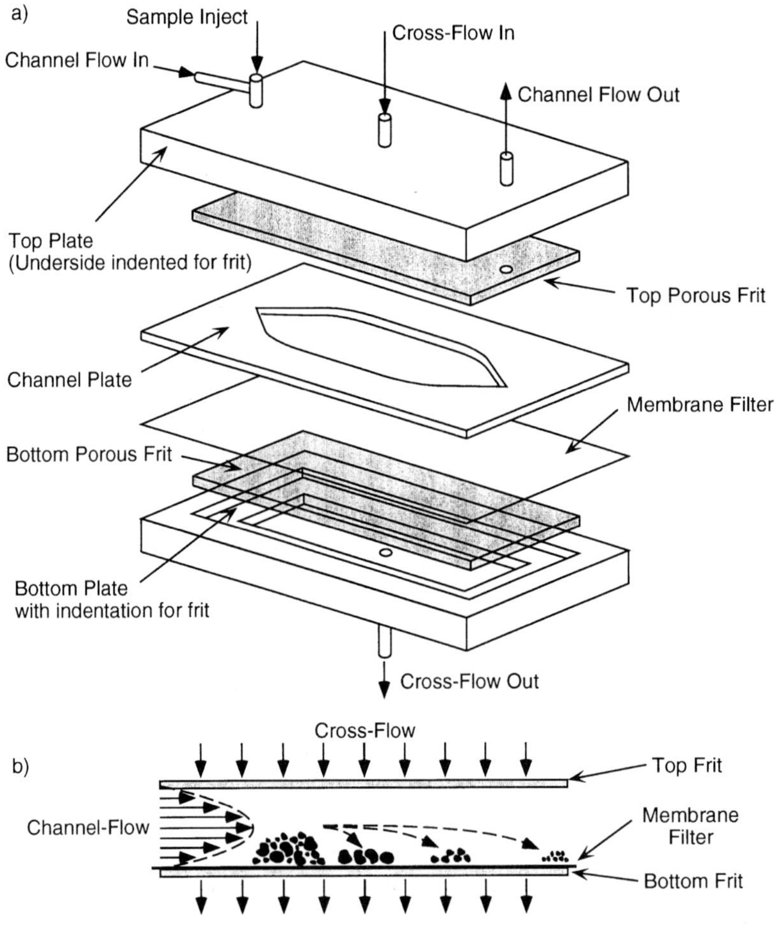

Figure 5.23. Construction and operation of a standard Field Flow Fractionation (FFF) channel. a) Internal structure of the FFF channel. b) Enlarged side view of the internal operation of an FFF channel [60].

sedimentation FFF's the force field is gravity. In this technique a ribbon like channel is cut out of a thin 0.050–0.5 mm plastic and the ribbon is sandwiched between two walls composed of materials that will seal the system and transmit the desired driving force. The system is shown in Figure 5.23. When the colloidal particles to be fractionated are injected at the entrance to the ribbon through which the liquid will flow, the wet stream spreads out into a fan across the channel as shown. Just after the sample has been placed in the channel the flow is usually stopped for a while to allow the fineparticle component to be settled on the floor of the ribbon by gravity. Essentially, in operating of the SFFF system, fractionation is achieved because of the competition between Brownian motion tending to lift the particles off the floor of the ribbon and the Gravity forces tending to return them to the floor. Potentially the larger colloidal particles are unable to lift very far from the floor of the ribbon and must travel along the floor of the ribbon under the influence of low flow velocities. On the other hand the smaller fineparticles are able, by Brownian motion, to move up into the higher velocity field of the parabolic velocity front and are carried further down the groove than the larger fineparticles. It follows that the smallest colloidal particles emerge first from the flow channel. To increase the resolution performance of the technique sometimes centrifugal force is applied to the ribbon of fluid using equipment illustrated in Figure 5.24.

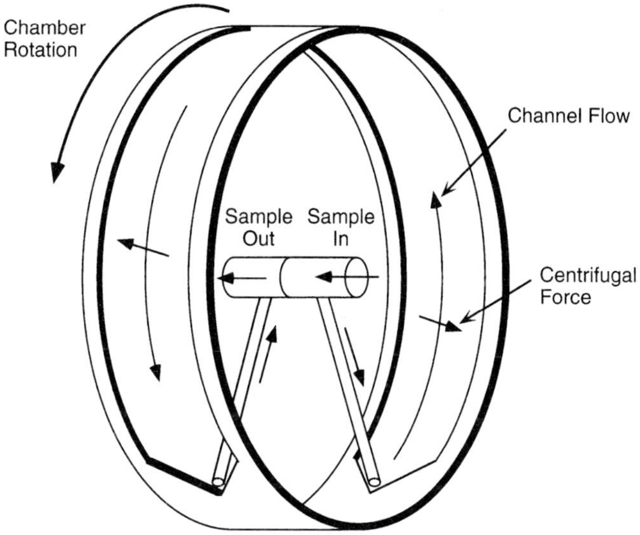

Figure 5.24. Sketch of a Field Flow Fractionation chamber which uses centrifugal force, induced by rotation of a cylindrical chamber about an axis, to act as the force driving the fineparticles to the wall of the chamber [60].

When the colloidal particles to be fractionated are larger than one micron then the physical size of the particles in the ribbon interferes with the competition between Brownian motion and the viscous flow essentially because, for such larger fineparticles, Brownian motion is much less than for smaller fineparticles and also the larger fineparticles cannot be confined to the lowest velocity stream in the parabolic flow profile. However, in such a situation one can exploit the fact that the larger fineparticles protrude into the higher velocity flow and are carried more quickly down the ribbon. For these larger fineparticles the reverse phenomenon occurs and the larger fineparticles emerge first from the SF^3 ribbon of fluid. This variation of the field flow technique exploiting the larger fineparticles is known as Steric FFF.

FFF has proved to be extremely useful in the study of colloid fineparticles and is being used to separate materials of high molecular weight and also the debris from

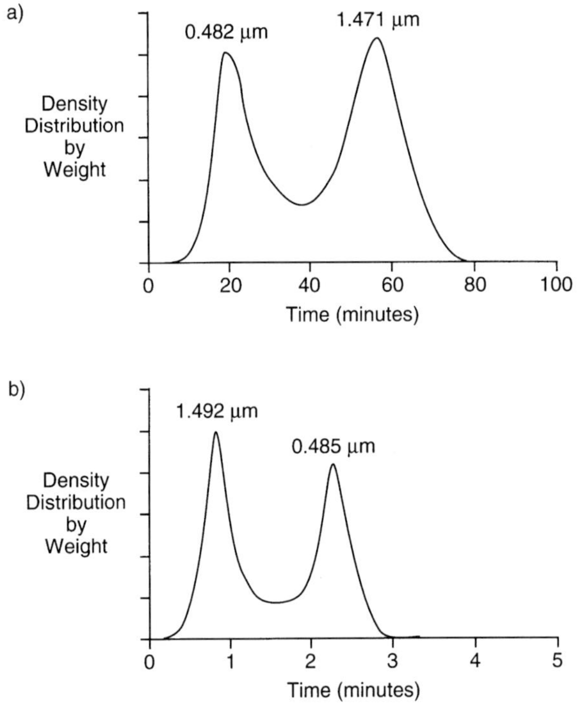

Figure 5.25. Field flow fractionation fractograms of an iron oxide pigment. a) Fractionation using normal flow conditions in the chamber. (Conditions in a 210 μm thick channel; channel flow 5.5 ml/min; cross-flow 0.8 ml/min;) b) Fractionation using Stearic flow conditions on the chamber [60]. (Conditions in a 100 μm thick channel; channel flow 7.4 ml/min; cross-flow 1.7 ml/min)

dissolved material from biological cells. Symposia on the use of Field Flow Fractionation have been held [60]. (See also chapters in reference 55.) The resolution achievable using Steric FFF is illustrated by the data shown in Figure 5.25. The output from the field flow fractionation by analogy with chromatography is called a fractogram. In figure 5.26 the fractogram of the separated sample of latex spheres is shown.

T. Schauer has used modified flow field flow fractionation to determine the particle size of each individual constituents of a three component mixture of paint binder, pigment and filler, characterized by relatively broad and overlapping distributions. He was able to follow the changes on mixing them [61].

In this chapter we have concentrated on laboratory scale air classifiers. Large-scale air classifiers for use in industry have been discussed by several workers [62, 63, 64].

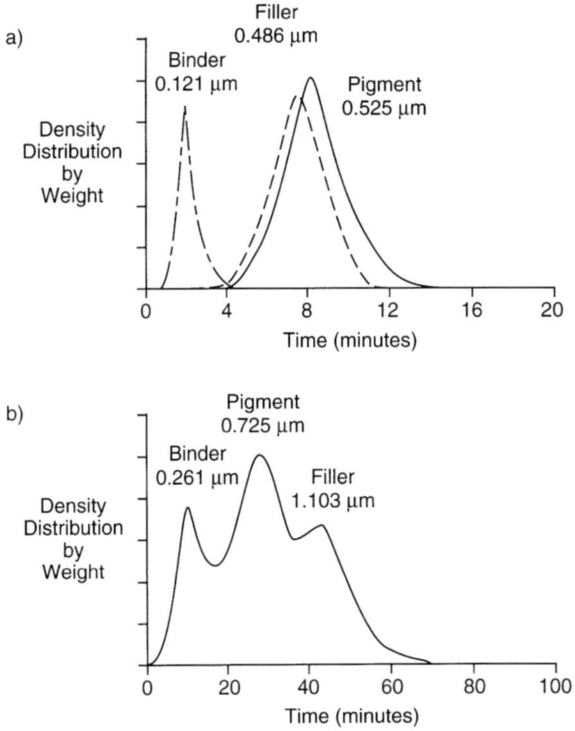

Figure 5.26. Field flow fractionation fractograms of paint pigment components [60]. a) Fractionation of the individual paint components superimposed. (Conditions in a 100 µm thick channel; channel flow 5 ml/min; cross-flow 1.5 ml/min); b) Fractionation of a mixture of the paint components. (Conditions in a 210 µm thick channel; channel flow 5.5 ml/min; cross-flow 0.8 ml/min;)

References

[1] See Chapter 6 in B. H. Kaye, Chaos and Complexity; Discovering the Surprising Patterns of Science and Technology, VCH, Weinheim, 1993.

[2] T. Scholz, D. R. Uhlmann, and B. Chalmers, "Elutriation Particle Separator," *Rev. Sci. Instrum.,* 36 (December 1965), 1813–1816.

[3] For a discussion of the use of Tromp's curve, see K. Leschonski and H. Rumpf, "Principle and Construction of Two New Air Classifiers for Particle Size Analysis," *Powder Technology.* 2 (March 1969), 175–186.

[4] H. W. Gonell, "Die Bestimmung der Kornzusammensetzung, Staubfor minger Stoffe ins Besondere von Zement," *Ton Industrie Zeitung* No. 13 (1929); *Ton Industrie Zeitlung,* No. 25 (1935).

[5] C. J. Stairmand, "Some Practical Aspects of Particle Size Analysis in Industry," in Symposium on Particle Size Analysis (1947), *Suppl. Transact. Inst. Chem. Eng.* 25.

[6] O. Lauer, Grain Size Measurements on Commercial Powders, Alpine AG. Augsburg, Germany (available in English and German).

[7] M. Weilbacher, H. Rumpf, *Aufbereitungstechnik* 9 (1968) 223–300.

[8] Available from Fritsch and Company, D-6581, Idar-Oberstein, 1, West Germany.

[9] Agents for Fritsch in the United States are Geoscience Instrument Corporation, 435 East Third Street, Mount Vernon, New York 10553.

[10] For a discussion of the use of Tromp's curve see K Leschonski, H. Rumpf, "Principle and Construction of Two New Air Classifiers for Particle Size Analysis", *Pow. Technology.,* 2(1969) 175–186.

[11] See Chapter 3 of B. H. Kaye, Powder Mixing, Chapman and Hall, London, 1997.

[12] P. S. Roller, *Ind. Eng. Chem.* (April 1931).

[13] "Subsieve Analysis of Granular Metal Powders by Air Classification," *ASTM Part B,* 293–360.

[14] "British Standard Methods for the Determination of the Particle Size of Powders," Part 3, "Elutriation Methods," *British Standards* No. 3406, Part 3, 1963.

[15] The Roller Elutriator is available from the American Instrument Company, 8030 Georgia Avenue, Silver Spring, Maryland, USA.

[16] The Aminco classifier is available commercially from the American Instrument Company; see reference 13.

[17] H. E. T. Haultain, *Transactions of the Canadian Institute of Mining and Metallurgy,* 40 (1937), 229. See also E. W. Price, *Ind. Eng. and Chem. Process Design and Development,* 1962, p. 79.

[18] Information on the Haultain Infrasizer is available from the Department of Mining, University of Toronto, Ontario, Canada.

[19] H. N. Blythe, E. G. Pryor, and A. Eldridge, "Recent Developments in Mineral Dressing," *Inst. Mining Met. (Lond.)* (Symposium 23 to September 5, 1952), p. 11

[20] Commercial information on the Hexhlet Aerosol Sampler is obtainable from the Casella Corporation, Regent House, Britannia Walk, London, UK.

[21] V. Timbrell, "The Terminal Velocity and Size of Airborne Dust Particles." Suppl. No. 3, *Br. J. Appl. Phys.* (1954). Commercial information is available from Flemming Instruments Ltd. Caxton Way, Stevenage, Hertfordshire England.

[22] K. T. Whitby and W. E. Clarke, "Electric Aerosol Particle Counters and Size Distribution Measuring System for the Range 0.015 to 1 Micron Size." *Tellus* 18 (1966), 573–585. Commercial information on the Whitby aerosol analyzer is available from Thermo Systems, Inc., 2500 Cleveland Avenue North, St. Paul, Minnesota, USA 55113.

[23] W. H. Marlow, "A Limitation on Electrical Measures of Aerosols," *NBS Publication* 464, 1977, p. 213.

[24] H. Rumpf, K. Leschonski, *Chemie-Ingenieur-Technik*, 21 (1967), 1231.

[25] K. L. Metzger, K. Leschonski, "Untersuchung Eines Quernstromsichters zur Online-Teilchengrossenanalyse," *Dechema Monographien,* 79 (1589–1615), Part B (1975), 77–94.

[26] K. Leschonski, K. L. Metzger, U. Schindler, "A New Online Particle Size Analyzer," proceeding of Salford Conference on Fineparticle Characterization, September 1977; Proceedings, M. J. Groves, Ed. Heyden & Son, London, 1978.

[27] Information on the Konimeter is available from Sartorius, 34 Weenderlandstrasse, Gottingen. West Germany 96102

[28] Sciotic Corporation, 4037 Robert Road. Co1umbus, Ohio, USA 43228.

[29] A. A. Anderson, "A Sample for Respiratory Health Hazard Assessment." *Am. Ind. Hyg. Assoc. J.*, 27 (1966), 160–165. Commercial information on the Anderson sampler is available from "2000, Inc.," 5899 South State Street Salt Lake City, Utah, USA 84107.

[30] K. R. May, *J. Sci. Instrum*, 22 (1945), 187. Commercial information on the May cascade impactor is available from Casella Corporation: see reference 21, or BGI Incorporated: see reference 42.

[31] D. A. Lundgren, *J. Air Pollution Control Assoc.*, 17(1967), 225. Commercial information is available from Sierra Instruments, Box 909, Vil1age Square Carmel Valley, California, USA 93924.

[32] S. M Lippmann. *Ind. Hyg. Assoc. J.*, 22 (1961), 348. Commercial information is available from Unico Environmental Instruments, Inc., P. O. Box 590, Fall River, Massachusetts, USA 02722.

[33] T. G. Dzubay, R. K. Stevens, C. M. Peterson, "X-ray Fluorescence analysis of environmental Samples", in Application of the Dichotomous Sampler to the Characterization of Ambient Aerosols, T. Dzubay (ed.), Ann Arbor Science Publishers, Ann Arbor, Michigan, 1978.

[34] B. W. Loo, J. M. Jaklevic, F. S. Goulding, "Dichotomous Virtual Impactors for Large-Scale Monitoring of Airborne Particulate Matter", in Fineparticles, B. H. Liu (ed.), Academic, New York, 1976, 311–350.

[35] R. F. Hounam and R. J. Sherwood. "The Cascade Centripeter, a Device for Determining the Concentration and Size Distributions of Aerosols." *Am. Ind. Hyg. Assoc. J.*, 26 (March/April 1965) 122–131.

[36] Commercial literature on this equipment is available from BGI Incorporated, 58 Guinan Street, Waltham, Massachusetts 02154 or Bird and Tole Ltd., Bledlow Ridge, High Wycombe, Buckinghamshire, England.

[37] J. G. Olin, F. J. Sem, and D. L. Christenson, *Am. Ind. Hyg. Assoc. J.*, 32 (1971), 209. Commercial information is available from Thermo-Systems Inc.; see reference 24.
[38] K. L. Williams and R. P. Vincent, "Evaluation of the TEOM Dust Monitor," *Bureau of Mines Information Circular*, 1986, United States Department of the Interior.
[39] H. Patashinck and G. Rupperсht, "Microweighing Goes on Line in Real Time," *Research and Development*, Technical Publishing, June 1986.
[40] Commercial information available from Ruppercht and Patashinck Inc., 17 Maple Road, P. O. Box 330, Voorheesville, NY, USA 12186.
[41] H. Patashinck and G. Ruppercht, "Advances in Microweighing Technology," Reprinted from *Am. Lab.*, July 1986.
[42] A. Goetz, "The Aerosol Spectrometer, A New Instrument for the Analysis of Airborne Particles in the Sub Micron Range," *Public Works Magazine* (February 1959), 91–93. Commercial information on the Goetz aerosol spectrometer is available from the Zimney Corporation, 1900 South Myrtle Avenue, Monrovia, California 91016.
[43] D. Hochreiner and P. M. Brown. *Env. Sci. Technol.* 3 (1969), 813.
[44] D. Hochreiner, "A new centrifuge to measure the aerodynamic diameter of aerosol particles in the submicron range," *J. Colloid, and Interface Sci.*, 36 (1971), 191.
[45] W. Stober and H. Flachsbart, *Env. Sci Technol*, 3 (1969), 1280.
[46] J. H. Burson, A. Y. H. Keng, and C. Orr, "Particle Dynamics in Centrifugal Fields", *Powder Technol.*, 1 (1968), 305–316. The elutriator developed by Burson and colleagues was made available commercially by the Micromeritics Corporation under the name "Micromeritics Aerosol Sampler Model 1001." Information is available from Micromeritics Instrument Corporation, 800 Goshen Springs Road, Norcroft, Georgia 30071.
[47] See Chapter 7 of B. H. Kaye, Direct Characterization of Fineparticles, John Wiley and Sons, New York, 1981.
[48] The BAHCO classifier was developed at the laboratories of Hjorth and Company, Bahco, Sweden. See H. Ashley, "Particle Size Analysis by Elutriation and Centrifugation, as Exemplified by the BAHCO Microparticle Classifier," in Particle Size Analysis, J. D. Stockham and E. G. Fochtman, Eds., Ann Arbor Science Publishers Incorporated, P. O. Box 1425, Ann Arbor, Michigan 48106. The BAHCO classifier is known in the United States as the Microparticle classifier and is available there from the Harry W. Dietert Company, 9330 Roselawn Avenue, Detroit, Michigan 48204.
[49] Committee Number TA5 of the Air Pollution Control Association (dust, fumes, mists, and fog collectors) has recommended that the BAHCO classifier be adopted as a standard instrument for determining the size characteristics of fineparticles when specifying and testing industrial dust collecting devices. A standardized operation procedure for use with the BAHCO for this purpose has been prepared by the American Society of Mechanical Engineers. The procedure is entitled "Determining the Properties of Fine Particulate Matter," Power Test Code PTC 28, American Society of Mechanical Engineers, New York, 1965.
[50] F. J. Colon, J. W. Van Heuven, and H. M. Van der Laan, "Centrifugal Elutriation of Particles in Liquid Suspension," in Proceedings of the Conference on Fineparticle Characterization, Bradford, England, 1970, British Society for Analytical Chemistry, Savil1e Row, London, 1971.

[51] D. F. Kelsal and J. C. H. McAdam, "Design and Operating Characteristics of a Hydraulic Cyclone Elutriator," *Transact. Inst. Chem. Eng.* 41 (1963), 84.

[52] Commercial information is available from Warman International Proprietary, Ltd., 18–26 Dixon Avenue, Artarmon, Sydney, Australia.

[53] A. L. Hinde and B. J. D. Lloyd, "Realtime Particle Size Analysis in Wet Closed Circuit Milling," *Powder Technol.* 12 (1975), 37–50.

[54] H. Small, F. L. Saunders, and J. Solc. "Hydrodynamic Chromatography a new approach to particle size analysis" *Advances in colloid and interface science* **6** (1976) 237–266.

[55] Particle Size Distribution II – Assessment and Characterization. Edited by T. Provder. ACS Symposium series #472. Proceedings of a Symposium held Boston April 22–27, 1990 Published by American Chemical Society Washington D. C. 1991.

[56] B. A. Buffham, "Model independent aspects of hydrodynamic chromatography theory" *Journal of Colloid and Interface Science*,. 67:1 (October 15, 1978) 154–165.

[57] See commercial literature of Matec Applied Sciences 75 South St. Hopkinton MA, USA 01748.

[58] J. G. DosRamos, R. D. Jenkins, and C. A. Silebi, "Efficiency of Particle Separation in Capillary hydrodynamic fractionation," Particle Size Distribution Assessment and Characterization II Edited by T. Provder in Symposium series #472 Published by American Chemical Society, Washington DC 1991.

[59] J. McHughag (1984) Particle Size Measurement u sing Chromatography *CRC Critical Review of Analytical Chemistry* 15 pp. 63–117.

[60] J. C. Giddings "Field Flow Fractionation." *C&EN,* October 10.1988 pp. 34–45. Report on the Second International Symposium on Field Flow Fractionation, *American Laboratory, March* 1992. 40t–40x.

[61] T. Schauer, Symmetrical and Asymmetrical flow field flow fractionation for particle size determination, Part. Part. System Characterization, 12, 1995, pp. 284–288.

[62] S. R. DeSilva, D. C. Walsh, S.T Johannson, T. Bergstrom, Air classification computer models and experimental results, KONA, No. 9, 1991, p. 131

[63] Air classifiers for use in industry are available from Sisshin Engineering Company Ltd., Tommen America Inc., 1000 Corporate Grove Drive, Buffalo Grove, Illinois, 60089-4507.

[64] For review articles on FFF and HDC see H. Barth, Modern Methods of Particle Size Analysis, Wiley, New York, 1984.

6 Stream Methods for Characterizing Fineparticles

6.1 Basic Concepts

In stream methods the fineparticles to be characterized are passed in a stream through an interrogation zone. Changes in the physical properties of the interrogation zone caused by a fineparticle are related to the size of the fineparticle. In this book, stream methods for characterizing fineparticles are classified according to the type of physical signal used to monitor the interrogation zone behavior. Thus, if the changes in electrical resistance of the interrogation zone are monitored, the method is described as a resistazone stream counter. The resistazone principle has been widely exploited in such instruments as the Coulter Counter® and other instruments described in Section 6.2.

Some stream methods require special preparation of the fineparticles before they are presented to the stream counter. For example, if a resistazone stream counter is used, the fineparticles have to be dispersed in an electrically conducting fluid. Inappropriate techniques at this stage can alter the measured size distribution of the fineparticles. This aspect of preparatory technique is usually specific to a given fineparticle system, and I do not feel that it is useful to give a discussion of the many different types of fineparticle that may be encountered and the problems of dispersing them in particular fluids. Many manufacturers of instruments for characterizing fineparticles by stream methods are happy to give advice and assistance in selecting appropriate fluids and dispersion technology prior to the use of their proprietary instrument to characterize a fineparticle system.

In many stream methods for characterizing fineparticles it is not economical to measure the physical properties of the interrogation zone over all the zone for all the time used in the study. One economic compromise that is widely used in the design of stream counters is to design the system so that the statistical probability of multiple occupancy of the interrogation zone is negligible. When this assumption is made, all changes in the physical properties of the interrogation zone are assumed to be caused directly by the presence of a single fineparticle. In practice, some stream methods do not always achieve single occupancy of the zone. Multiple occupancy of the interrogation zone can give rise to three different errors in the measured size distribution of the fineparticle system: (1) primary count loss, (2) secondary count gain, and (3) shadow loss.

"Primary count loss" means that two or more fineparticles in the zone together results in the underestimation of the population frequency of the smaller fineparticles. Sometimes this primary count loss is accompanied by the fact that the multiple occupancy results in a false signal of a pseudolarge fineparticle, thus resulting in the measured frequency of these larger fineparticles being higher than in the real population. This apparent enrichment of the larger fineparticle population is known as secondary count gain. Sometimes the presence of one fineparticle desensitizes the interrogation zone for a short time with a subsequent failure to register the presence of a closely following fineparticle. This type of error is described as shadow loss. For example, in the acoustical stream counter described in Section 6.3, there appears to be strong evidence that during the period when the primary sonic signal from the passage of a fineparticle is decaying, the system is unable to register the presence of any other fineparticles entering the interrogation zone. Most manufacturers of stream counters for characterizing fineparticles recommend carrying out measurements at a series of increasingly diluted suspensions until further dilution leaves the measured size distribution of the fineparticle system unchanged. At this stage of dilution it is assumed that multiple-occupancy effects including close-following shadow effects, are no longer significant. Some workers have used statistical theory to predict the potential primary count loss and secondary count gain when using dilutions above those at which single occupancy can be assumed [1]. However, since the predictive statistics require a primary knowledge of the distribution function of the system being measured, this can only be a procedure of successive approximations utilizing initial measurements to refine the estimates of the underlying distribution function by statistical manipulation of data. Wherever possible, the technologist should opt for extreme dilution since this greatly simplifies the transformation of the appropriate experimental data to the desired size characteristic distribution.

Fractional occupation of the interrogation zone can also distort the measured fineparticle distribution function. For example, in some light scattering stream counters, the interrogation zone is not restricted physically but is a region defined by the optical illumination system. In such systems it is possible to have partial occupancy of the zone by a fineparticle traveling near the outer limits of the optically defined region. Designers of aerosol stream counters using optical interrogation zones have attempted to deal with this problem by using a sheath of clean air to confine the fineparticles well within the optically sensitive region of the interrogation zone.

When interpreting the data from a stream method, one must be concerned about the fact that the physical properties of the interrogation zone may not be isotropic. For example, in the literature on the Coulter Counter, a well-known resistazone stream counter, it was assumed for a long time that the signal generated in the instrument was independent of the location of the fineparticle within the zone. More recent work, however, has established that, if the fineparticle is close to the wall of the cylinder of fluid in the interrogation zone, the signal will differ from that generated by the same fineparticle moving down the axis of the interrogation zone. For high-resolution and accuracy work, the Coulter Electronics Corporation has introduced what it calls an "electronic editor" to reject signals from fineparticles not traveling down the center of the interrogation zone [2].

6.2 Resistazone Stream Counters

The physical change in the electrical resistance of a well-defined cylinder of conducting fluid caused by the presence of a fineparticle is the basis of the resistazone stream counter. Commercially available resistazone stream counters are the Coulter Counter, the Electrozone® Counter. and the Nanopar® Analyzer [2–4]. The basic system used in a resistazone stream counter is shown in Figure 6.1. The fineparticles to be characterized are suspended in an electrolyte and made to pass through a cylindrical orifice placed between two electrodes. The presence of a fineparticle in the orifice between the electrodes causes a voltage pulse. This pulse is amplified and recorded. Commercial instruments based on this basic principle became available in the early 1950's. Since then, the sophistication of the electronics used to evaluate and record the electronic pulses generated by a resistazone counter has increased tremendously. Current technology is able to investigate the shape and the duration of a series of sequential pulses, thus enabling the technologist to obtain information on the shape and size characteristics of a fineparticle system [5–8]. Research publications dealing with the applications of resistazone counters number many hundreds. Summaries of this technical literature are available from the manufacturers of the instruments. In this presentation we do not discuss special applications of the instruments, but confine the discussion to a review of the evolution of significant instrument features.

Consider the ideal case of a single smooth sphere present in the center of a cylindrical orifice. It is possible to deduce a simple relationship relating resistance changes in the orifice to the dimensions of the fineparticle and the orifice. Early workers with resistazone counters established that this relationship held for spherical fineparticles provided that the diameter of the fineparticle did not exceed 40 % of the diameter of the orifice. Other experimental studies show that, in general, using the simple orifices available in the early instruments, fineparticles with dimensions of less than 3 % of the diameter of the orifice could not be characterized easily. This lower limit has been reduced by modern techniques such as hydrodynamic focusing and the use of contoured orifice equipped with flow straighteners of the type discussed later in this section [6–8].

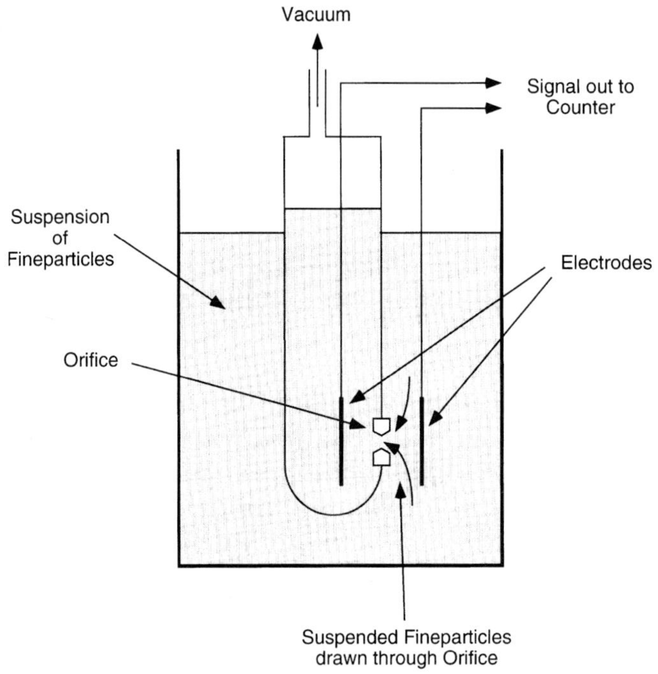

Figure 6.1. A basic resistazone counter uses the change in resistance of the volume of fluid between the two electrodes to count fineparticles as they are drawn through the orifice.

Because of the success in correlating the volume of spheres with the electrical signals generated by their passage through the orifice, it is often claimed that resistazone counters measure the volume of a fineparticle. It must be stressed that this is an unwarranted generalization. It has been shown that the signal generated by such counters is dependent on the shape of the fineparticle and hence cannot be related in a simple manner to the volume of an irregularly shaped fineparticle [5].

In practice, resistazone counters are calibrated by using standard spherical fineparticles. Often the assumption that the pulse height is proportional to the volume of the fineparticle is used to extend a one-point calibration by using spheres of known size to calibrate the instrument. Therefore, data from most resistazone characterization studies are not always absolute. and the sizes quoted are often for equivalent spheres. Any report of data from a resistazone counter study should clearly state any use of a calibration standard.

Early workers using resistazone counters anticipated that the behavior of metal fineparticles in the counter would differ from that of dielectric fineparticles. In practice, it was found that all fineparticles investigated behaved as if they had insulating properties. This has been variously attributed to the fact that many fineparticles have oxide layers on them or chemisorbed layers of nonconducting species that in the electrolyte serve to give the fineparticles insulating properties under the operating conditions of the resistazone counter.

Earlier models of the resistazone counters were found to be particularly sensitive to the presence of any chips or irregularities on the periphery of the orifice delineating the interrogation zone. Studies showed that this arose from the fact that the Reynolds numbers of the flow dynamics through the orifice were of the order of 100–1000. This is sufficiently close to the critical Reynolds number for the changeover from laminar to turbulent flow that the presence of a chip or an irregularity was likely to initiate turbulent flow. Eddies in the turbulent flow passing through the sensitive zone of the instrument were recorded as false fineparticles. It has been shown that a contoured orifice leads to a smoother flowline configuration at the entry into the orifice, with a consequent lessening of the effect of any problems due to chipping or any slight protuberance. Also, in general, the contoured orifice decreases noise signals in a characterization study and improves the resolution achievable with the instrument [6–8]. Since the operative size range of a characterization study is 3–40 % of the orifice diameter, the characterization of a wide-size-distribution fineparticle system can require the use of a series of orifices of different diameters.

The presence of fineparticles larger than the orifice will clog it. An instrument innovation that prevents problems arising from a blocked orifice is to protect the entrance to the orifice with an electrode made of metal screen with the apertures of the screen 40 % of the diameter of the orifice [6]. If this type of electrode is combined with a contoured orifice plus a flow-straightening collar, the resolution of the measurements is increased and background noise decreased. The orifice system incorporating these various modifications, all developed at IITRI (the Illinois Institute of Technology Research Institute) is illustrated in Figure 6.2.

Early discussions of the performance of resistazone counters assumed that the electric field in the orifice of the resistazone counter was isotropic and that the exact location of the fineparticle within the zone was unimportant. However, later investigations showed that the shape of the pulses generated by fineparticles off center in the orifice were different from those generated by those exactly in the center of the interrogation zone. Thus Grover in Israel [9] described the occurrence of four different types of pulse generated by fineparticles following different paths through the orifice. Similar results were reported by Davies et al. [7]. At about the same time, Thom [10] at the medical clinic of the University of Berlin published studies on the pulse shape obtained when spheres were drawn through an orifice on nylon threads. They showed that when the spheres moved along the axis of the orifice, one obtained smooth peaked pulses. Different-shaped pulses were found to be associated with the passage of the sphere on an off-axis trajectory, with the shape depending on the proximity of the sphere

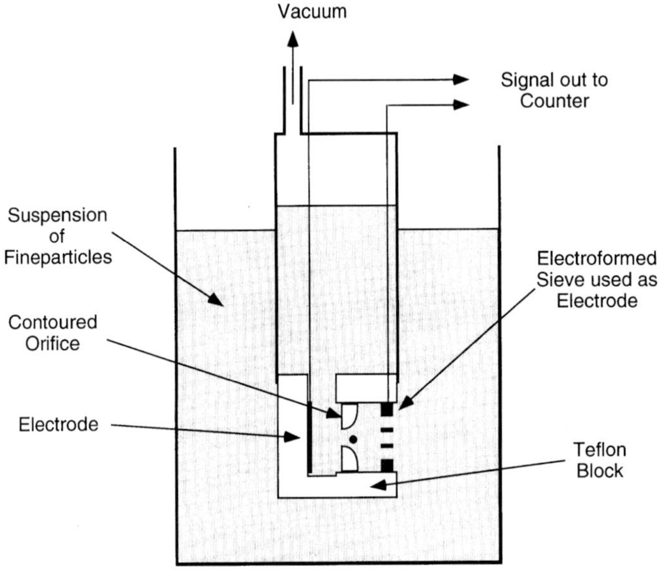

Figure 6.2. In the IITRI improved resistazone counter, a metal electroformed sieve acts as one electrode and prevents oversize fineparticles from entering the sensing zone or extremely large fineparticles from blocking the orifice. The hole through the Teflon block between the electrodes acts as a flow straightener.

to the wall of the orifice. It was found empirically that by cutting the edges of the orifice to create a 45° angle plane to the axis of the orifice, the distortion of the pulses could be reduced, but that the pulse magnitude was greater for peaks created by off-axis fineparticles than for those for on-axis fineparticles [11]. To avoid distorted pulses that result in error in the measured size, various approaches have been adopted. Thom used a capillary tube to fire the fineparticles down the axis of the orifice. This approach is effective with systems such as blood cells, where the range of sizes present is relatively small, but is difficult with a wide-size-range powder. Davies et al. concluded that their flow straightener plus contoured orifice in conjunction with the protective mesh electrode in effect had structured the flow into the orifice, thus enabling the fineparticles to move down the axis of the system [7]. The system developed by Thom and coworkers was exploited commercially by the German company Telefunken. The physical principles employed in the Telefunken system are illustrated in Figure 6.3(a). When suction is applied on the downside of the orifice, clean electrolyte is drawn through the orifice along with a stream of suspension from the capillary. The clean electrolyte is said to focus the suspension stream, and the technique is described as hydrodynamic focusing [12]. The Telefunken system has been licensed by the Coulter Electronics

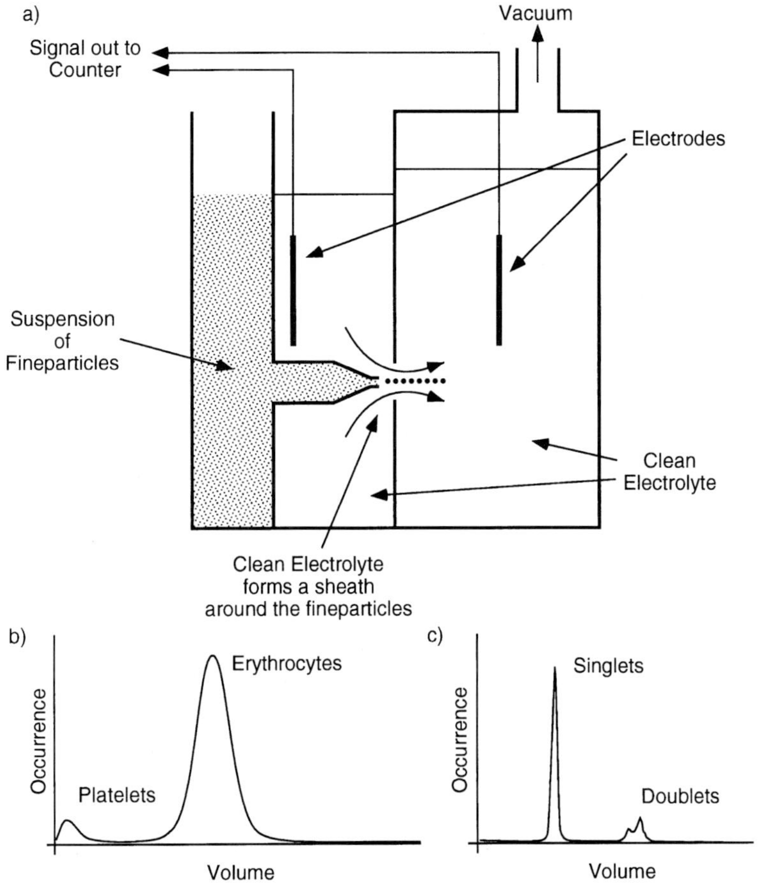

Figure 6.3. Hydrodynamic focusing is used in the Coulter Counter TF resistazone stream counter to ensure that fineparticles travel down the centre of the interrogation zone. a) Hydrodynamic focusing as used in the Telefunken system. b) Size distribution of whole blood by the Coulter TF system. c) Size distribution of standard 2.02 mm latex spheres.

Organization and sold under the name Coulter Counter TF®. In the commercial literature provided with the Coulter Counter TF® system, it is stated that the clean flow of electrolyte narrows down the stream of suspension leaving the capillary throat by a factor of 10. The high-resolution data obtainable with the Coulter Counter TF® system is illustrated in Figure 6.3(b) [2].

The Coulter Electronics Organization has developed an electronic editor that looks at the structure of a pulse and rejects all types except those characteristic of fineparticles traveling down the axis of the orifice [11]. Karuhn and Berg described the use of a combination of hydrodynamic focusing plus electronic pulse inspection to measure fineparticle shape [13].

Davies et al. have used a resistazone orifice shielded with a mesh electrode to measure fiber length [14]. They protected the orifice with an electrode having apertures of a given size. The registration of fineparticles having equivalent volumes greater than this aperture size is then interpreted as being due to the presence of fibers. To enable a whole range of fiber lengths to be investigated, a resistazone orifice was equipped with a steel wheel on which six electroformed screens of different aperture sizes were mounted. The wheel could be rotated to bring any given disk into position in front of the orifice. Davies et al. report that this system was rather difficult to operate in a leak-free manner but that fiber length data obtained were comparable to that obtained by microscope examination for the systems investigated [14].

DeBlois and Wesley used a Nuclepore® filter to create a resistazone counter of very small aperture that could be used to measure the size of small objects such as bacteria and viruses [4]. Typical pulses obtained by DeBlois are shown in Figure 6.4. The equipment is available commercially as the Nanopar analyzer® [4].

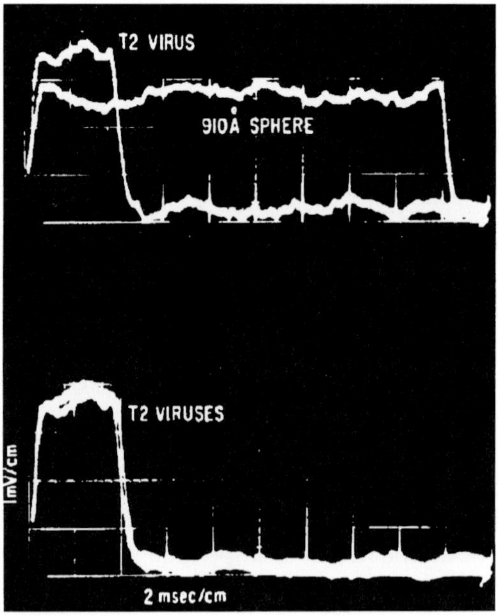

Figure 6.4. Using a Nuclepore membrane filter, DeBlois and co-workers made a resistazone counter with an orifice small enough to measure the size of viruses. (Reprinted from [38], John Wiley and Sons.)

Using modern electronic systems, and computer generated graphics, resistazone counters can generate highly resolved data. This is illustrated by the data of Figure 6.5. It was reported by Kahrun, using a sophisticated resistazone counter known as the Elzone™ [15]. In Figure 6.5(a), the number count for a phosphor powder used in the manufacture of television screens is reported, the geometric mean size is 1.327, and the arithmetic median size is 1.625. In Figure 6.4(b), the same data, transformed into a volume distribution histogram is shown. It can be seen that the bulk of the powder is log-Gaussian with a geometric mean diameter of 6.9 microns and an arithmetic mean size of 10.55.

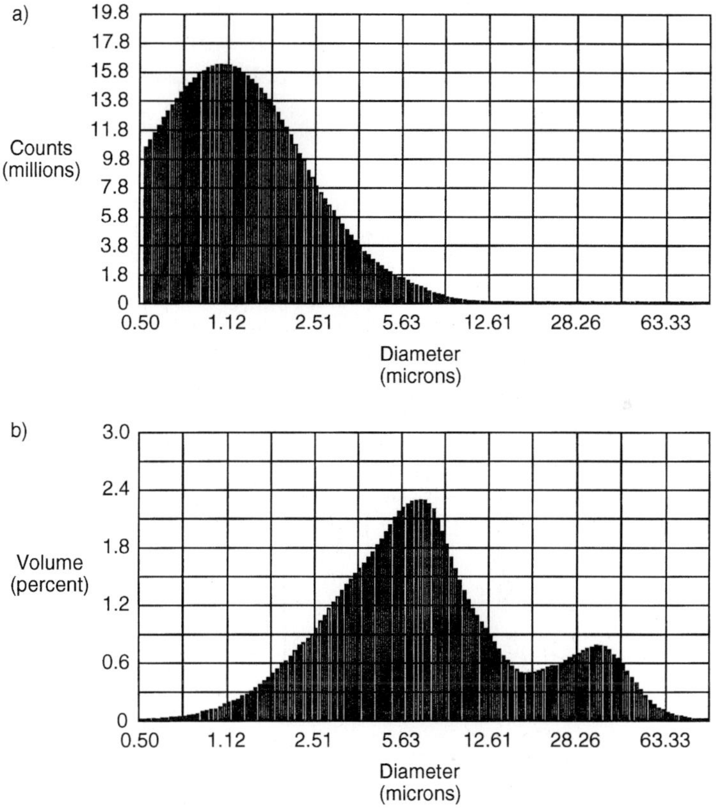

Figure 6.5. High resolution data from the Elzone resistazone counter for a phosphor powder used in the manufacture of television screens, as reported by Kahrun. a) Differential number data for the phosphor powder. b) Differential volume data for the phosphor powder. (Used by permission of R. Kahrun.)

The way in which the data from a resistazone counter is affected by shape of the powders involved is demonstrated by the data shown in Figure 6.6 generated by Alliet and Switzer [16]. The powders studied are shown in Figures 6.6(a) and (b). (The shape and surface structure of these powders were discussed in Chapter 2.) In Figures 6.6(c)

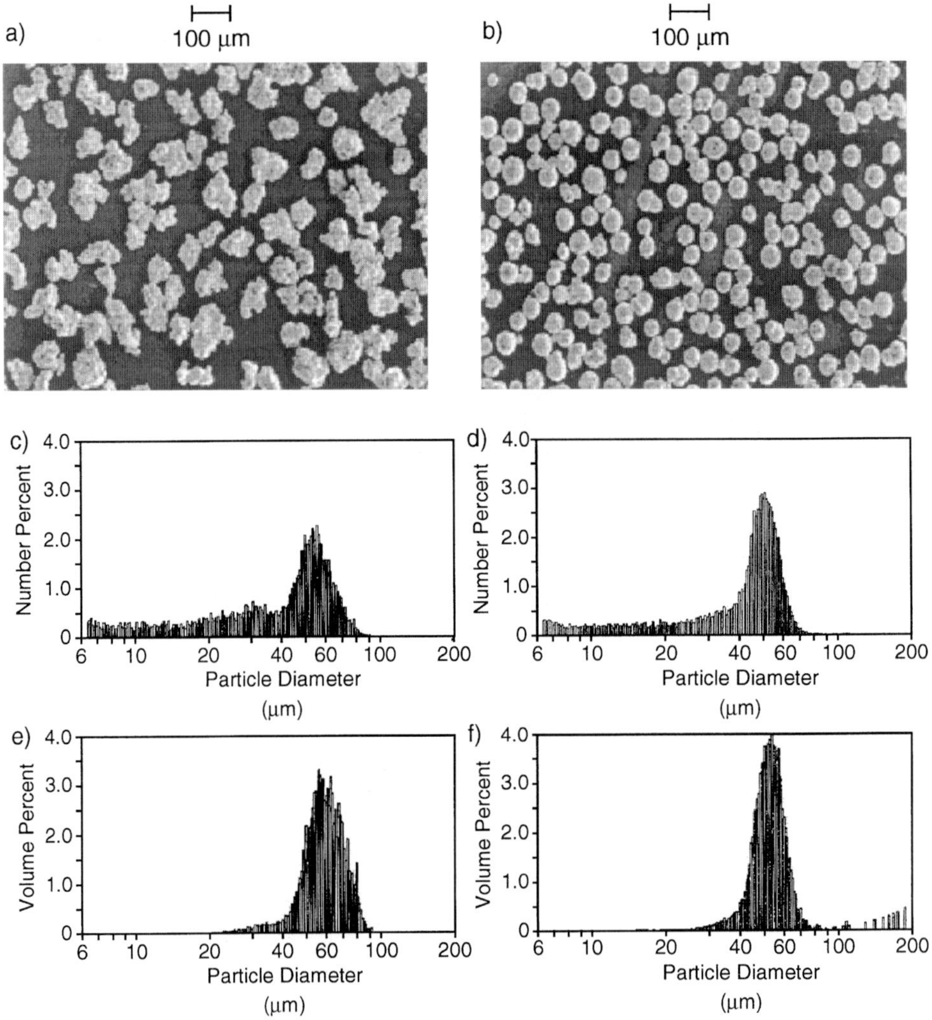

Figure 6.6. Size distributions of two metal powders by the Coulter® Multisizer Accucomp® show differences between the two powders due to the method used to produce them. a) Micrograph of an iron powder. b) Micrograph of a spray dried ferrite powder. c) and d) Number size distributions for the powders of (a) and (b). e) and f) Volume size distributions for the powders of (a) and (b). (Micrographs provided by D. Alliet of Xerox Corporation.)

and (d), the number distribution of the two powders as measured by the Coulter Counter is shown. It can be seen that the number distribution is essentially log-Gaussian with a small fine tail as shown in the diagram. The irregular shaped powder of (a) has been prepared by sieving a large amount of powder, whereas the powder in (b) had been prepared by spray drying. The volume size distribution of the powders as measured by the Coulter Counter are shown in Figures 6.6(e) and (f). From 6.6(f), it can be seen that the main size distribution is essentially log-Gaussian, with a lower geometric standard deviation demonstrating that the spray dried powder is a narrower ranged powder system. A very interesting feature of the volume distribution of the spray dried powder is the presence of minor populations of size 55 and 105 microns, which indicates the presence of doublets formed by collision of the falling spray droplets in the spray drying process. Note in particular, that two 52 μm droplets can combine to form a 105 μm agglomerate and that there is a subsidary peak at this point. The peak at 85 μm probably represents combinations of smaller and larger droplets in the original spray used to produce the droplets. Out in the larger sizes it can be seen that there is the odd larger agglomerate produced by the spray drying process. The graphs of Figures 6.5 and 6.6 indicate the different kind of information that is available on the size distribution of the powders using resistazone counters and even on the formation dynamics of the powder being studied using resistazone counters.

6.3 Stream Counters Based on Accoustic Phenomena

A stream counter in which the size of a fineparticle present in an interrogation zone is measured with the use of ultrasonic waves was reported by Albertson [17] and evaluated by Gayle and co-workers [18]. In this instrument, shown in Figure 6.7, a piezoelectric crystal mounted in the base of a fluid column functioned alternately as an emitter and receiver of ultrasonic waves. At the start of a measurement cycle the piezoelectric crystal emitted a short, converging pulse of ultrasonic waves. The crystal was then switched to a neutral mode for a time α, after which it became a receiving crystal for a very short period, $\Delta\alpha$. As the pulse of ultrasonic waves moved up the column, whenever a fineparticle was encountered a signal was reflected back to the crystal. However, only pulses reflected from a height equal $\alpha/2V$; where V is the velocity of the ultrasonic waves in the fluid containing the fineparticles to be characterized, are recorded. Thus the location and the dimensions of the interrogation zone and its height above the crystal were varied by altering the periods α and $\Delta\alpha$. The resolution of the technique was also limited by the duration of the burst of ultrasonic waves emitted from the crystal. In theory, the reflected signal is proportional to the area of

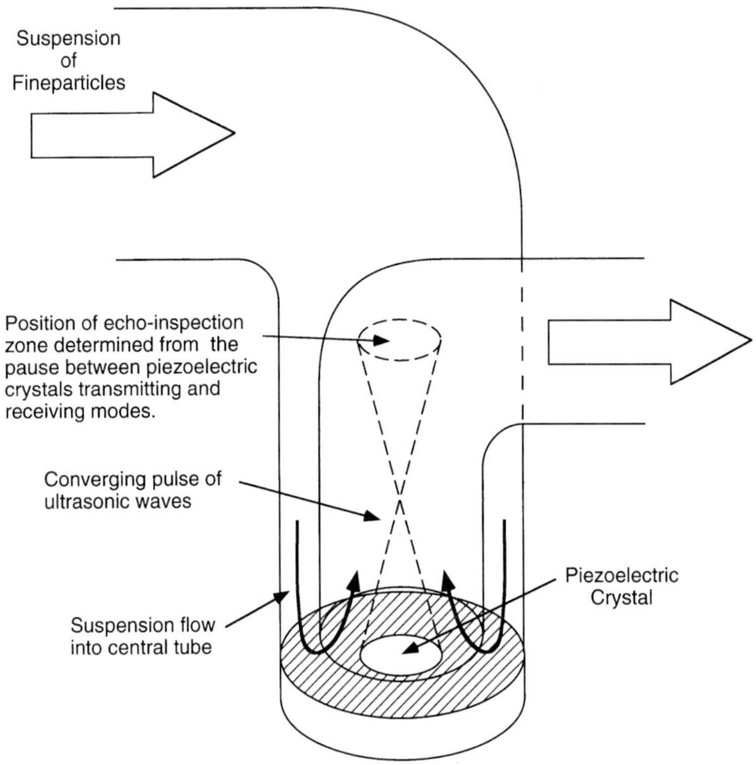

Figure 6.7. A sonizone stream counter, as suggested by Albertson, uses ultrasonics to inspect a suspension of fineparticles by operating a piezoelectric crystal in two alternating modes as a transmitter and a receiver [17].

the fineparticle; in practice, extensive calibration procedures had to be employed in order to make the instrument functional. It was found that turbulence and bubbles in the fluid moving through the instrument gave false signals during the measurement procedure. The instrument was originally developed for monitoring the cleanliness of opaque fluids since the procedure did not require that the fluid be either conducting or transparent. However, the instrument did not receive further development, presumably because of the difficulty of calibrating it. Future developments in ultrasonic technology may perhaps make it possible to reexamine this type of accoustic based stream counter. The developers of the instrument indicated that they thought that it could be developed to monitor fineparticles as small as 10 μm, but that at the stage of development reported in the literature, a more realistic limit was 25 μm.

The Autometrics Corporation has developed a system for determining both solids concentration in a suspension and the average size of the fineparticles in a stream counter that uses ultrasonic sound waves [19]. A beam of ultrasonics passing through a slurry of fineparticles suffers power loss from two sources [20]. The first source of power loss occurs because the fineparticles attempt to move with the wavefront of the ultrasonic wave. As the fineparticles attempt to move with the wavefront they experience viscous drag in the fluid, which leads to power dissipation. The second type of power loss occurs when sonic energy is diffracted out of the ultrasonic beam. It has been shown that for solids that are less than 20 % by volume of fineparticles, it is possible to find two different ultrasonic frequencies such that at one frequency the change in power is basically due to the mass of fineparticles present in the suspension, whereas at a different frequency, small changes in size distribution function of the suspended fineparticles will cause a large power loss. For any given slurry of fineparticles, these two different frequencies have to be determined by experimental investigation. The system used in the Autometrics device is illustrated in Figure 6.8(a). Typical performance data are summarized in Figure 6.8(b) [21]. The equipment is widely used in the process in industry for on line monitoring of the performance of crushing and grinding equipment.

The acoustic fineparticle counter was originally developed by Langer [22]. Langer observed that fineparticles moving down a vacuum hose created audible clicks as they moved through a constriction made by a clamp. Langer built and tested various-shaped capillary tubes and established that the sound made by a fineparticle could be heard only upstream of the capillary constriction and that the walls of the tube connecting the capillary to the aerosol stream had to be smooth and gradually contoured down to the capillary orifice size. He also showed that the shape of the exit from the capillary was immaterial and that no sound could be heard downstream of the capillary. He showed that the performance of the instrument could be greatly improved by making the entrance to the capillary section bell shaped. Later work by various investigators has established that the sound produced in this type of instrument is generated in the following manner. The Reynolds number of the airflow through the capillary has to be greater than or equal to 2000, and for this type of flow one would normally anticipate turbulent flow. Indeed, if rough surfaces are present in the throat leading to the capillary tube or in the capillary itself, the flow appears to break down into turbulent motion with consequent continuous noise generation. In the absence of a fineparticle, the contoured throat leading to the capillary section maintains laminar flow conditions (see discussion of Reynolds number in reference 23). Thus Scarlett showed by using Doppler shift measurements in laser light that the air jet emerging from the capillary tube maintained its integrity for a considerable distance down the tube and that the conditions inside the jet were laminar [24].

By measuring the pressure fluctuations down the capillary pipe section of the instrument, Karuhn postulated that the mechanism producing the sound was as follows. The fineparticle entering the capillary section of the instrument penetrates the boundary

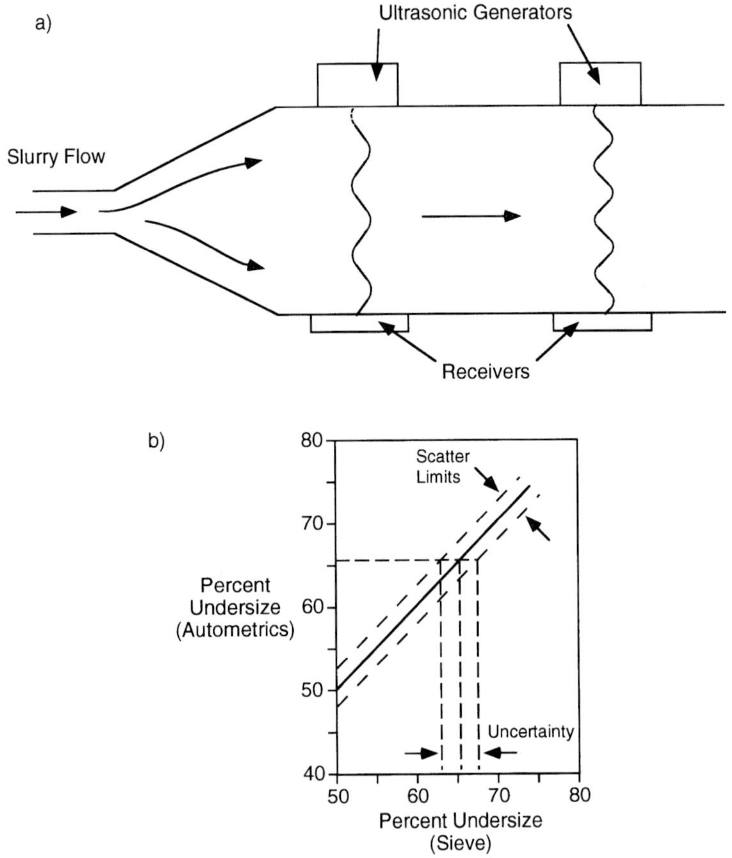

Figure 6.8. A sonizone stream counter by Autometrics uses two different ultrasonic generators to monitor the fineparticles passing the instrument [19]. a) Schematic of the Autometrics system. b) Comparison of results from the Autometrics stream counter with sieve analysis of crushed rock from a mine.

layer of the laminar flow initiating the shedding of a vortex. This vortex will travel at a much higher speed than the fineparticle and would be reflected by the end of the pipe to create a sound wave in the capillary [25]. Scarlett and Sinclair showed that the frequency of the sound generated by the fineparticle was dependent on the length of the capillary tube, and the length was related to half the wavelength of the emitted sound wave. Thus the capillary tube acts as an organ pipe with one end effectively closed and the bell-shaped end open. This explains the effectiveness of the bell-shaped entrance in both contouring the streamline flows into the capillary to maintain laminar flow, and at the same time acting as a resonator to enhance the note emitted by the capillary pipe when stimulated into acoustic emission by the vortex shed from the boundary layer by the fineparticle.

Until the acoustic signal has faded away and laminar flow conditions have been reestablished in the pipe, the presence of a subsequent fineparticle cannot be detected. It has been suggested that this type of acoustical counter could be very useful in measuring the gross level of dust fineparticles in industrial working areas. The extreme simplicity and robustness of the equipment makes it attractive for such situations. It should be noted that Scarlett and Sinclair established that the fineparticles emerging from the capillary were still accelerating and were traveling at velocities well below those of the gas. In general, the strength of the acoustic signal emitted from the counter does not seem to depend on the size of the fineparticle to within any achievable sensitivity at the present state of the art in acoustical engineering. However, Karuhn has pointed out that since the basic mechanism involved in the generation of the acoustical signal was an interruption in the type of flow present in the emergent jet issuing from the capillary section of the instrument, it should be possible to measure pressure changes in this jet and relate them to the size of the fineparticle causing the perturbation in the jet flow [25].

The use of ultrasonic measurements to study submicron particles has been reported by Scott, et al. [26]

6.4 Stream Counters Using Optical Inspection Procedures

In many stream counters the presence of a fineparticle is detected in an interrogation zone by means of an optical signal. The characteristics of the fineparticle are deduced from the properties of that signal. In passive photozone counters the optical signal is generated external to the interrogation zone and the fineparticle characteristics deduced from the magnitude and spatial distribution of the reflected signal. In an active photozone counter the fineparticles emit light when stimulated by means of some external energy beam. Thus fineparticles could fluoresce in the optical region when stimulated by ultraviolet light or laser light. Thus in an instrument known as a flow cytometer, used to study living cells such as phytoplankton in seawater (cytology is the biological study of cells), the cells being studied are suspended in a rapidly moving, laminar stream of liquid and made to flow in single file past an array of sensors which measure optical properties. At the point of interrogation, each cell is illuminated by laser or arc lamp. The resulting scattered light and fluorescence emitted, are detected by the sensors [27, 28, 29]. Active photozone counters have received little attention; however, they do have the advantage of being species specific and could pick out certain types of fineparticles present in a general aerosol or slurry stream. The need to detect specific air pollutants could give stimulus to the development of this type of stream counter.

There is a vast body of literature on the theory and design of optically based stream counters. In the space available, it is possible to give only a brief review of the various instruments that are available commercially. The light-scattering properties of fineparticles depend in a complex way on their refractive index, physical dimensions, and the wavelength of light incident on the fineparticles. Entire books have been written on the theory of light scattering; however, only in rare sample systems, such as spherical fineparticles, is one able to relate the size of the scattered light signal to the magnitude of the fineparticle by theoretical considerations [30, 31]. In practice, most of the widely used optically based counters are calibrated empirically by using standard fineparticles of known dimensions. Variations between the different optical designs of the various instruments makes it difficult to correlate their performance except on an empirical basis. When the fineparticles are considerably larger than the wavelength of light, the fineparticles can be considered as generating a geometric shadow so that the energy removed from the light beam is a simple function of the diameter of the fineparticle. In such situations it has become commercial practice to refer to the instrument as being based on the measurement of "light blockage."

As will be discussed in more detail in Chapter 7, when the dimensions of a fineparticle start to approach the wavelength of light, the complex scattering behavior is studied by use of scattering theory developed by a German scientist, Mie, whereas when the fineparticles are much smaller than the wavelength of light, the appropriate light-scattering theory is called the Rayleigh light-scattering theory. Other specialized light-scattering theories are known as Rayleigh-Gans and Lorentz-Mie theories [30, 31].

The first optically based stream counters were developed in the United States in the late 1950's and early 1960's. They were used to measure the size distributions of aerosols [32, 33]. The stream of aerosol to be characterized was injected into an illuminated area by using hydrodynamic focusing to provide a well-defined stream of aerosol passing down the middle of the optically defined interrogation zone. To minimize problems that could arise from using gases of different constitution and/or different temperature to that of the aerosol stream, the stream of gas used in the hydrodynamic focusing is usually obtained by filtering a diverted portion of the aerosol stream. For the same reason, any dilution of the aerosol required to achieve single occupancy of the interrogation zone should be carried out with a portion of the filtered aerosol stream.

The HIAC Royco Instrument Corporation manufactures a whole range of photozone stream counters [34, 35]. The optical system used in the different instruments depends on the information requirements of the particular study. For example, in Royco Model 203, shown in Figure 6.9(a), right angle scattered light is used to measure the fineparticles. This instrument is similar to a family of photozone counters developed at IITRI known by the family name Aerosoloscope [36]. In Royco Model No. 218, shown in Figure 6.9(b), a near-forward scattered light is used. The company recommends that the near-forward scattering instrument be used when one must monitor large volumes of gases in which the suspended fineparticles can vary widely in composition size and optical properties. The Royco Model 345, shown in Figure 6.9(c), is used to characterize fineparticles in a liquid suspension. In this instrument the energy removed from a forward traveling beam is measured. Royco instruments

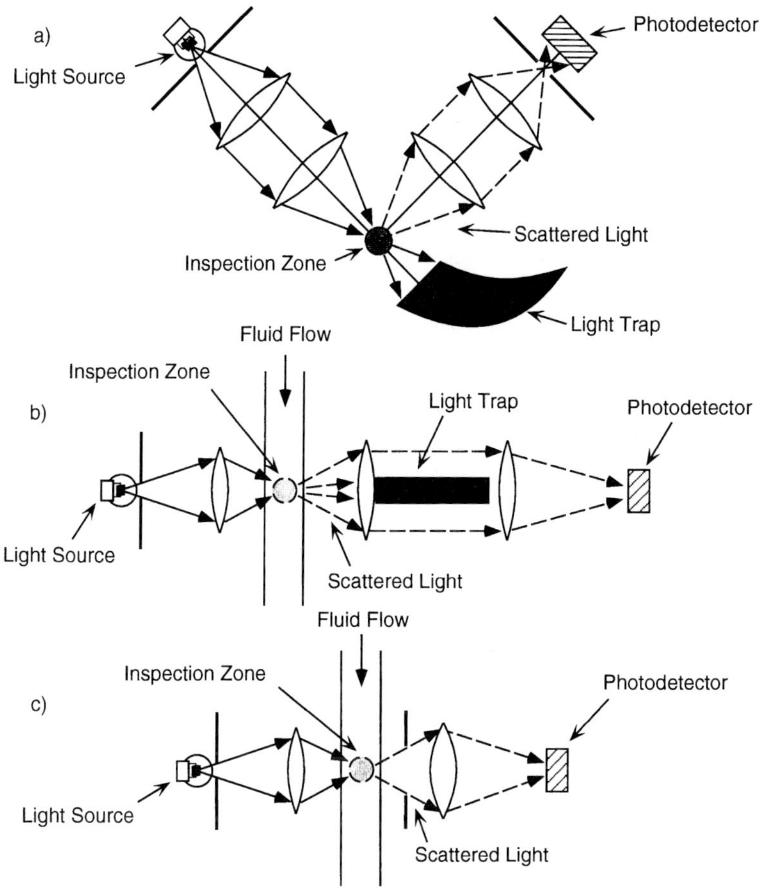

Figure 6.9. Royco corporation manufactures several different photozone stream counters which exploit different types of light scattering. a) Top view of the optical arrangement of the Model 203 photozone stream counter. b) Optical arrangement Model 218 photozone stream counter. c) Optical arrangement Model 345 photozone stream counter.

are calibrated using fineparticles of known characteristics. If one is interested in research studies of specific aerosols and/or fineparticles in a liquid suspension, the optics of the various instruments are well defined, and one can carry out more complex interpretation of experimental data by using the appropriate light-scattering theory. Note that the Bausch and Lomb Corporation used to manufacture a photozone stream counter for the characterization of aerosols [37, 38].

The Climet stream counter, shown in Figure 6.10, characterizes the fineparticles by collecting and measuring most of the forward scattered light and a high proportion of the back-scattered light with the aid of an elliptical mirror [39]. The interro-

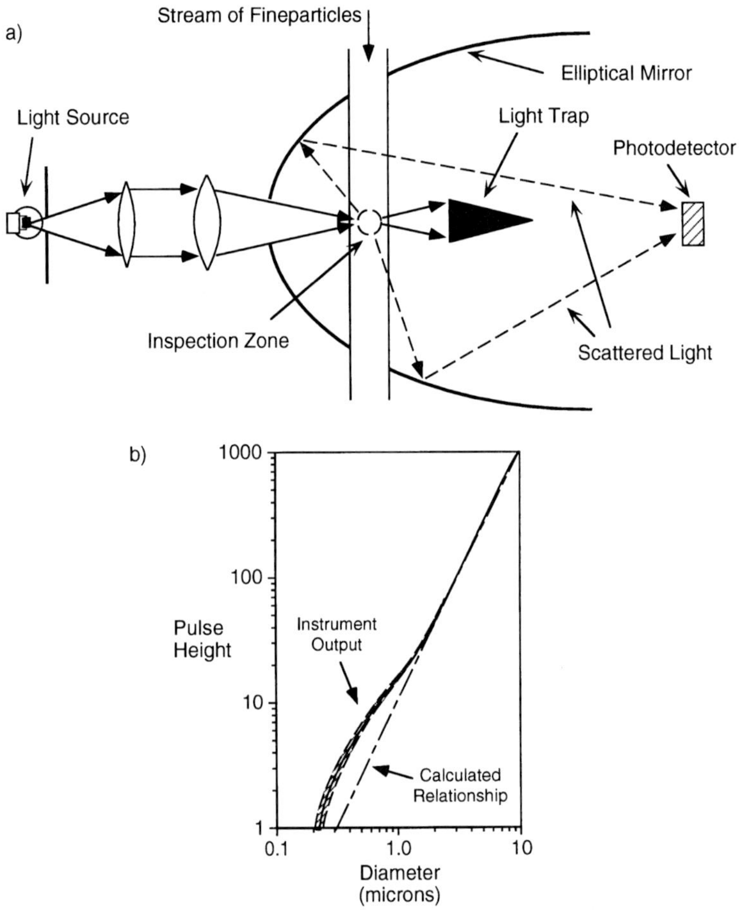

Figure 6.10. The Climet photozone stream counter uses an elliptical mirror to collect scattered light and focus it to the photodetector. a) Schematic of the Climet photozone stream counter. b) Typical calibration curve for the Climet counter.

gation beam enters through a small aperture at the front end of the elliptical mirror and is focused onto the interrogation zone. This zone is at one focus of the elliptical mirror. Light not scattered from the interrogation beam is absorbed by a light trap. If a source of light is placed at one focus of an ellipse, the energy is focused at the other. Therefore, all the back-scattered and forward scattered light within the confines of the elliptical mirror are reflected to the other focus of the elliptical mirror where the photoelectric device is placed. It can be seen from the calibration curve shown in Figure 6.10(b) that by measuring the scattered light over a much larger solid angle, the calibration curve is sensitive to changes in size below 1.5 μm.

In the HIAC (an acronym for HIgh Accuracy Counter), fineparticles in a liquid stream are characterized by the power drop they cause in a forward-traveling beam of white light [34]. The system along with calibration data are shown in Figure 6.11. It

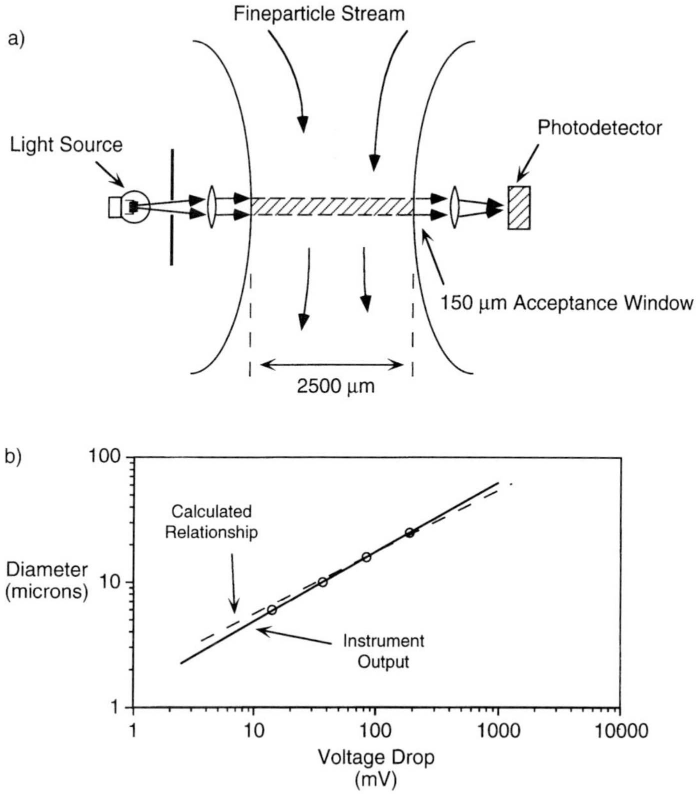

Figure 6.11. The HIAC photozone stream counter characterizes fineparticles by the power drop they cause as they pass through a beam of white light. a) Arrangement of the inspection zone of the HIAC counter. b) Calibration curve for the HIAC counter showing that the curve deviates from the predicted relationship for power blocked by a geometric shadow.

can be seen that the calibration curve deviates slightly from the square relationship predicted for straightforward geometric shadow power removal. The deviation is not unreasonable in the sense that the signals for less than 7.3 μm are slightly higher than they should be, which would indicate that, for these fineparticles, the instrument is beginning to pick up some forward scattered light. Behringer et al. adapted the HIAC system to study relatively large fineparticles suspended in a gas stream [40]. They showed that for fineparticles in the size range 20–100 μm, the geometric shadow power loss relationship can be used to interpret the experimental data. A range of stream counters using optical techniques for evaluating a stream of fineparticles is available from Particle Measuring Systems Inc. [41]. A modern, high resolution version of an optical particle stream counter is available from Particle Sizing Systems [42]. The high resolution obtainable with this system for a mixture of latex spheres is shown in Figure 6.12.

An instrument developed for the characterization of submicron fineparticles is shown in Figure 6.13. Fineparticles too small to be resolved as images can still be seen in dark-field-illuminated microscope systems by the scintillations of light that they create as they move into the plane of the microscope. Techniques for studying objects in this way are described as ultramicroscopy. The instrument shown in Figure 6.13 exploits ultramicroscopy and the measurement procedure is known as "flow ultramicroscopy". In the original system developed by Davidson, Collins and Haller, a suspension stream of latex was made to flow up and through the plane of a dark-field-illuminated microscope and the scintillations created by the presence of the latex fineparticles counted

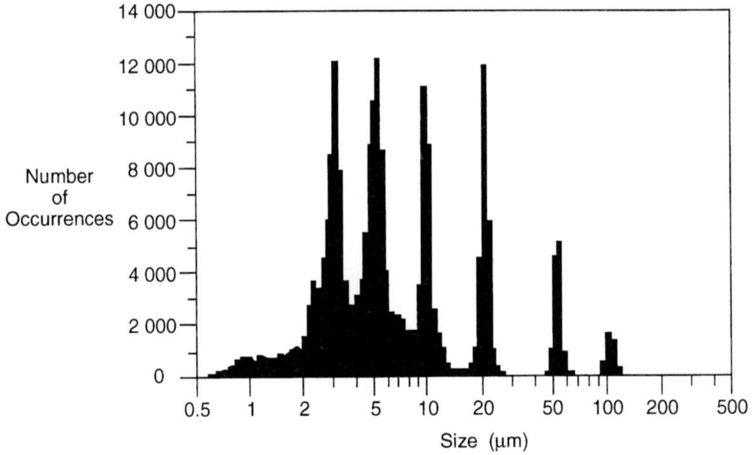

Figure 6.12. The resolution of the particle size information which can be obtained with photozone stream counters is illustrated by data for a mixture of latex spheres generated by the Particle Sizing Systems Accusizer® 770.

[43]. In the automated version of their equipment, shown in Figure 6.13(a), the microscope is replaced by a high-amplification photoelectric device linked to data processing equipment. A laser light shines through the latex suspension welling up from a feed system and flowing over into a discharge container. As the latex spheres move into the laser light, they scintillate and generate a signal at the photomultiplier. In the experiments reported by Davidson and co-workers, the magnitude of the spheres was related to the energy scattered into the receiving angle of the photoelectric device using the Lorentz-Mie theory of light scattering of monochromatic radiation [43]. From the data shown in Figure 6.13(b), it can be seen that measurements made by this technique compare well with those obtained from electromicrographs.

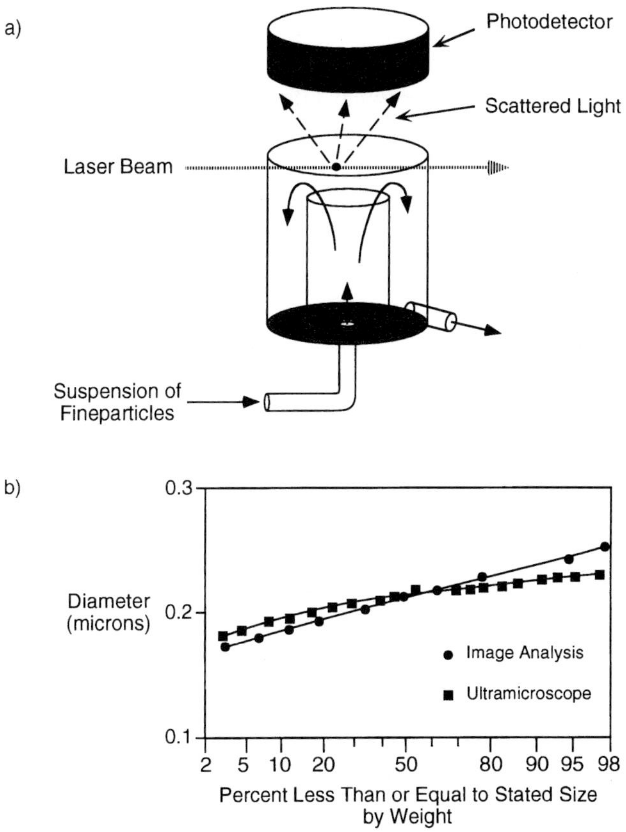

Figure 6.13. In the flow ultramicroscope, fineparticles welling up the centre of the column scintillate as they pass through the laser beam and are counted. a) Sketch of the ultramicroscope system. b) Comparison of results from the flow ultramicroscope to image analysis data for a suspension of latex spheres.

The Prototron fineparticle counter is a specialized photozone system used to inspect parenteral solutions for tramp fineparticles [44]. Parenteral solutions are pharmaceutical preparations injected into the body [45].

In the Prototron the contents of a sealed vial containing liquid is scanned in a circular pattern by a laser beam. The unscattered laser light is absorbed by a trap. Any fineparticle lying within the scan pattern will scatter light. The light scattered in the forward direction is picked up by a lens and sent to a photoelectric device. It is claimed by the manufacturer that the instrument can size fineparticles in the range 1–100 μm. Presumably the instrument is calibrated empirically by using spherical fineparticles.

Heidenreich and co-workers, have investigated the calibration of an optical stream particle counter using the Aerodynamic Particle Sizer manufactured by TSI Instruments Inc which will be described in the next section [46].

Vonnegut and Neubauer have described an instrument for sizing droplets. The basis of their instrument is the fact that when a droplet strikes a hot wire, the energy taken from the wire as the droplet evaporates causes a temperature drop in the wire, thus increasing the electrical resistance of the wire. Monitoring the electrical resistance of the wire results in a series of pulses that can be related to the size of droplets hitting the wire [47].

6.5 Time-of-Flight Stream Counters

A new type of stream counter in which the particles to be characterized enter, at random, a zone in a fluid suspension inspected by a scanning laser was developed by the Galai Corporation [48]. (Note: for many years the Galai instrument was sold in the United States by the Brinkmann Instrument Company. In many publications on particle size the instrument is referred to as the Brinkmann Size Analyzer.) The basic system of the Galai instrument is shown in Figure 6.14. A narrow, focused beam of laser light explores an area of a suspension and the size of the particles in suspension are measured by the time it takes for the laser beam to track across the profile of the fineparticle. Sophisticated electronic editors are used to generate the size distribution data from the information generated by the scanning laser. The laser is made to scan by means of a rotating optical wedge, as shown in Figure 6.14(b), and the time the light is blocked is related to the size of the fineparticle, as shown in Figure 6.14(c). The system also incorporates a video camera for inspecting the actual fineparticles being measured. The company also provides logic modules for advanced image analysis using the video camera data collection system. A typical set of data generated using the

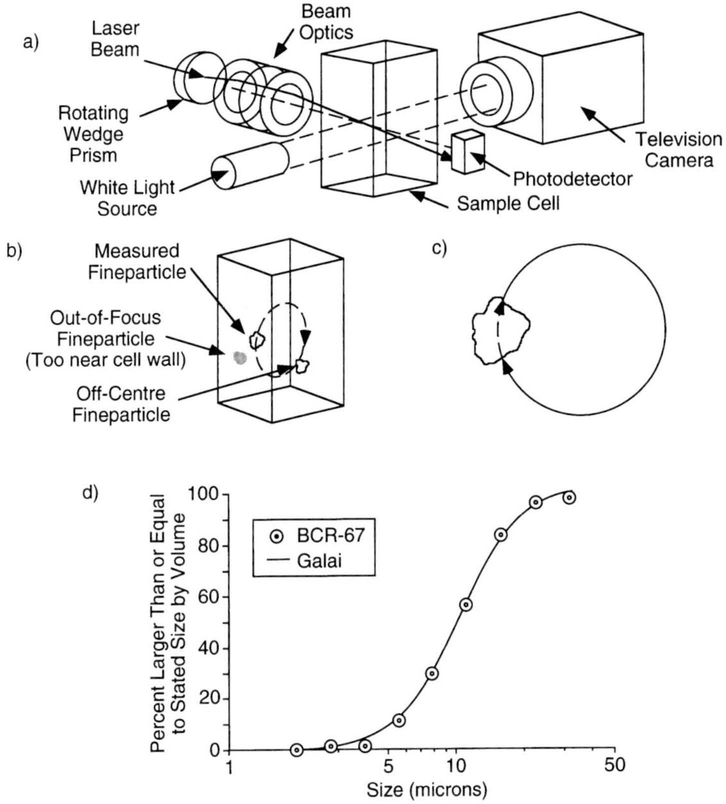

Figure 6.14. The Galai particle size analyzer uses a laser beam and a rotating wedge prism to measure the size of fineparticles. a) Basic layout of the Galai instrument. b) The laser beam traces a circular path with in the sample cell and the logic of the instrument rejects any fineparticles which are off-centre or out of focus. c) The length of time that the laser beam is blocked from reaching the photodetector is related to the size of the fineparticle. d) Data for BCR-67, standard quartz test powder, from the Galai particle size analyzer compared with it's known size distribution.

Galai instrument is summarized in Figure 6.14(d). This type of instrument is known as a time-of-flight or time-of-transit analyzer.

Another type of time-of-flight analyzer which uses a similar type of system to the Galai instrument is known as the Lasentec instrument [49]. This system is portable and has been used for on-line monitoring of particles in a slurry or suspension, as well as for size analysis in the laboratory. The basic system used in the Lasentec instrument is illustrated in Figure 6.15.

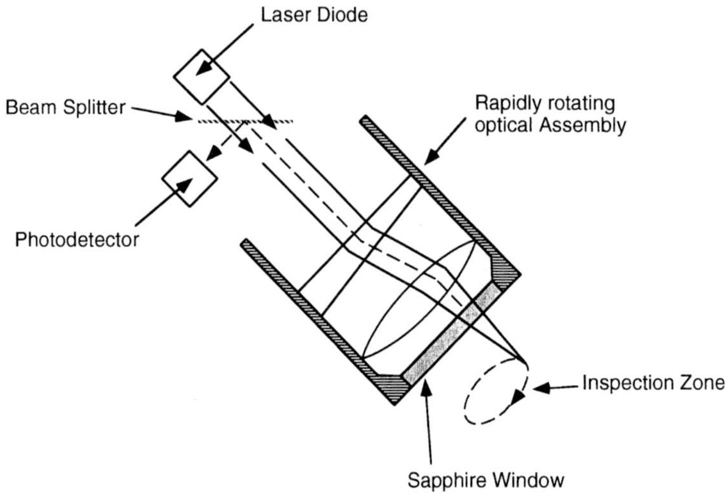

Figure 6.15. In the Lasentec Focused Beam Reflectance Measurement system the probe is inserted in a stream of fineparticles and the rotating optics scan the laser beam across fineparticles that pass in front of the probe's window.

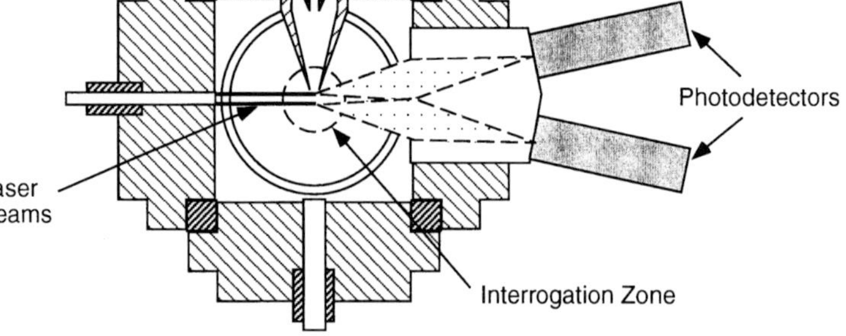

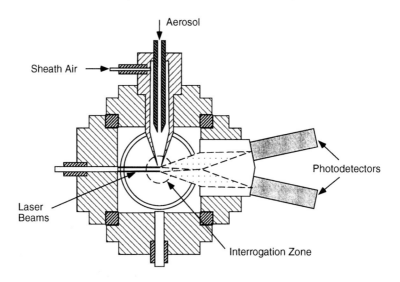

Figure 6.16. Schematic diagram of the inspection zone of the Amherst Process Instruments Inc. Aerosizer® time-of-flight particle sizer.

In a time-of-flight instrument, the fineparticles to be characterized are studied in an airstream of aerosols. The instruments are variously referred to as aerosol spectrometers, or aerosol size analyzers. In this class of instruments, one of the major problems is to ensure that the particles to be characterized are efficiently dispersed as an aerosol [50]. The Aerosizer®, manufactured by Amherst Process Instruments Inc. is a time-of-flight aerosol spectrometer based upon original investigations by Dahneke [51, 52, 53]. The basic system of the Aerosizer® is shown in Figure 6.16. A stream of aerosol surrounded by clean sheath air is directed across a gap between two laser beams. The aerosol fineparticles are accelerated across this gap. It can be shown that because of the force operating on the fineparticles, the smaller fineparticles travel across the gap between the laser light faster than the larger fineparticles which are accelerated more slowly by the projecting airstream. The equipment has a sophisticated electronic editor to correlate which pulse generated on leaving the second beam correlates with the pulse in the first beam [54, 55]. The aerosol fineparticles to be characterized are sucked into the expansion zone operating at a partial vacuum. A stream of clean air focuses the aerosol stream, and can operate at very high fineparticle flow densities. The equipment can measure fineparticles at a rate of 10,000 per second. The equipment is calibrated using standard fineparticles. In Figure 6.17(a), a typical calibration curve generating what is known as the geometric diameter of the particles being studied. In essence, for spheres, the geometric diameter is the Stokes Diameter, generated by converting the aerodynamic diameter measured directly in the instrument using the density of the material. In the other parts of this figure, typical data generated by studying various systems are shown. The machine can be used for particles as small as 0.3 microns up to 700 microns. The parameter measured by the Aerosizer® is the aerodynamic size, which is the diameter of the sphere of unit density that has the same aerodynamic behaviour as the particle. In the pharmaceutical industry, and occupational health and hygiene, the aerodynamic particle size governs the dynamics of inhalation of the particle [54, 55].

In Figure 6.18, the Stokes Diameter distribution of the metal powder in Chapter 2 is shown, both as a number distribution and a volume distribution. In the transformation to the volume distribution, the small particles with a mean diameter of 0.8 microns virtually disappear from the display graph. This demonstrates the importance of studying some powders by more than one technique since the diffractometer studies of this powder, to be described in Chapter 7, do not usually indicate the presence of the small fineparticles, which would constitute a health hazard as compared to the major portion of the powder. It can be seen that the size distribution of this powder is essentially log-normal and that when one converts to the volume based assessment, the geometric mean shifts from 30 μm to 38 μm because of the more significant contributions of the larger fineparticles. The geometric mean diameter of 38 μm as compared to the value of 59 μm generated by the Coulter Counter® study, reported in

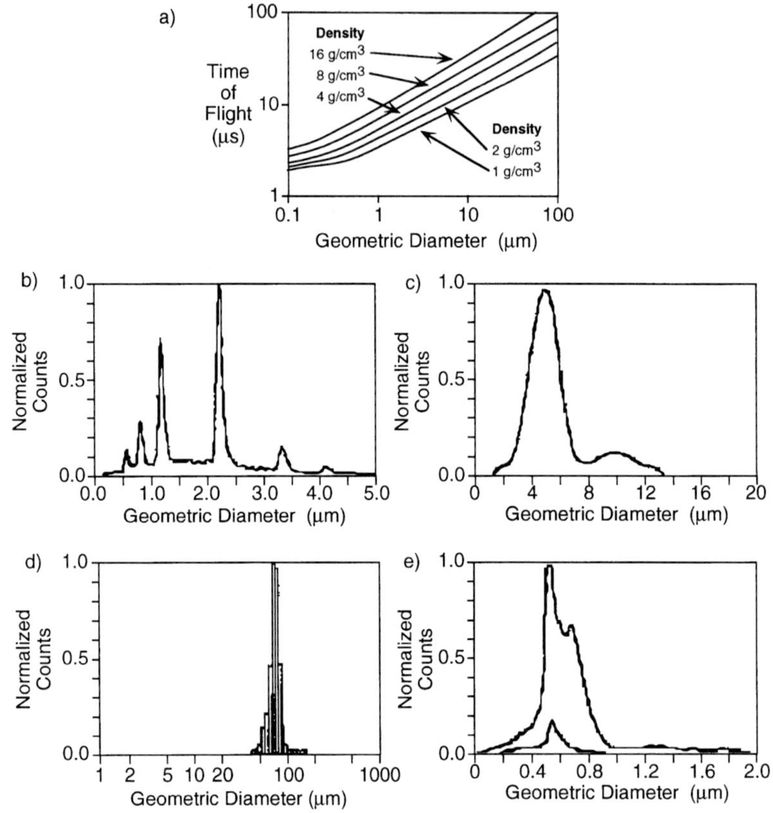

Figure 6.17. Typical data generated by the Amherst Process Instruments Inc. Aerosizer® for aerosol systems, and powders aerosolized prior to characterization studies. a) Calibration using aerosols of standard latex spheres of known size. b) Characterization of a mixture standard of polystyrene latex spheres. c) Characterization of a sample of 5 micron silica microspheres mixed with a small number of 10 micron microspheres. D) Glass spheres used in reflective paint (ballotini). E) Two oil mists characterized by direct injection into the Aerosizer®.

Figure 6.6, yields a ratio of the two geometric mean diameters of 59/38 = 1.6, which is of the same order as the average shape factor of 1.72, found in our discussion of the shape of the powder grains of these powders given in Chapter 2. This confirms that the particles are being characterized by an aligned aerodynamic diameter in the Aerosizer®. This shape factor ratio cannot be an exact correlation because of the ranges of shapes in the powder.

The importance of measuring the aerodynamic diameter directly in occupational health and pharmaceutical inhalation technology is illustrated by the data of Figure 6.19, in which sets of isoaerodynamic uranium dioxide and thorium dioxide

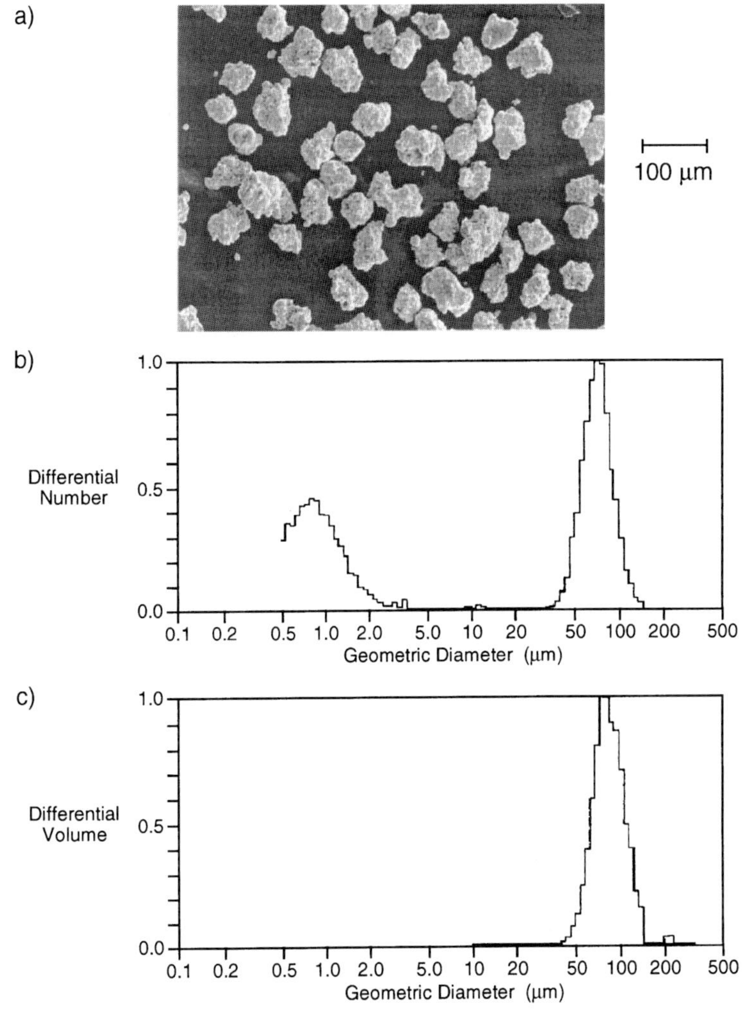

Figure 6.18. Data generated by the Amherst Process Instruments Inc. Aerosizer® for an irregular iron powder clearly demonstrates the effect of interpreting the data as a number size distribution as compared to a volume size distribution. a) Micrograph of the irregular iron powder. (Micrograph courtesy of D. Alliet, Xerox Corp.) b) Number size distribution of the iron powder of (a). c) Volume size distribution of the iron powder.

fineparticles are shown. These isoaerodynamic diameters were isolated and characterized using the Stöber centrifuge [56]. It can be seen that there is no way that one could deduce the aerodynamic diameter from the image analysis data.

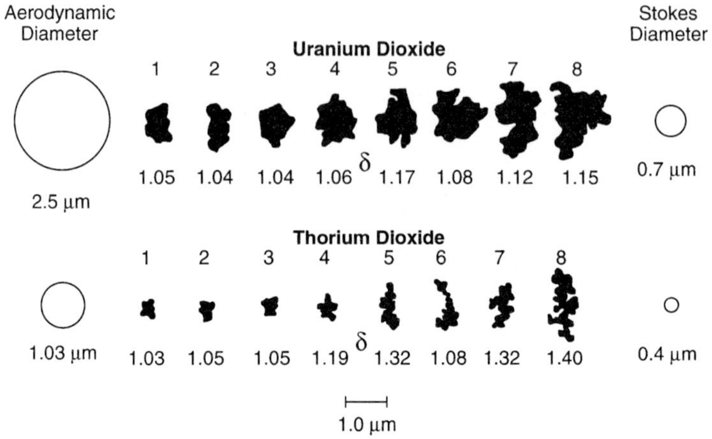

Figure 6.19. The Stöber centrifuge was used to prepare samples of fineparticles having the same aerodynamic diameter, called isoaerodynamic fineparticles. The two samples shown above demonstrate how the shape of a fineparticle can affect it's measured aerodynamic diameter.

A piece of ancillary equipment, which can be used with the Aerosizer® to study the efficiency of the dispersion of the powder under study into the aerosol form is the Aero-Disperser®. This equipment is shown in Figure 6.20(a). The powders under study are dispersed in a primary air stream, which moves into a second area in which the primary aerosol is moved through a gap, the dimensions of which can be altered to change the shear forces experienced by the aerosol moving through the gap A. In Figure 6.20(b), the effect of dispersing a pharmaceutical powder used in a dry powder inhaler, for treating asthma, is shown. It can be seen that when dispersed by an applied pressure across the dispersing gap of 5 pounds per square inch, the powder was all below the size considered to be the upper limit for use in therapeutic aerosols. A further increase in applied pressure did not change the measured size distribution [53].

Pigments, such as titanium dioxide, are deliberately manufactured so that their geometric mean diameter is of the same order as the wavelength of light. In Figure 6.21(a), characterization of a titanium dioxide pigment dispersed in the Aero-Disperser® and fed into the Aerosizer® are shown. In Figure 6.21(b), for comparison, the data from a centrifugal sedimentation study using x-rays to measure the concentration changes (see Chapter 4), manufactured by Brookhaven Instruments [57]. Other comparisons of the Aerosizer® generated data with different instruments has been published by Etzler and Sanderson [58].

Another aerosol spectrometer is manufactured by TSI Incorporated. Their instrument is known as the Aerodynamic Particle Sizer (APS). This system operates at subsonic flow conditions, and has a lower rate of count than the Aerosizer®. The performance of the instrument has been documented in various publications [59, 60, 61]. In

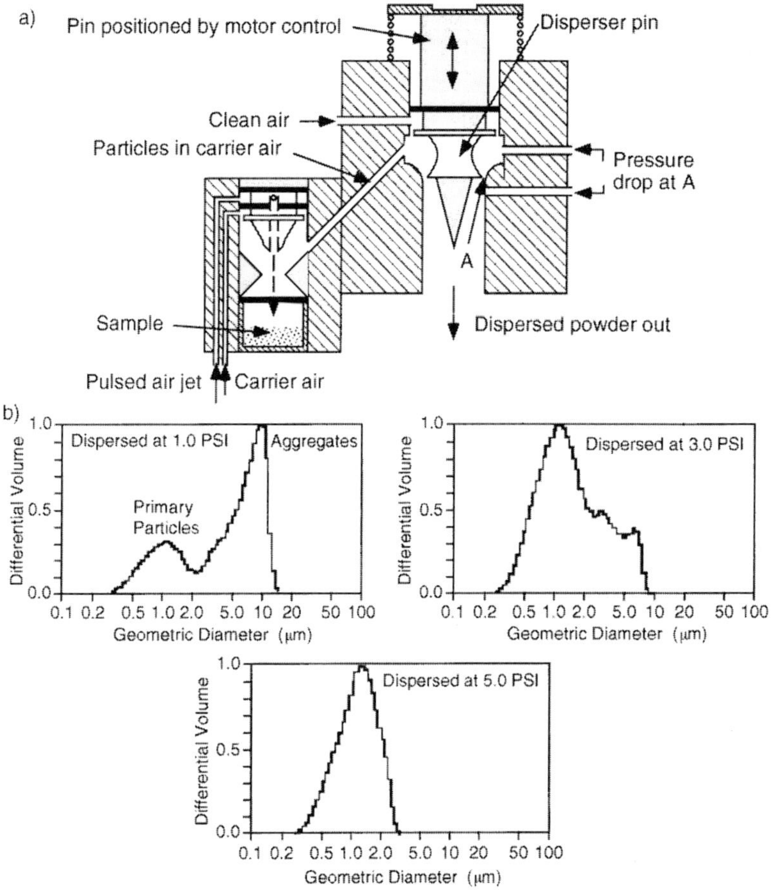

Figure 6.20. The Aero-Disperser™ attachment for the Aerosizer®, enables aerosol preparation at controlled shear rates. a) Internal structure of the AeroDisperser. b) The effect of various shear rates can be seen from the size distributions generated for a pharmaceutical powder intended for use as a therapeutic aerosol.

early 1997, the TSI company introduced a new design of their instrument, shown in Figure 6.22(a). In this device, known as the model 3320 Aerodynamic Particle Sizer, what is known as double crested optics is used. This system is shown in Figure 6.22(b). In literature announcing the new developments, it is stated that

> *"in the previous design of the equipment, the time-of-flight spectrometer used two tightly focused laser beams resulting in two distinct signals for each particle. This required that the two signals be correlated to a start and stop time for every*

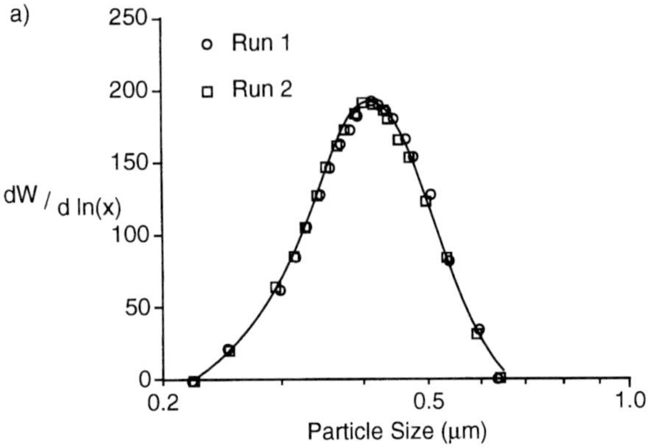

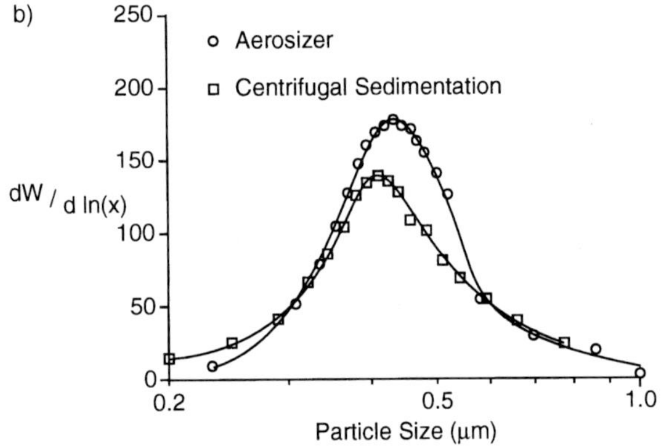

Figure 6.21. The high shear forces available in the AeroDisperser enables one to size cohesive powders such as titanium dioxide. a) Repeatability of Aerosizer® characterization studies. b) Comparison of Aerosizer® data with data generated in centrifugal sedimentation studies.

particle. Significant errors often occur due to coincidence, or due to phantom measurements, which arise when particles near the detection threshold produce only one signal."

With the new equipment it is claimed that both types of errors are eliminated by using overlapping laser beams, producing one double crested beam profile. Each particle then creates a single, continuous signal that has two crests. Particles with only one crest are phantom particles, or with more than two crests are coincident errors and are logged for concentration calculations, but are not used in building the size distribution. The commercial development of aerosol spectrometers is a very active area of research, and one should check for the latest information with the manufacturers of such instruments.

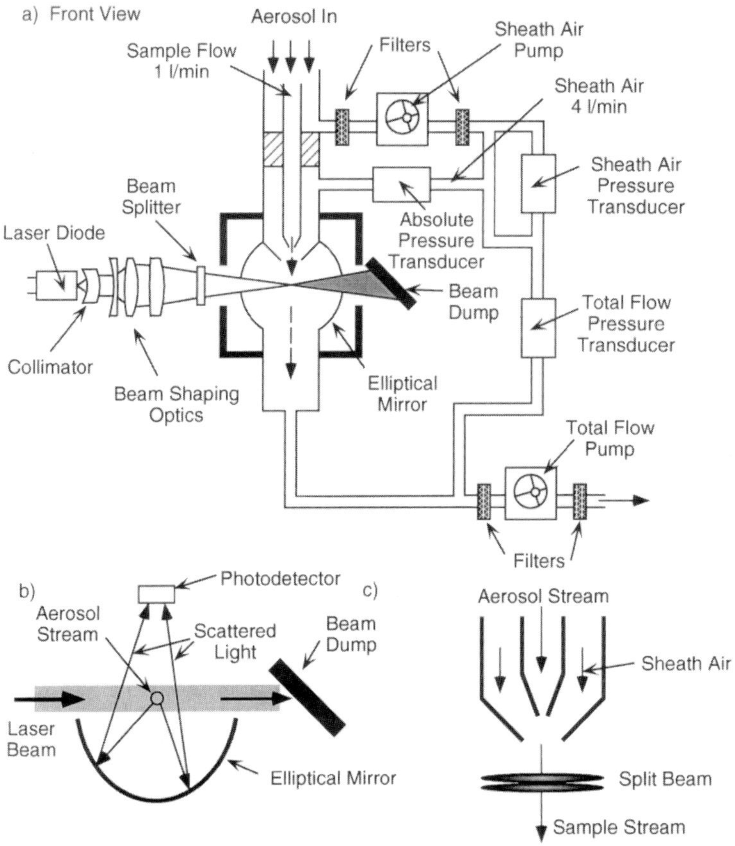

Figure 6.22. The TSI Aerodynamic Particle Sizer Model 3320 measures the velocity of fineparticles passing through a so-called "double crested" laser beam in order to determine their size from the acceleration they experience at the nozzle. (Larger fineparticles accelerate more slowly and so have a lower velocity.) a) Layout of the TSI Aerodynamic Particle Sizer. b) View of the inspection zone from above. c) Enlarged side view looking into the laser beam.

References

[1] E. J. Wynn, M. J. Hounslow, "Coincidence Correction for Electrical-Zone (Coulter Counter) Particle Size Analyzers", *Powder Technol.*, 93 (1997) 163–175
[2] Information on the Coulter Counter is available from Coulter Particle Characterization Electronics Inc., P. O. Box 169015, M/C 195-10, Miami, Florida, USA, 33116-9015.
[3] Information on Celloscope and Electrozone resistazone counters can be obtained from Particle Data, Inc., P. O. Box 265. Elmhurst, Illinois 60126.
[4] R. W. DeBlois, K. A. Wesley, "Size Concentration and Electrophoretic Mobilities of Several C-Type Encornaviruses, and T-even Bacteriophages by the Resistive-pulse Technique," reprint available from the General Electric Corporation, Research and Development Distribution, P. O. Box 43, Building 5, Schenectady, New York 12301. Commercial information on the Nanopar analyzer is available from the same address.
[5] R. Davies, R. Karuhn, L. Townsend, "The Effect of Particle Shape on the Response of an Electrical Resistance Zone Counter," in *Proceedings of the Symposium on Particle Size Analysis, Nuremburg, Germany, 1975*; published as Dechema Monographien, 79 (1589–1615), Teil B (1975).
[6] R. Karuhn. R. Davies, B. H. Kaye, M. J. Clinch, "Studies on the Coulter Counter, Part One: Investigations into the Effect of Orifice Ceometry and Flow Direction on the Measurement of Particle Volume", *Powder Technol.* (1975) 157–171: see also references 6 and 7.
[7] R. Davies. R. Karuhn, J. Craf, "Studies on the Coulter Counter, Part Two: Investigations into the Effect of Flow Direction and Angle of Entry of a Particle on Both Possible Volume and Pulse Shape", *Powder Technol.*, (1970) 157–166.
[8] R. Davies, R. Karuhn, J. Graf, J. Stockham, "Studies on the Coulter Counter. Part Three: Applications of the IITRI Sensing System", *Powder Technol.* 3 (1976) 193–201.
[9] N. J. Grover et al., "Electrical sizing of particle in suspensions III Regid spheroids and red blood cells," *Biophys J.*, 12 (1972) 1099.
[10] R. Thom, *Zeitschrift Gesamte Exp. Med.*, 151 (1969) 331.
[11] R. W. Lines, "A Short Contribution to the Proceedings of the First European Particle Size Analysis Conference", Nurnberg, September 1975. pp. 424–431. Also see comments on this contribution by R. Davies, pp. 432 and 433 in reference 4.
[12] L. Spielman, S. I. Goren, "Improving Resolution in Coulter Counting by Hydrodynamic Focussing", *J. Colloid Interface Sci.,* 6 (1968) 175–182.
[13] R. F. Karuhn, R. H. Berg, "Particle Shape Analysis Via Electrolytic Sensing Zone", presented at the International Symposium on Powder and Bulk Solids. Chicago, May 1978: proceedings published by the International Powder Institute, Chicago, Illinois.
[14] R. Davies, R. Karuhn, J. Graf, J. Stockham, *Bull. Parenteral Drug Assoc.,* 29 (2) (March/April 1975) 98–109.
[15] Personal Communication R. Karuhn, Particle Technology Labs, 1705 West 61st Street, D? Grove, IL 60516.

[16] Personal Communication from D. Alliet and L. Switzer, Xerox Corporation, Webster, NY, 14580.

[17] C. Albertson, "Detection of Particles in Liquids by the Use of Ultrasonics", in *Proceedings of the First National Symposium on Non destructive Testing of Aircraft and Missile Components, San Antonio, Texas*, February 16, 1960.

[18] J. B. Gayle, "An Evaluation of the Sperry Ultrasonic Particle Counter", *J. Am. Assoc. Contamination Control*, 2 (10) October 1963) 16–21

[19] Commercial information is available from the Denver Autometrics Inc., 6235 Lookout Rd., Boulder, Colorado, 80301.

[20] The basic physics of the attenuation of ultrasonic beams passing through a slurry of fineparticles was studied in depth by G. H. Flammer, "The Use of Ultrasonics in the Measurement of Suspended Sediment Size Distribution and Concentration," Ph. D. thesis, University of Minnesota, 1958.

[21] A. L. Hinde, P. J. Lloyd, "Real time Particle Size Analysis in Wet Closed Circuit Milling", *Powder Technol.*, 12 (1975) 37–50.

[22] G. Langer, "An Acoustic Particle Counter Preliminary Results", *J. Colloid Sci.*, 20 (1965) 602–609.

[23] B. H. Kaye, Chapter 6 in *"Chaos and Complexity"*, VCH Publishers, Weinheim, Germeny, 1993.

[24] B. Scarlett, I. Sinclair, "Operation of the Langer Acoustic Counter", contribution to the proceedings of the Nuremberg Conference on Fineparticle Measurements, September 1975 (see reference 4).

[25] R. F. Karuhn, "The Development of a New Acoustic Particle Counter for Particle Size Analysis", in *Proceedings of the Conference on Powders, IIT Research Institute, September 1974;* proceedings published by IITRI. 10 West 35th St., Chicago, Illinois.

[26] D. M. Scott, A. Bauzmann, C. Jochen, "Ultrasonic measurement of submicron particles", *Part, Part. Syst. Charact.*, 12 (1995) 269–273.

[27] B. Li, "Analysis of phytoplankton autoflourescence in size by flow cytometry," *Canadian Research,* Feb (1988) 18–22.

[28] H. M. Shapiro, "Practical Flow Cytometry", Alan Arliss Inc., New York, 1985.

[29] M. A. Van Diller, P. N. Dean, O. D. Laerum, M. R. Melamed, "Flow Cytometry Instrumentation and Data Analysis", Academic Press, London, 1985.

[30] M. Kerker, "The Scattering of Light and Other Electromagnetic Radiation," Academic, New York, 1969.

[31] H. C. Van der Hulst, "Light Scattering by Small Particles", Wiley, New York, 1957.

[32] D. Sinclair, "A New Photometer for Aerosol Particle Size Analysis", *J. Air Pollution Control Assoc.*, 7 (2) (February 1967) 105–108.

[33] J. K. Channell, J. Hanna, "Experience with Light Scattering Particle Counters", *Arch. Env. Health,* 6 (March 1963) 386–400.

[34] Information is available from HIAC Instrument Division, P. O. Box 3007, 4719 West Brook Street, Mont Clair, California 91763.

[35] Information is available from Royco Instruments, Inc., 141 Jefferson Drive, Menlo Park, California 94025.

[36] Information on the various versions of the Aerosoloscope is available from Fineparticle Section, IITRI, 10 West 35th St., Chicago, Illinois.
[37] A. E. Martens, "An Electro-optical Dust Counter", paper presented to the American Association for Contamination Control, Albany, New York. February 1966. Commercial information is available from Bausch and Lomb. Rochester, New York.
[38] See discussion of stream counters in Chapter 8 of B. H. Kaye, *"Direct Characterization of Fineparticles"*, Wiley, New York, 1981.
[39] Information is available from the Climet Corporation, 1320 Colton Avenue, Redlands, California 92373.
[40] A. J. Behringer, T. S. Mika, D. H. Wood, T. W. Nash, "Use of a Modified HIAC Criterion PC 320 as an Airborne Particle Size Analyzer", paper presented at the Conference on Particle Size Analysis, September 1977, Salford, England; proceedings to be published by the Analytical Chemistry Section of the Royal Institute of Chemistry, Great Britain.
[41] See trade literature Particle Measuring Systems Inc., 5475 Airport Boulevard, Boulder, Colorado, 80301.
[42] See trade literature Particle Sizing Systems, 75 Aerocamino, Santa Barbara, CA 93117.
[43] J. A. Davidson, E. A. Collins, H. S. Haller, "Latex Particle Size Analysis, Part 3: Particle Size Distribution by Low Ultra Microscopy", *J. Polymer Sci.*, Part C 34 (1971) 235–255.
[44] Commercial information is available from Prototron, Inc., 1020 Corporation Way, Palo Alto. California 94303.
[45] M. J. Groves, "Parenteral Products", William Heineman, Medical Books Limited. London 1973.
[46] S. Heidenreich, H. Buttner, F. Ebert, "Investigations in the behaviour of an aerodynamic particle sizer and its applicability to calibrate an optical particle counter", *Part. Part. Syst. Charact.*, 12, 1995, pp. 304–308.
[47] B. Vonnegut, R. Neubauer, *Anal. Chem.*, 24 (1954) 100.
[48] Galai Instruments Inc., 577 Main Street, Islip, NY, USA, 11751.
[49] Lasentec, 15224 NE 95th Street, Redmond, WA, USA, 98052.
[50] B. H. Kaye, "Generating Aerosols," KONA, 15 (1997) 68–81.
[51] B. E. Dahneke, "Aerosol Beams Spectrometry", *Nature Phys. Sci.*, 244 (1973) 54–55.
[52] B. E. Dahneke, "Aerosol Beam Specrometers in Recent Developments in Aerosol Science", Ed. E. T. Shore. John Wiley & Sons, New York, 1978, 187–223.
[53] Amherst Process Instruments, The Pomeroy Building, 7 Pomeroy Lane, Amherst, MA, USA, 01002-2941.
[54] R. W. Niven, "Aerodynamic Particle Size Testing Using the Time of Flight Aerosol Beam Spectrometer", *Pharm. Technol.,* 17 (1993) 72–78.
[55] M. Hindle, P. R. Byron, "Size Distribution Control of Raw Materials for Dry Powder Inhalers Using the Aerosizer® with the Aero-Disperser®", *Pharm. Technol.,* June 1995, 64–78.
[56] W. Stöber, H. Flachsbart, *Environmental Science and Technology,* 3 (1969), 1280. Recent information is available from the Institute for Aerobiology, Grafschaft, Sauerland, West Germany.

[57] Brookhaven Instruments, Brookhaven Corporate Park, 750 Bluepoint Road, Holtsville, New York, USA, 11742.
[58] F. N. Etzler, M. S. Sanderson, "Particle Size Analysis: A Comparative Study of Various Methods", *Part. Part. Syst. Charact.* 12 (1995) 217–224.
[59] B. T. Chen, Y. S. Cheng, H. C. Yai, "Performance of a TSI Aerodynamic Particle Sizer", *Aerosol Science Technology,* 4 (1985) 89–97.
[60] See Trade literature on the Aerodynamic Powder Sizer (APS Spectrometer) manufactured by TSI Incorporated, Particle Instruments, P. O. Box 64394, St. Paul, MN, USA 55164.
[61] P. A. Baron, "Calibration and Use of the Aerodynamic Particle Sizer – APS 300," *Aerosol Science Technology,* 5 (1986) 55–67.

7 Light Scattering Methods for Characterizing Fineparticles

7.1 The Basic Vocabulary and Concepts of Light Scattering

The nature of light remained controversial into the 1800's. Newton had proposed that light traveled in small packets, like small balls, whereas experimentally discovered phenomena indicated that the nature of light was a wave motion. Today we know that when it is being given out or absorbed, light energy exists as photons (discrete packages of light energy). As it travels around microscopic objects it is necessary to treat light energy as waves. In undergraduate studies of physical phenomena students are introduced to light diffraction, a phenomenon that occurs when light passes though a slit of the same order of magnitude as the wavelength of light. The wavelengths of visible light range from 0.4 to 0.7 microns. Alternatively this amount is often quoted in nanometres – a nanometre being one billionth of a meter. In Figure 7.1 the type of diagram shown in undergraduate textbooks demonstrating the optical diffraction of monochromatic light through a narrow slit is shown. The series of light and dark bands in the diffraction pattern spread out as the width of the slit decreases. Basic theory of light diffraction is covered in many undergraduate physics textbooks see for example Halliday and Resnick [1].

An ISO standard procedures for using light scattering methods to characterize fineparticles has been formulated [2]. The suggested definitions for this area of particle size characterization, as set out in the standard, are summarized in Table 7.1. The second and third part of this standard procedure deal with the design and operational requirements for equipment used to characterize fineparticles by light scattering. A surprising omission from Table 7.1 is the term optical interference which occurs when the light from two adjacent slits each produce their own diffraction patterns which interact to generate a pattern described as an interference pattern. The typical textbook illustration of this phenomena is shown in Figure 7.1(b). From the structure of the interference pattern, the size of the diffracting slits and their separation can be deduced. This is the same type of information which can be obtained by shining the monochromatic beam of light through an array of fineparticles which diffracts light and creates an interference pattern.

The subject of studying the size, shape, and the constitution of a population of fineparticles from their diffraction patterns is complex. Many books have been written on the topic, see for example references 3, 4, 5, 6, and 7. International confer-

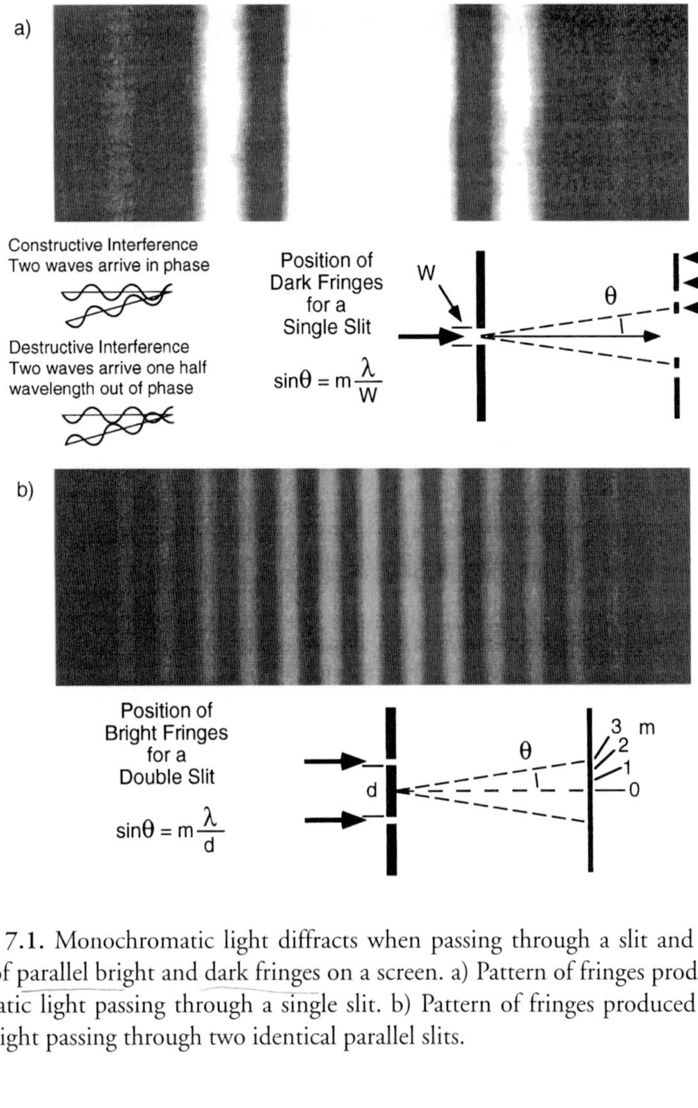

Figure 7.1. Monochromatic light diffracts when passing through a slit and will produce a series of parallel bright and dark fringes on a screen. a) Pattern of fringes produced by monochromatic light passing through a single slit. b) Pattern of fringes produced by monochromatic light passing through two identical parallel slits.

ences have been held on the topic of optical particle sizing; for example the 5th International Congress on Optical Particle Sizing was held in August 1998 [8]. Venturing into the specialized books on light scattering theory can be an intimidating experience. However, the powder technologist wanting to use techniques based on light scattering to study powder systems only needs to understand the basic concepts involved in light scattering studies of fineparticles. For more advanced information should they wish to carry out research work and more specialized studies using light scattering technology, various advanced text books can be consulted. The technologist who needs

Table 7.1. Definitions of the terms used in light scattering characterization of fineparticles as set out in the ISO standard.

Absorption: reduction of intensity of a light beam traversing a medium throughenergy conversion in the medium.

Complex Refractive Index (m): refractive index of a fineparticle relative to that of the medium, consisting of a real and an imaginary (absorption) part.

$$m = \frac{n_p - ik_p}{n_m}$$

Deconvolution: mathematical procedure whereby the size distribution of a fineparticle ensemble is inferred from measurements of the scattering pattern produced by the ensemble.

Diffraction: spreading of light around the contour of a fineparticle beyond the limits of its geometrical shadow with small deviation from redtilinear propagation.

Extinction: attenuation of a light beam traversing a medium through absorption and scattering.

Model Matrix: matrix containing light scattering vectors for differential size classes, scaled to the detector" geometry, as derived from model computation.

Obscuration: percentage or fraction of incident light that is attenuated due to scattering and/or absorption, also known as optical concentration also known as Optical Concentration.

Optical Model: theoretical model used for computing the model matrix, viz, optically homogeneous sphere with specified complex refractive index.

Reflection: return of radiation by a surface, without change in wavelength.

Refraction: Change in direction of propagation of light determined by change in the velocity of propagation in passing from one medium to another, in accordance with Snell's law.:

$$n_m \sin\theta_m = n_p \sin\theta_p$$

Scattering: general term describing the change in propagation of a beam of radiation at the interface of two media.

Scattering Pattern: angular, c. q. spatial pattern of light intensities (*I*(*q*) and *I*(*r*) RESPECTIVELY) originating from scattering.

Single Scattering: scattering whereby the contribution of a single member of a fineparticle population to the scattering pattern of the entire population is independent of the other members of the population.

to use commercially available characterization instruments based on diffraction phenomena finds that these relatively expensive instruments have all of the necessary interpretive theorems built into the logic of the machine. For commercial security reasons it is often very difficult to know the assumptions used when turning a measured diffraction pattern into a size distribution. In our discussion we can point out the questions one should ask if an investment is to be made into such machines and we will note the caveats to be raised when one is interpreting data from a commercially available machine.

The key considerations leading to the acceptance of the wave nature of light were carried out by a English physicists Thomas Young (1773–1829) [9]. For this reason the type of interference fringes shown in Figure 7.1(b) are called Young's interference fringes. Prior to the advent of the laser, generating a diffraction pattern was a complex and difficult task involving equipment necessary to generate monochromatic light as well as equipment to study the diffraction pattern. Today the generation of patterns such as those in Figure 7.1 is a simple matter and laser light based diffractometers are available from many different companies. [10–17].

As noted in a reference book on the biographies of scientists, Young's position on the wave nature of light was unacceptable during his lifetime because it was regarded anti-Newtonian. Young became famous in history as one of those contributing to the deciphering of inscriptions of the Rosetta stone which gave historians the ability to read Egyptian hieroglyphics [9]. It is rather interesting to note that Young actually exploited the diffraction of light to develop a commercial instrument for measuring the fineness of wool (the diffraction pattern of a single hair is the same as the diffraction pattern of the slit of Figure 7.1(a)). I was intrigued on a recent visit to New Zealand (December 1996) to see in the Wellington newspaper comments on the prices being paid for wool in which the price varied by a significant amount for its fineness. Thus, in 1996 18.3 micron wool sold for $ 8/kg where as a 17 micron wool cost considerably more. The fineness of the wool can be studied to determine which part of the sheep the wool came from and also variations in the diet of the sheep over a period of time.

Diffraction patterns have also been used to study the fineness of asbestos fibres and other fabric fibres. Young called his instrument an eriometer from the Greek work for wool erios. He also showed that one could measure the diffraction pattern of an array of blood cells to determine the size of the cells [18]. In the days before fast, cheap computers were available to tackle the complex theory of the light scattering of fineparticles, simplified theories were used to interpret diffraction patterns. Thus diffraction patterns in which it was assumed that the optical light source was infinitely distant from the diffracting object and the detection equipment also at an infinite distance from the scattering object are known as the Fraunhofer Diffraction Patterns [1, 9]. (It should be appreciated that, when discussing infinite distances with regard to fineparticles, it is the number of diameters away that is significant. If the fineparticle is only 10 microns then a few hundred centimeters is, in effect, an infinite distance.)

Young's diffraction and interference fringes shown in Figure 7.1 are Fraunhofer Patterns. In some instruments used for studying fineparticles by diffraction pattern, Fraunhofer optical theory is used to interpret the diffractometer data because, as pointed out in the ISO standards, for these calculations one does not need to know the complex refractive index of the diffracting objects making the calculation of the pattern relatively simple and fast. Fraunhofer patterns of some regular and irregular geometric shapes are shown in Figure 7.2 [19].

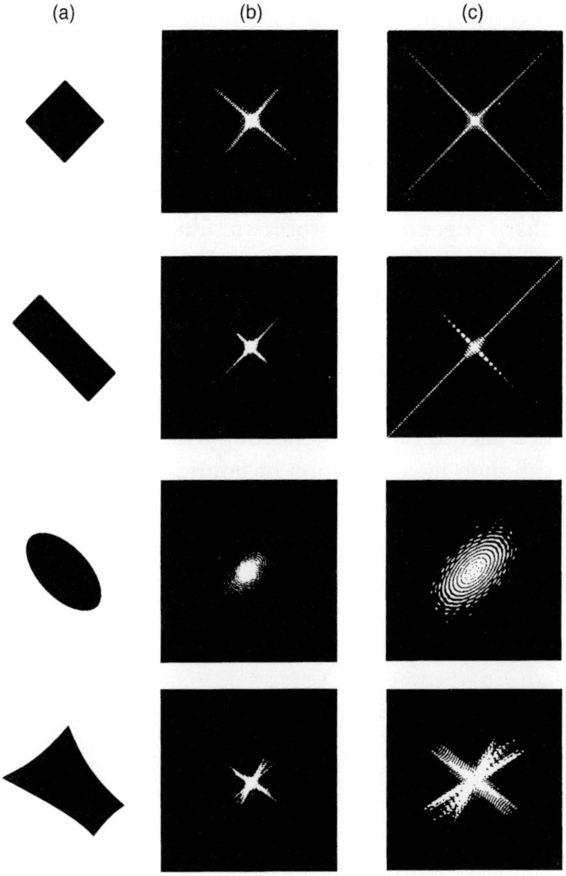

Figure 7.2. A two dimensional Fourier transform of a shape is equivalent to using the shape to produce a diffraction pattern. a) Some simple shapes used in this comparison. b) Optical diffraction patterns produced from the shapes of (a) using a laser and recorded on a photographic negative. c) Two dimensional Fourier transforms calculated mathematically from the shapes of (a)

Fraunhofer diffraction patterns are named after Joseph Fraunhofer, a German physicist (1787–1826) who invented the diffraction grating. It can be shown that the Fraunhofer diffraction pattern of a fineparticle is the same as the two-dimensional Fourier Transform. (One-dimension Fourier Transforms were discussed in the chapter on shape.) It is difficult to explain in simple terms what is meant by the Fourier Transform in two and three-dimensional space but computer programs are readily available to calculate the two-dimensional Fourier Transform. In Figure 7.2 the calculated two-dimensional Fourier Transforms of the four shapes shown are given along with the Fraunhofer diffraction patterns. One feature to note in the patterns shown in Figure 7.2 is that the presence of a sharp edge results in a diffraction of energy a long way from the center of the pattern. This is an important fact when considering the accuracy of size determinations by means of diffracted energy since virtually all commercial machines currently used for determining the size of fineparticles interpret the diffraction pattern in terms of light scattering from equivalent spheres. In essence the appearance of sharp edges on the profiles results in the spherical equivalent size being smaller than the physical size.

Sometimes it is useful to calculate the Fourier Transforms of particle grains in a population to anticipate what kind of problems one will encounter when trying to interpret the significance of size distribution data generated by diffractometers. Thus in Figure 7.3 we show several profiles of two different powders, which will be used in experiments discussed later in this chapter, along with their two-dimensional Fourier Transforms [20]. Although the calculation of Fourier Transforms is a complex problem it is not necessary for the powder technologists to involve themselves in a detailed

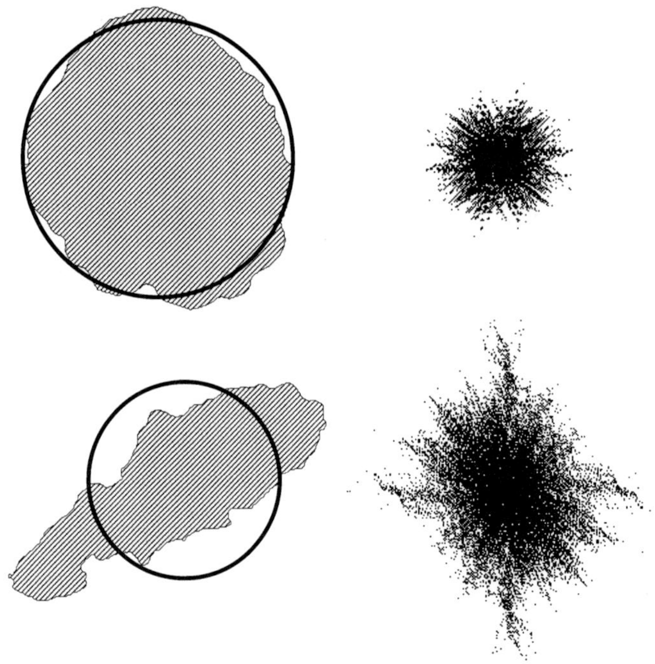

Figure 7.3. Calculated Fourier transforms of two different actual fineparticle profiles, shown with their circle of equal area imposed over them, demonstrate the effect of shape on the diffraction pattern. This indicates that caution should be exercised when using diffractometers to measure the size distribution of an irregular powder.

study of the techniques of Fourier analysis because of the availability of computer programs. The reader who wishes to pursue the concept of Fourier analysis can find a useful discussion in reference 21.

Another simplified diffraction theory used when interpreting diffraction patterns is known as Fresnel theory. In this theory one still considers an infinitely distant light source but the diffraction rays converge on a detector screen at a finite distance. This approach to diffraction is named after the French physicist Fresnel (1788–1827) who developed a mathematical theory of wave motion of light.

For fineparticles greater than 20 microns in diameter Fraunhofer calculations are often adequate, however, for fineparticles approaching the wavelength of light calculations are much more complex and it is necessary to use a generalized theory of light scattering developed by the scientist Gustav Mie. His original work was published in German in 1908. His theory is discussed in detail in several of the books [3–7]. When discussing light scattering according to Mie theory it is necessary to use the concept of complex optical refractive index of the substance. In elementary texts one considers the refractive index of a substance such as water as being the ratio of the velocity of light in substance compared to the velocity of light in a vacuum. Many real substances absorb light as the electromagnetic wave passes through the material. To allow for this, one can regard the refractive index as a complex number consisting of two parts,

$$\mu = \tau + ik\rho$$

where τ is the real component and $ik\rho$ is the imaginary component (i is the square root of -1). This takes into account the absorption properties of the material. Mie theory deals with the energy removed from a forward traveling beam at an infinitely distant detector. To describe the energy removal from a beam by a spherical particle the parameter known as the extinction coefficient of the material is used. The extinction coefficient defines the combined effects of scattering and absorption in the removal of energy from a beam. In Figure 7.4(a) the energy removed by spheres of different size and different refractive index are summarized. As can be seen by this diagram, the extinction coefficient oscillates about the value of 2 for small particles with differences which vary with the refractive index value and approaches the value of 2 for larger particles of almost any refractive index. Sometimes people are surprised that this limit's value is 2. It must be remembered that the blocking fineparticle removes energy by absorbing the incident beam and an equal amount of energy diffracted. In general it can be stated that the effect of a complex component indicating absorption in the media is to suppress the oscillations in the scattering power of the fineparticle. This is shown for the data in Figure 7.4(b).

Another way to summarize the scattering patterns developed by light when the fineparticles are of the order of the wavelength of light is to draw a polar energy pat-

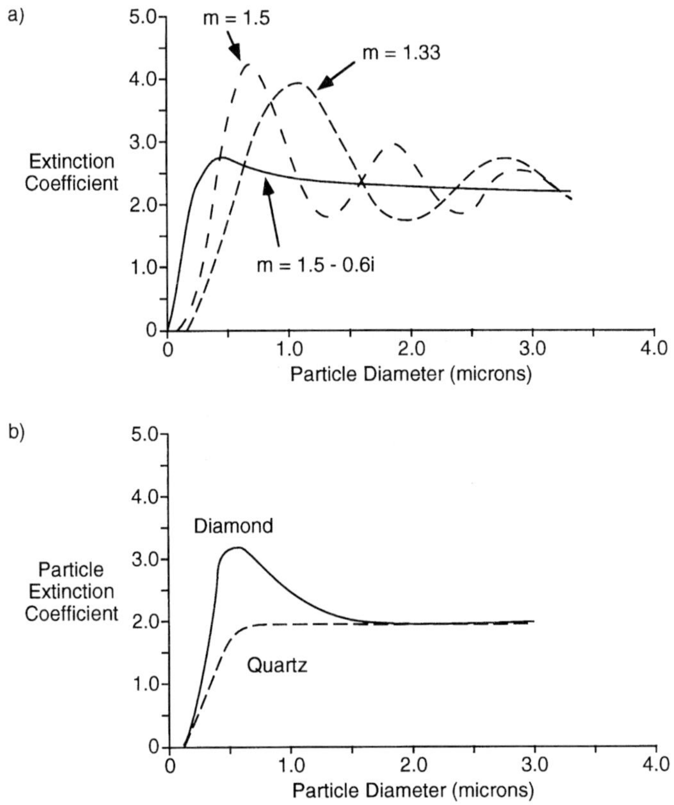

Figure 7.4. As the size of the fineparticle interacting with a light beam approaches the wavelength of the light, the extinction coefficient begins to oscillate around a value of 2.00 with the oscillations becoming larger as the fineparticles become smaller. a) Theoretical calculation for fineparticles of various refractive index. b) Measured data for two particular types of fineparticles.

tern. In Figure 7.5 the complex polar radial energy distribution in the scattering pattern for various sized fineparticles is shown. As the particle size approaches the wavelength of light, an increased amount of light is scattered at right angles the direction of the incident beam. This is one of the reasons why diffractometers intended to study particles smaller than 1 or 2 microns use side scattered light in their measurements.

When the particles are less than the wavelength of light the appropriate theory of light scattering is known as the Rayleigh Theory. It was developed by Lord Rayleigh an English Physicist (1842–1919). Before he acquired the title from his father, his family name was John William Strutt. Scientists interested in the optical properties of very small spherical particles, such as the latex emulsions used in the paint industry and in

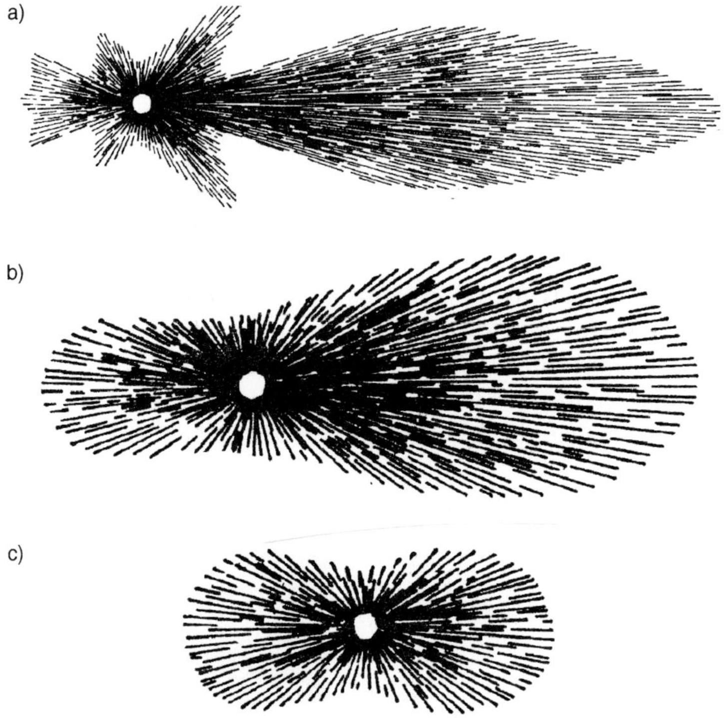

Figure 7.5. Light scatters in different ways depending on the size of the fineparticle with respect to the wavelength of the incident beam of light. In all the above cases the fineparticles are illuminated by a light beam incident from the left. a) Scattering when the fineparticle is larger than the wavelength of the incident light is predominantly in the forward direction. b) Scattering from a fineparticle which is about one quarter the size of the wavelength of the incident light. c) Scattering from a fineparticle less than one tenth the size of the wavelength of the incident light.

the droplets of fat in milk, have published many articles on the light scattering properties of such fineparticles [22, 23, 24]. In Section 7.2 we will discuss the work on determining the structure of individual fineparticles from light scattering studies.

Before we can understand the basic theory of commercially available diffractometers for characterizing the size distribution of a powder it is necessary to consider what happens when the light generating the diffraction pattern passes through a cloud of fineparticles. Basically the Fraunhofer diffraction pattern of a sphere is a set of circles consisting of a bright center, and the rings of energy around the center which are known as the first order minimum, first order maximum etc. The formula for the first order minimum in a Fraunhofer pattern from a sphere is

$$\frac{y}{x} = \frac{1.22\lambda}{2}$$

Where **y** is the distance from the axis and **x** the distance between the sphere and the plane where the pattern is generated, and $\boldsymbol{\lambda}$ is the wavelength of the light.

An important concept is that if one has a regular array of monosized spheres, the diffraction pattern is a series of points as shown in Figure 7.6(a). On the other hand, if the scattering spheres are a random array then the pattern is the same as a single particle's but the energy is multiplied by the number of scattering spheres as shown in Figure 7.6(b). Therefore, for a random array of monosize spheres one can calculate the size of the spheres from the dimensions of the diffraction pattern. The advantage of using a diffraction pattern is that the dimensions of the diffraction pattern is orders of magnitude larger than the diffracting spheres. When one has a random array of

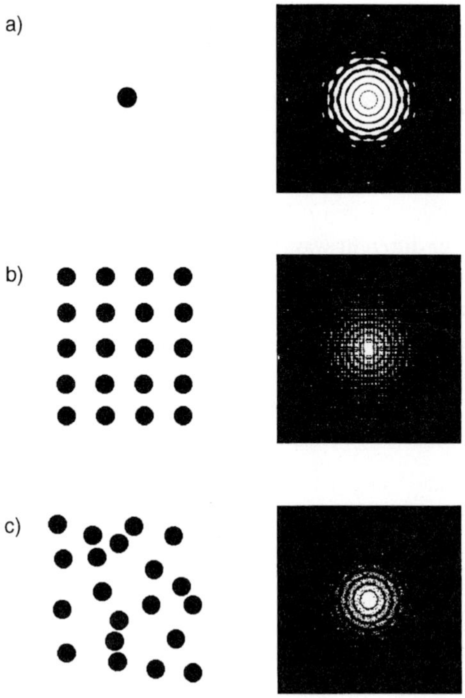

Figure 7.6. The diffraction pattern produced by a set of round (spherical) profiles can provide information about the arrangement of the profiles in the array. a) A single round profile and its diffraction pattern. b) A regular array of profiles like that of (a) results in a diffraction pattern which looks like a regular grid of dots of light. c) A random group of profiles like that of (a) produces a diffraction pattern similar to that of the single profile.

different sized spheres each of the different sized groups contribute their own diffraction pattern to the group diffraction pattern. The basic idea used in diffractometers for measuring the size distribution of fineparticles is to measure the energy distribution from the center outwards in a group diffraction pattern and then, by making assumptions on how the different sizes would contribute to different regions of the diffraction pattern, one can work out the size distribution of the scattering material. This process of breaking down the diffraction pattern into its constituent energy bands with the ultimate calculation of the size distribution is known as deconvolution of the scattering pattern. For an interesting introduction to the calculations, using computers, of the diffraction pattern of arrays of small circles see reference 25.

7.2 Studies of the Light Scattering Properties of Individual Fineparticles

Kerker has written an interesting review of the use of light scattering to study aerosols [26]. One section of this review entitled "Single Particle Instruments" describes how scientists adapted Millikan's methods of suspending an oil droplet in space to study aerosol products. Millikan, in his research of the magnitude of the charge of an electron, suspended an electrically charged oil drop between condenser plates in order to measure the charge [1, 27]. Apparently there is some conflict as to who really carried out this research and Kerker points out that another scientist, Harvey Fletcher, was involved in the original work and that the aerosol particle used to measure the charge was monitored by observing scattered light. In 1926 Whytlaw-Gray and Pattison used the basic systems used by Millikan and Fletcher to measure the radius of the particle suspended electrically by measuring the light intensity scattered at 90 degrees to the direction of the light. They were able to develop an empirical expression relating the intensity of scattered light to the size of the droplets. In 1961 Gucker and Egen built an advanced version of the Millikan and Fletcher equipment and they were able to measure light scattering over a range of angles from 40 to 140 degrees. They used a spray of an oil, dioctylphthlate solution, with droplet sizes ranging from 0.7 to 1.5 microns in radius. They were able to demonstrate the validity of calculations based upon Mie Theory [27]. In recent years Colbeck has used the same basic concepts to isolate a single soot fineparticle from a combustion chamber [28]. By means of electric fields he is able to rotate and/or invert a soot fineparticle to study its detailed structure by means of light scattering and optical microscopy. He has been able to measure the fractal dimension of the soot fineparticle and its change in structure as it ages. Ashkin and co-workers have used optical pressure to levitate single fineparticles for detailed light scattering studies [29]. Kerker's review contains some interesting details on the biographies of people who have contributed to aerosol light scattering studies and also reviews the development of optical type stream counters discussed in Chapter 6.

7.3 Light Scattering Properties of Clouds and Suspensions of Fineparticles

Kerker, in his review quoted in the previous section, records that one of the first scientists to discuss the optical properties of clouds was Leonardo da Vinci (1452–1519). In da Vinci's notebooks published in 1890 it is recorded,

> *"if you produce a small quantity of smoke from wood and the rays of the sun fall on this smoke and if you place behind it a piece of black velvet in which the sun does not fall you will see that the black stuff will appear of a beautiful blue colour – water violently ejected in a fine spray and in a dark chamber when the sun beams are emitted produces thin blue rays – hence it follows as I say that the atmosphere assumes this as you view by region of the particles of moisture which catch the rays of the sun" [26].*

The study of clouds in the atmosphere by recording the way in which they scatter light is one of the few tools available for studying clouds in situ. The use of light scattering to study atmospheric clouds has been reviewed by Harris [22]. In Harris's review he makes the following comments,

> *"if there are many particles close together in a cloud, or a fog, a photon may interact many times with different particles before escaping from the cloud, the result is that the characteristics of single scattering are smooth with respect to direction, for example the inability to detect the suns position in fog."*

He goes on to state that multiple scattering is common in nature and it is difficult to calculate because of mathematical complexities. Studies indicate that in a hazy atmosphere multiple scattering is much more important in the visible light range than in the infrared. In the laboratory studying the light scattering behaviour of a suspension of fineparticles, or an aerosol cloud in a chamber, is described as studies in turbidometry and nephelometry. The word turbid comes from a Latin word meaning disturbed and obviously refers to systems which are similar to stream water when it becomes cloudy from the sediment stirred up from the bottom of the stream. The use of light scattering to obtain average particle sizes is useful in industries where the dispersed particles are spherical and have a known type of size distribution. For example, Goulden studied the size of fat globules in milk using light scattering techniques [23].

Fiber optics have been used to study the light scattering from concentrated dispersions [30]. In studies of concentrated suspensions by light-scattering it is sometimes more convenient to study back scattering rather than forward or side scattered light (see for example the study by Heffels et al [31]). Overall, light-scattering of a smoke cloud is a powerful technique for characterizing a rapidly changing system such as aging cigarette smoke [32]. By making certain assumptions, one can measure the surface area and median aerodynamic diameter of an aerosol cloud [33]. Light scattering to study the structure of aerosol clouds and suspensions is an active area of research, and one should consult the research literature for recent developments [34].

7.4 Diffractometers for Characterizing Particle Size Distributions of Fineparticles

As mentioned in Section 7.1, Young pioneered the use of diffraction methods to study fiber width, and the size distribution of systems such as blood cells or pollen, in the 19th century. The modern diffractometers for sizing fineparticles in a powder became possible with the development of the laser, for generating the diffraction pattern, and the availability of lower cost powerful computing methods. In modern diffractometers a laser is used to generate the group diffraction pattern of the fineparticle population under study and this pattern falls on a special photocell detector which is a series of concentric circular photocells. A typical arrangement of a laser diffraction particle size analyzer and a sketch of the typical detection system is shown in Figure 7.7. The energy from the beam which is not diffracted from the particles being inspected is absorbed by a central auxiliary photocell as shown. Using this device, one can calculate the energy removed from the forward traveling beam by the fineparticles to obtain a measure of the overall strength of the suspension's scattering power. One of the first diffractometers for characterizing a powder system was developed by Cornillaut who was interested in sizing cement powders. His instrument has developed into the modern diffractometer available from CILAS [13]. By the late 1990's, there were seven commercially available diffractometers for sizing fineparticles and more will probably become available in the very near future [10–17]. (In the quickly changing market it is difficult to keep track of the companies when they change their names and those of their instruments. For example a very well known instrument, one of the first to be developed, known as the Microtrac was developed and marketed by Leeds and Northrop. In the mid–1990's, this instrument began to be marketed by Honeywell Inc. [11].)

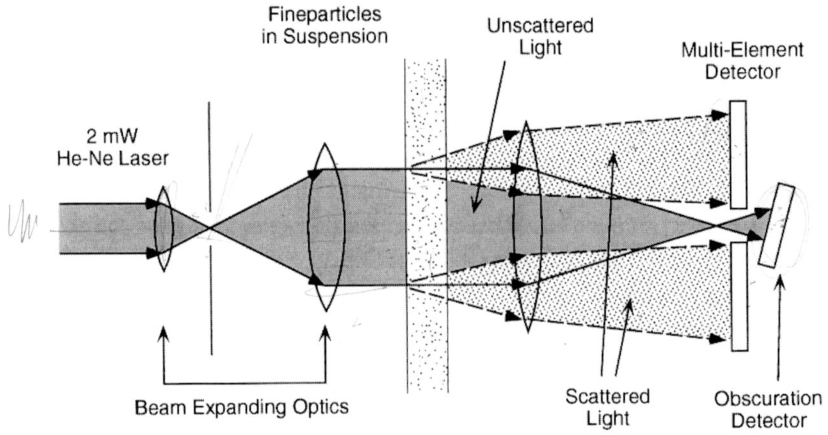

Figure 7.7. Optical arrangement of a typical laser diffraction particle size analyzer.

The several instruments vary somewhat in their optical configuration and also in the software used to transform the diffraction pattern data into a size distribution curve. An early discussion of the data processing used for such instruments was published by Swithenbank and his co-workers [35, 36]. In theory, one can use basic physical theory to interpret the diffraction pattern energy distribution, but in fact, this can be extremely tedious and, even with modern computers, time consuming. Many commercial machines actually take short cuts in their data processing by assuming that the data will fit a given size distribution. (See discussion of light diffraction data in References 37 and 38.)

The development of the software associated with the deconvolution procedure and built into the logic of a commercial diffractometer represents a major investment by the instrument companies and is often the essential commercially valuable part of the instrument. Therefore it is not surprising that the companies are reluctant to divulge the exact details of the physical basis of their logic. Before purchasing an optical diffractometer, one should inquire as to whether there are built in assumptions in the processing of the logic and how the diffractometer will handle multi-mode distributions of sizes in their distribution function. At the time of writing, inter-laboratory inter-equipment studies of standard powders had resulted in remarkably wide differences in the assessed particle size distribution, as demonstrated by the data reported by Merkus and co-workers, summarized in Figure 7.8 [39]. The reason for the differences between the assessed size distributions, using the same powder and between the various instruments, could arise from several factors. One of the main factors in comparing analytical data is the problem of uniform technique in the dispersion of the powder prior to the characterization study, particularly if one is using a dry aerosol in the equipment, and differences in built in software in the various instruments [40].

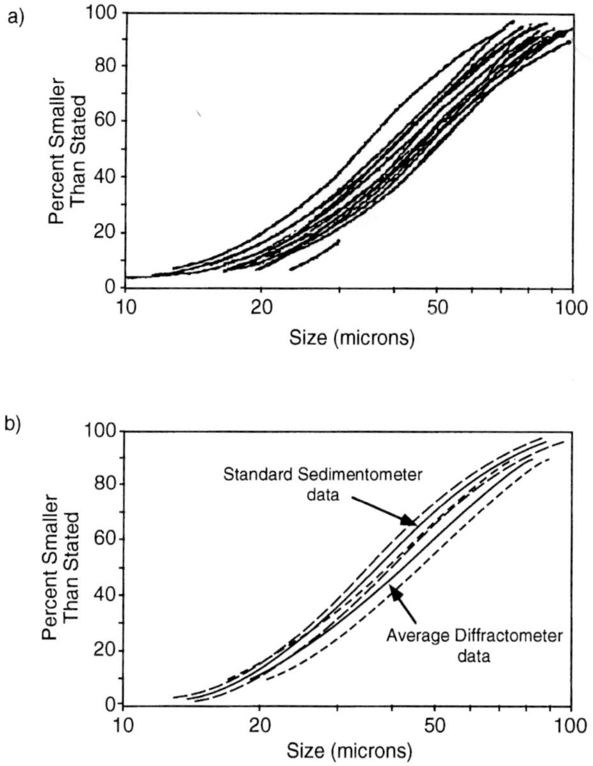

Figure 7.8. Although optical diffractometers can generate data very quickly, discrepancies between the data from various instruments or different laboratories can be quite large. a) Comparison of size distributions for various laboratories using the same type of instrument to size the Community Bureau of Reference of the European Economic Community standard reference material BCR69 as presented by Merkus et al [39]. b) Average and standard deviation of (a) compared with Standard sedimentometer data.

The software used by different manufacturers is in flux and one can only give some general guidance on what to look for when comparing data from different instruments. An example of how one particular instrument was using curve-fitting in the deconvolution algorithm is illustrated by the data shown in Figure 7.9 and 7.10. The four powders shown in Figure 7.9(a) were characterized using a commercially available diffractometer [37]. As described earlier, the first three powders were fractionated using sieves, whereas the fourth powder was prepared from a slurry by spray drying. The powders were first sized by image analysis. In Figure 7.9 the sizing data along with the images of the powder are shown. It can be seen that the agreement between the two methods is best for the spray-dried powder which is more spherical (see shape data

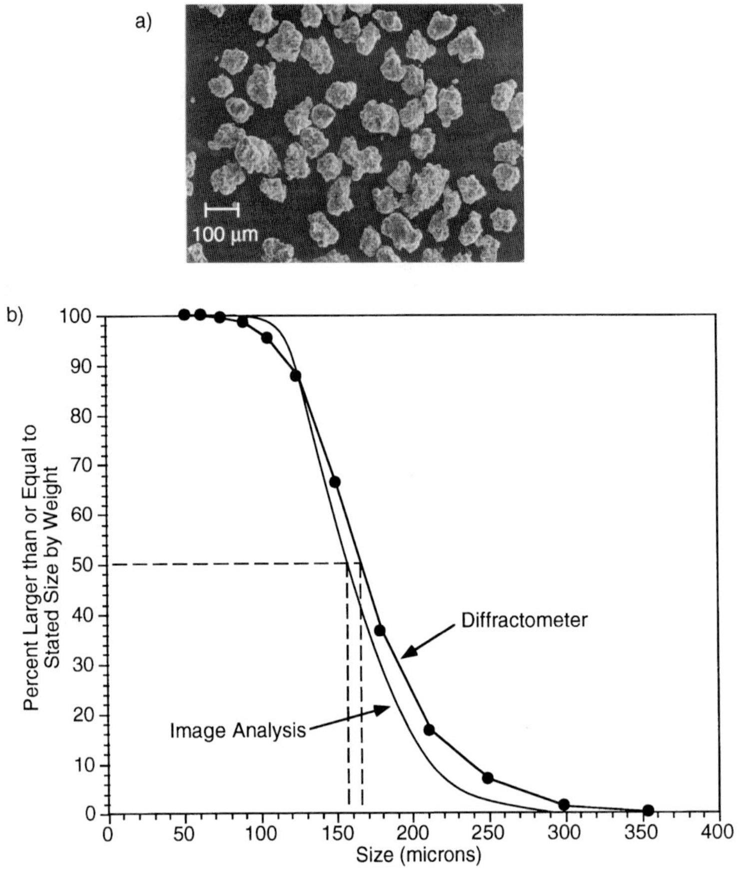

Figure 7.9. Size distributions of an irregular iron powder as determined by image analysis and diffractometer show significant differences. a) Electronmicrograph of an irregular iron powder. (Micrograph courtesy of D. Alliet, Xerox Corp.) b) Comparison of size distributions of the iron powder of (a) as measured by a diffractometer and by image analysis.

on these powders given in Chapter 2). In Figure 7.10(a), four measured distributions of the powders by diffractometer are plotted on log-probability graph paper in cumulative format. It can be seen that the data indicates that all four powders have the same type of distribution function, which is rather surprising in view of the fact that one of them was prepared by a very different method. In Figure 7.10(b), the image analysis data is plotted on log-probability paper and it can be seen that, particularly the spray dried powder does not generate a straight line relationship on such a graphical scale. From the way in which it was prepared, one would anticipate that the spray dried powder would be Gaussianly distributed, and when data for particular powder is plotted

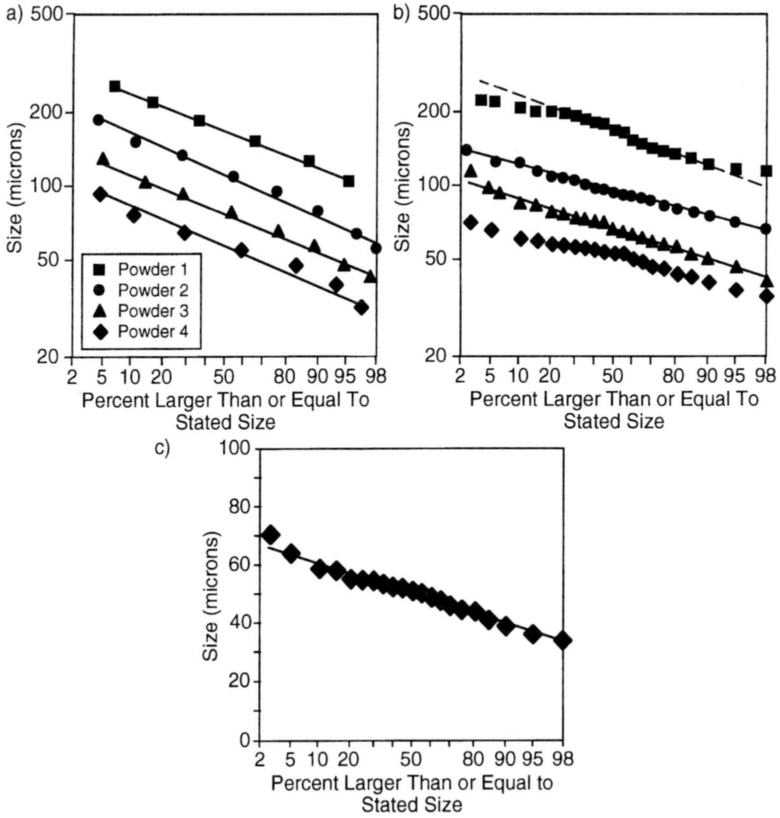

Figure 7.10. Size distributions of four irregular iron powders as determined by image analysis and diffractometer plotted on Gaussian probability graph paper. a) Size distributions obtained from an optical diffractometer plotted on log-Gaussian scales yields parallel lines, indicating that the algorithm used assumes a log-Gaussian distribution. b) Log-Gaussian plot of image analysis data, indicates that Powders 1 and 4 do not fit a log-Gaussian distribution. c) Image analysis data for Powder 4 plotted on arithmetic-probability paper shows that the distribution follows a Gaussian distribution.

on arithmetic-probability paper from the image analysis data in Figure 7.10(c), it can be seen that indeed this spray dried powder is describable by a Gaussian distribution. The data summarized in Figure 7.9 and 7.10 indicate that the diffractometer used curve fitting to an anticipated log-Gaussian distribution function in order to save computation time and cost. The particular diffractometer used in this study is not quoted since the situation differs from diffractometer to diffractometer and companies are continually adjusting and upgrading their deconvolution algorithms. The data for Figures 7.9 and 7.10 are presented to warn the technologist to ask questions as to the deconvolution algorithm in any size distribution measurements they are using, particularly if the data is required for research purposes. For quality control situations it is not particularly relevant that the deconvolution algorithm may incorporate a time saving strategy; curve fitting [37].

Palmer and co-workers have compared the data obtained using light-scattering diffractometers to data generated by a resistazone stream counter (these were discussed in Chapter 6) [41]. The powders that Palmer et. al. used in their study are summarized in Table 7.2. Typical fields of view of the powders that they used, along with the data summaries of the size distributions of the powders as measured by different techniques are summarized in Figure 7.11. It can be seen that the effect of fineparticle shape is quite pronounced, and that, especially if the fineparticles are porous, there can be considerable difference in the measured size distributions. Sometimes, a simple correlation factor can be used to allow for shape when carrying out intermethod comparisons. Thus, in Figure 7.12, the comparison for an essentially spherical powder versus an irregularly shaped powder as measured by the Microtrac diffractometer and the Sedigraph equipment (described in Chapter 4) is shown. However, such a correlation

Table 7.2. Data for Palmer's measurements of the size of the four types of fineparticles shown in Figure 7.11 as determined by diffractometer and resistazone measurements.

	Spherical Latex		
	Mean (microns)	**Median (microns)**	**Mode (microns)**
Diffractometer	20.19	20.29	20.56
Resistazone	20.11	20.27	20.44
	Spherical Diamond Dust with Cubes		
	Mean (microns)	**Median (microns)**	**Mode (microns)**
Diffractometer	4.543	4.811	4.792
Resistazone	4.237	4.139	4.193
	Boron Carbide Plates		
	Mean (microns)	**Median (microns)**	**Mode (microns)**
Diffractometer	28.93	29.71	29.60
Resistazone	19.45	20.41	21.11
	Difloxacin HCl Rods		
	Mean (microns)	**Median (microns)**	**Mode (microns)**
Diffractometer	9.090	9.436	10.870
Resistazone	6.566	6.509	6.574

can only be regarded as a powder specific correlation factor. One would have to establish this type of correlation between methods for any particular powder used in any given study.

Depending on the system used, the diffractometer can size fineparticles over a wide range. The literature of specific manufacturers should be consulted for their stated range of measurements.

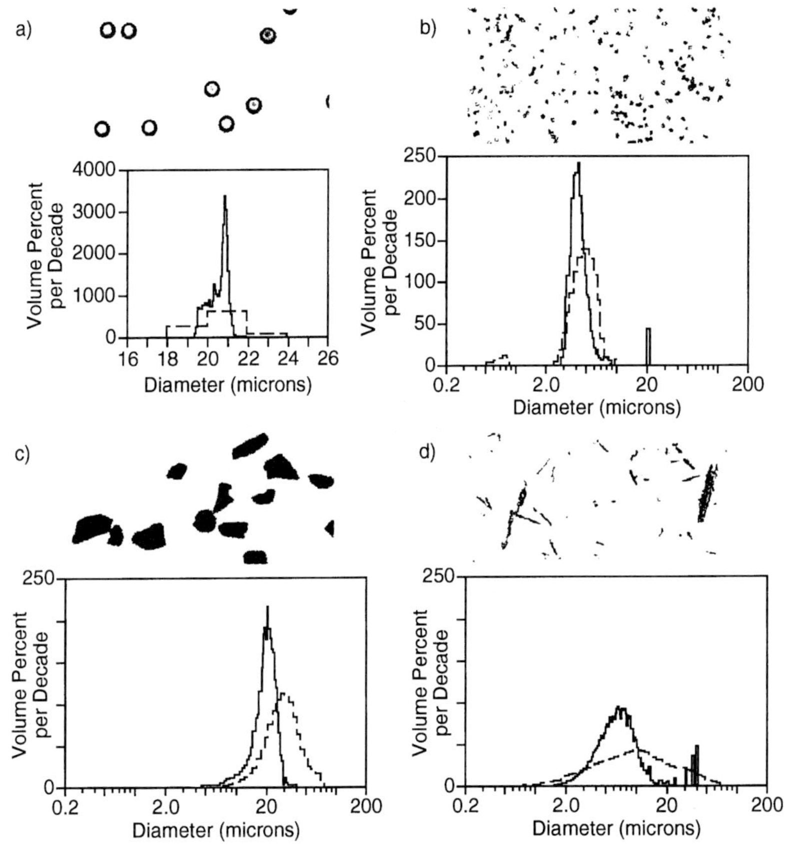

Figure 7.11. Comparison of results obtained by Palmer, using laser diffraction and resistazone stream counter for fineparticles of various shapes. a) Results for spherical latex fineparticles. b) Results for spherical diamond dust with some cubic fineparticles. c) Results for plate-like fineparticles. d) Results for rod-shaped particles. (Micrographs and data used by permission of American Laboratory and, J. Cowley, T. Palmer and P. J. Logiudice [41].)

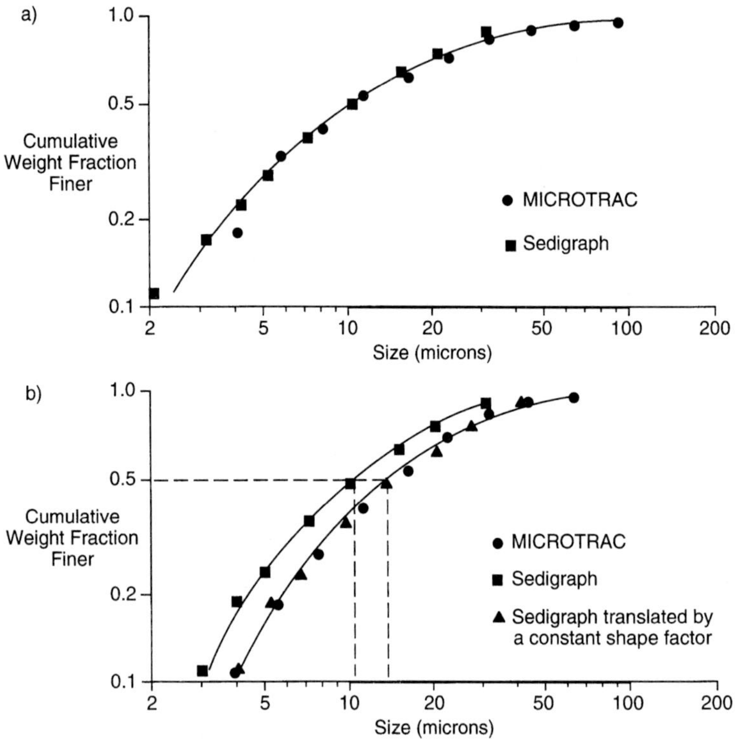

Figure 7.12. Comparison of sizing data obtained by sedimentation and laser diffraction for a smooth and an irregularly shaped powder demonstrates how fineparticle shape can affect results obtained by laser diffraction. a) Comparison of data for a rounded quartz powder. b) Comparison of data for an irregular limestone powder.

7.5 Measuring the Fractal Structure of Flocculated Suspensions and Aerosol Systems Using Light-Scattering Studies

Many precipitates and fumed aerosol systems have fractal structure. The reason for this is that they generally grow by a process known as diffusion limited aggregation [42–46]. This process can be modeled on a computer, as illustrated by the system shown in Figure 7.13. A single particle shown at the center of the matrix of squares is assumed to be a nucleating center, around which the precipitate or agglomerate grows. In diffusion limited aggregation it is assumed that the nucleating center grows by the ran-

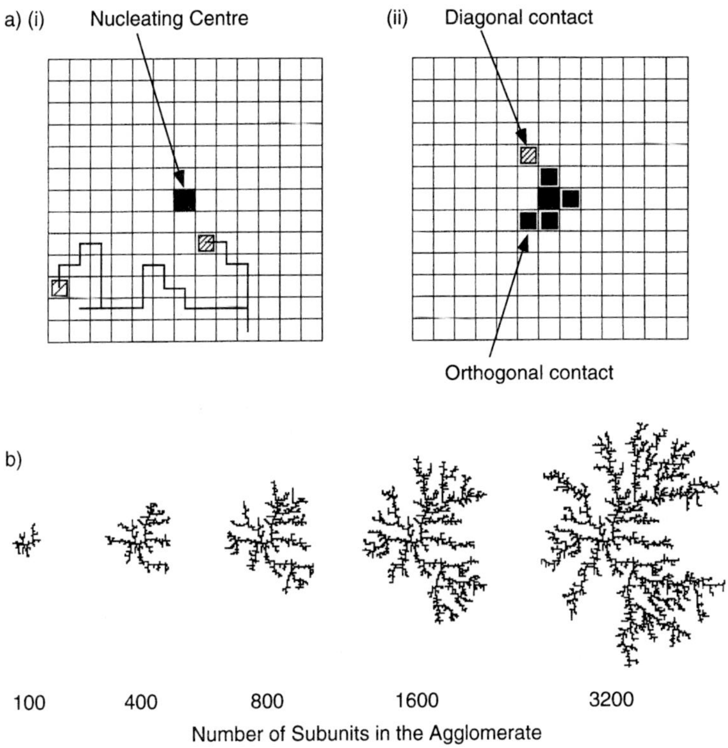

Figure 7.13. Diffusion Limited Aggregation (DLA) can be used to model the growth of fumes, clusters and flocs. a) Simplified steps in the DLA procedure. (i) A subunit enters the grid at random and undergoes a random walk (Brownian motion). (ii) If the random walking subunit contacts the cluster, and the probability of sticking is 100 percent, a subunit contacting orthogonally joins the growing cluster. If the contact is diagonal, the subunit continues to wander until it again contacts the cluster and the same joining rule is applied once more. b) Stages in the growth of a cluster using the method of (a) with 100 % probability of joining on orthogonal contact.

dom collision of wandering subunits in the area around the nucleating center. When modeling this on the computer, the pixel representing a subunit capable of joining the growing agglomerate approaches the nucleating center by undertaking a random walk (for details of such modeling see reference 42). When setting up this type of model one has to decide if the incoming wandering subunit is going to stick to the nucleus when it encounters it. It is usual in such a model to permit orthogonally touching subunits to attach themselves to the growing aggregates. One also has to specify the probability of the encounter forming a permanent union. In Figure 7.13 the successive stages in the growth of a simple agglomerate, grown according to the rule that a orthogonal encounter results in a permanent bond with a 100 % probability if it touches the growing agglomerate are shown. This type of agglomerate was one of the earliest to be studied by pioneers in subject of fractal geometry and agglomerates grown according to this rule are known as Whitten and Sander fractal aggregates [43]. If you were to be given a picture of a portion of the Whitten and Sander aggregate you would not know which part of the overall aggregate it came from demonstrating that the structure has the property known as statistical self-similarity.

To describe a fractal such as that of Figure 7.13(b)(iii) one uses what is known as a density of mass fractal. This is a measure of how the system occupies space. This fractal dimension is different from the boundary fractal dimension used in Chapter 2 to describe the rugged boundary of a dense fineparticle. If one grows several agglomerates of the Whitten Sander type they may look different but are of the same mass fractal

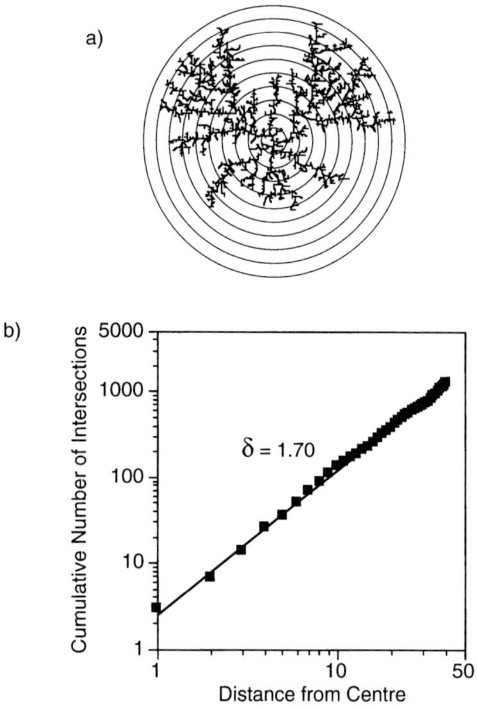

Figure 7.14. The fractal dimension of a DLA cluster can be determined by plotting the cumulative total of the number of times the cluster intersects a series of concentric circles drawn about its nucleating centre against the radius of the circle. a) The largest cluster of Figure 7.13(b) with concentric circles. b) Log-log plot of the cumulative number of intersections against the radius of the circle in subunit diameters. The fractal dimension is equal to the slope of the best-fit line to the data.

dimension. This is shown in Figure 7.14(a). One way to measure the mass fractal dimension of this type of agglomerate is to count the frequency of intersection on a set of standard circles. By moving out from the center of this type of study one eventually arrives at the situation in which the branches of the agglomerates are no longer reaching out and the number of intersections falls off. For such a situation it has been shown that the slope of the cumulative frequency of intersections plotted on log-log scale has a slope equal to that of the mass fractal dimension as shown in Figure 7.14(b). It has been shown by extensive studies that the theoretical value of the Whitten and Sander mass fractal dimension is 1.7. If the probability of adhering of a random walking subunit is different from 100 %, the resulting agglomerate has a much more dense

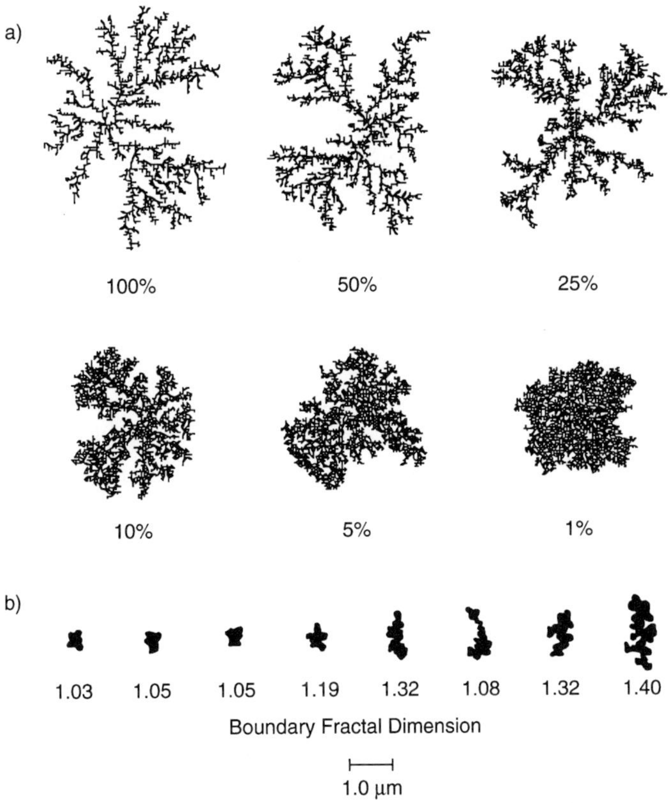

Figure 7.15. The way in which a cluster grows determines the overall ruggedness of the profile and that, in turn can affect the aerodynamic behaviour of the cluster. a) Simulated clusters grown using various probabilities that a wandering subunit will join the cluster. b) Real fumed thorium dioxide agglomerates, all having the same aerodynamic diameter, but different boundary fractal dimensions as shown.

structure, as shown by the fractal agglomerates of Figure 7.15. Each agglomerate was grown using the probability of a pixel joining its agglomerate on collision shown.

It can be seen that, particularly for the denser structures, one can also draw a boundary fractal around the shape and that, as the mass fractal dimension goes up, the boundary fractal dimension goes down. At 1 % probability of joining, the agglomerate grown is very dense. In the study of the colloidal precipitates one can have structures anywhere intermediate in the series shown in Figure 7.15. As demonstrated by the profiles of a set of thorium dioxide fumes, the boundary fractal dimension of profiles that all have the same aerodynamic diameter vary dramatically as shown in Figure 7.15(b). Measured boundary fractal dimensions of these profiles are shown under the profiles.

The structure of precipitates can be studied using light scattering techniques. However in such studies what is being measured is the internal structure of the aggregate which determines the interaction of the various parts of a scattered electromagnetic wave passing through the aggregate [46]. In particular in the mineral processing and water purification industries, residual solids in water that has been purified are brought together in a loosely structured precipitate which is called a floc and the process is called focculation. Schaefer has described structures of various types of fractal agglomerates and has used light scattering to study the fractal dimensions of the internal structure of silicate precipitates [45]. Raper and co-workers have made extensive studies of flocculants used in mineral processing and water purification [46]. It should be noted that light scattering studies in this case are used to look at the internal structure and that an independent method has to be used to measure the physical size of the floc scattering the light.

Many of the manufacturers of commercial diffractometers provide extensive tutorial literature for developing an understanding of the physics of diffractometers. I found some notes provided by Wedd of Malvern Instrument Company particularly useful [47]. (See also references 13, and 48.) Some workers have used x-ray scattering to study particle size. An ISO standard for this technique is in preparation (ISO – TC24-SC4). In the technique a narrow beam of x-rays passes through a powder layer containing ultrafine particles and the energy is dispersed around the incident beam at small angles by the electron scattering from the fineparticles. The distribution of scattered intensity is closely related to the particle size distribution. The draft of the ISO standard notes that this method is suitable for ultrafine powders containing particles in the size range of 1 nm to 300 nm it should not however be used for powders containing particles whose morphology is far from spherical, powder containing micropores, or mixtures of powders. Details are to be found in the ISO standard procedure [2].

References

[1] D Halliday, R. Resnick, *Fundamentals of Physics,* Second Edition, John Wiley and Sons, New York, 1981.

[2] ISO-CD13323-1 "Determination of Particle Size Distribution, Part 1 – Light Interaction Considerations" Information from Building Division of DIN. Burggrafenstrasse 6 Berlin-Tiergarten 10772 Berlin, Germany. In Canada information is available from the Standards Council of Canada, 1200-45 O'Connor Street, Ottawa, Ontario K1P 6N7.

[3] M. Kerker, *The Scattering of Light and other Electromagnetic Radiation,* Academic Press, New York, 1965.

[4] H. C. Van der Hulst, *Light Scattering by Small Particles,* Wiley, New York, 1957.

[5] C. F. Bohren, D. R. Huffman, *Adsorption and the Scattering of Light by Small Particles,* Wiley, New York, 1983.

[6] B. Chu, *Laser Light Scattering – Basic Principles and Practice,* Second Edition, Academic Press, 1991.

[7] L. P. Bayvel, A. R. Jones, *Electromagnetic Scattering and it's Application,* Applied Science Publishers, London, 1986.

[8] Inquiries regarding the proceedings of this conference obtainable from the Mechanical Engineering Department, University of Minnesota, 111 Church Street S. E. Minneapolis, MN 55455.

[9] See entry in *Larousse Dictionary of Scientists,* Larousse Publishing House, Anadale Street, Edinburgh and in New York, 1994.

[10] Horiba Instruments Incorporated, 17671 Armstrong Avenue, Irving, CA, 92714, USA.

[11] See literature from Honeywell Inc. Microtrac. RX100 16404 N. Black Canyon Hwy, Phoenix, AZ 85023, USA.

[12] Malvern Instruments Inc., 10 Southville Road, Southborough, MA 01772, USA. Insitec Inc. 2110 Omega Rd. San Ramon CA 94583 USA. Now a division of Malvern.

[13] CILAS U.S.A., agents: Denver Autometrics Inc., 6235 Lookout Rd., Bolder, Colorado 80301, USA.

[14] Sympatec Inc., Systems for Particle Technology Division, 3490 U.S. Route 1, Princeton, NJ 08540, USA.

[15] Coulter Corporation, Division of Scientific Instruments, P. O. Box 169015, Miami, FL 33116-9015, USA.

[16] Shimadzu Scientific Instruments Inc., 7102 Riverwood Drive, Columbia, MD 21046, USA.

[17] A Diffractometer intended for on-line monitoring of aerosol suspensions using a fiber optic detector has been developed by Combustion Engineering Incorporated, 100 Prospect Hill Road, Windson, Connecticut 06095. This machine was developed to measure pulverized coal being fed to an electrical power generating unit.

[18] E. Young, *An Introduction to Medical Literature,* Underwood and Blacks, London, 1818, p. 548.

[19] B. H. Kaye, *Direct Characterization of Fineparticles,* John Wiley & Sons, New York 1981.

[20] Data generated by Cherie Turbitt-Daoust, Laurentian University, personal communication.
[21] R. N. Bracewell, *The Fourier Transform and it's Applications,* Second Edition, McGraw-Hill, New York, 1978.
[22] S. Harris Jr., "Physics of Light Scattering," *Optical Spectra,* July-August (1970) 52–56.
[23] J. D. S. Goulden, "The Use of Turbometric Methods for Testing Homogenized Milk," *Journal of Dairy Research*, 27 (1960) 67–75.
[24] The Diffractometer methods are used extensively to study the size of latex suspensions in the coating industry. For example, T. Kourti, A. F. MacGregor, A. E. Hamielec, "Turbidometric Techniques -Capability to Provide the Full Size Distribution," Chapter 1 in the *Symposium Series 472, Particle Size Distribution II, Assessment and Characterization,* T. Provder (ed.), Published by the American Chemical Society, Washington, D. C., 1991. This book is the proceedings of a Symposium held in April, 1990.
[25] B. Hayes, "Digital Diffraction," *American Scientist,* 84 (May-June) 210 – 214.
[26] M. Kerker, "Light Scattering Instrumentation for Aerosol Studies and Historical Overview," *Aerosol Science and Technology*, 27 (1997) 522–540.
[27] Millikan's experiment is discussed in most first year physics textbooks, the original experiment is described in R. A. Millikan, "A New Modification of the Cloud Method of Determining the Elementary Electric Charge and the Most Probable Value of that Charge," *Philos. Magazine*, 19 (1910) 209 – 228.
[28] I. Colbeck, B. Atkinson, Y. Johar, "The Morphology and Optical Properties of Soot Produced by Different Fuels," *J. Aerosol Science,* 28 (1997) 715–723.
[29] A. Ashkin and J. M. Dziedzik, "Optical Levitation by Radiation Pressure," *Applied Physics Letter*, 19 (1971) 283–285.
[30] J. C. Thomas, "Fibreoptic Dynamic Light-Scattering from Concentrated Dispersions: Potential for On-line Particle Size" in book quoted in Ref. 24.
[31] C. Heffels, A. Willemse, B. Scarlett, "Possibilities of Near Backward Light Scattering for Characterizing Dense Particle Systems", *Powder Technology*, 86 (1996) 127–135.
[32] I. P. Chung, D. Dunn-Rankin, "Light Scattering Measurements of Main Stream and Side Stream Cigarette Smoke," *Aerosol Science and Technology,* 24 (1996) 85–101.
[33] S. Sato, J. H. Pinsent, D. Y. H. Pui, "The Characterization of Particle Size for Polydispersed Powders Aerosolized and Detected by Light-Scattering", *Part. Part. Syst. Charact.*, 13 (1996) 234–237.
[34] See for example, N. Whang, J. Wei, C. Hong and H. Zhang, "The Influence of Refractive Index on Particle Size Analysis Using Turbidometric Spectrum Methods", *Part. Part. Syst. Charact.*, 13 (1996) 238–244.
[35] J. Swithenbank, J. M. Beer, D. S. Taylor, D. Abbot, C. G. McCreath, "A Laser Diagnostic Technique for the Measurement of Droplet and Particle Size Distribution," Presented as paper 76–69 at the AIAA Aerospace Sciences Meeting, Jan. 1976. Published in "Experimental Diagnostics in Gas Phase Combustion Systems," *Progress in Astronautics and Aeronautics*, Ed. B. T. Zinn, Vol. 53, 1977.

[36] P. G. Felton, D. J. Brown, "Measurement of Crystal Growth Rates by Laser Diffraction," Presented at the International Symposium on In-stream Measurements of Particle Solid Properties, in Bergen, Norway, August 1978.
[37] B. Kaye, D. Alliet, L. Switzer, C. Turbitt-Daoust, "The Effect of Shape on Intermethod Correlation of Techniques for Characterizing the Size Distribution of Powder. Part I: Correlating the Size Distribution Measured by Sieving, Image Analysis, and Diffractometer Methods," *Part. Part. Syst. Charact.*, 14 (1997) 219–225.
[38] Nathier-Dufor et al. "Comparison of Sieving and Laser Diffraction for the Particle Size Measurements of Raw Materials used in Food Stuffs," *Powder Technology*, 76 (1993) 191–200.
[39] H. G. Merkus, O. Bischof, S. Drescher, B. Scarlett, "Precision and Accuracy in Particle Sizing – Round Robin Results from Sedimentation, Laser Diffraction, and Electrical Sensing Zone," PARTEC 95, 6th European Symposium on Particle Characterization, 21–23 March 1995, Proceedings published by Nürnberg Messer GmbH, Nürnberg, Germany.
[40] B. H. Kaye, "Generating Aerosols," *KONA,* 15 (1997) 68–81.
[41] A. T. Palmer, P. J. Logiudice, J. Cowley, "Comparison of sizing results obtained with electrolyte volume displacement laser light scattering instrumentation," *American Laboratory*, November, 1994. Reprints of the paper are available from Coulter Corp., Scientific Instruments Applications Laboratory, 1950 W 8th Ave., Hialeah, Florida, 333010.
[42] B. H. Kaye, *Chaos and Complexity – Discovering the Surprising Patterns of Science and Technology,* VCH, Weinheim, 1993.
[43] C. A. Whitten and L. M. Sander, "Diffusion-Limited Aggregation, a kinetic critical phenomenon," *Phys. Rev. Lett.* 47 (1981) 1400.
[44] B. H. Kaye, "Applied Fractal Geometry and the Fineparticle Specialist," *Part. Part. Syst. Charact.*, 10:3 (1993) 99–111.
[45] D. W. Schaefer, "Fractal Models and the Structure of Material," *Materials Research Society, M. R. S. Bulletin,* 13:2 (February 1988) 22–27.
[46] S-J Jun, R. Amal, J. A. Raper, "The use of Small Angle Light Scattering to Study the Structure of Flocs," *Part. Part. Syst. Charact.,* 12 (1995) 274–278.
[47] Personal communication from M. Wedd, Malvern Instruments, Spring Lane South, Malvern Works, WR14 1AT, United Kingdom.
[48] P. A. Webb, C. Orr, "Analytical Methods in Fineparticle Technology," Micrometrics Corporation, Norcross, Georgia, 1997.

8 Doppler Based Methods for Characterizing Fineparticles

8.1 Basic Concepts Used in Doppler Methods for Characterizing Fineparticles

The Doppler Shift of a wave motion is the perceived shift in frequency of a source of waves when either the source and/or the receiver system are in relative motion. Until the development of modern electronics, lasers, and computer processing systems the interest in the shifting of electromagnetic waves due to the Doppler effect was confined to looking at the red shift in light from the stars and the calculation of their relative motion due to the expansion of the Universe. Using modern technology one can measure the Doppler shift of small fineparticles in an air stream or ones which are undergoing Brownian motion in a liquid. The Doppler shift is named after the Austrian physicist Christian Johan Doppler (1803–53). He studied the change in the sound of trumpets being played on a wagon pulled by a locomotive moving to and away from his location. Undergraduate textbooks usually discuss the Doppler effect with respect to the sound of train whistles and or ambulance sirens moving toward and away from an observer.

There are two types of size characterization equipment based on laser Doppler shift measurement. In the first technique, the fineparticles under study move through a region illuminated by a laser beam. The Doppler shifts generated by the moving fineparticles are evaluated and related to the velocity of the fineparticles under study. The method used to study fineparticles less than 1 micron in size utilizes the Doppler shift created by Brownian motion as monitored with a laser beam [1]. This second type of technology is variously known as photoncorrelation spectroscopy (PCS) or dynamic light scattering (DLS).

Important concepts involved in the use of Doppler shifted laser light to characterize fineparticles can be understood by discussing an instrument known as FODA from the acronym Fibre Optic Doppler Anemometer [2]. The basic optical system of the FODA uses light from a helium neon laser which goes through a small hole in a mirror. The light is then focused down onto the aperture of a fibre optic cable as shown in Figure 8.1. This laser light moves along to the other end of the cable where it is used to illuminate fineparticles moving towards and past the fibre optic pipe. A certain percentage of the light arriving at the end of the fibre optic is reflected back down the pipe. Light bounced off the moving fineparticles, which is Doppler shifted by the

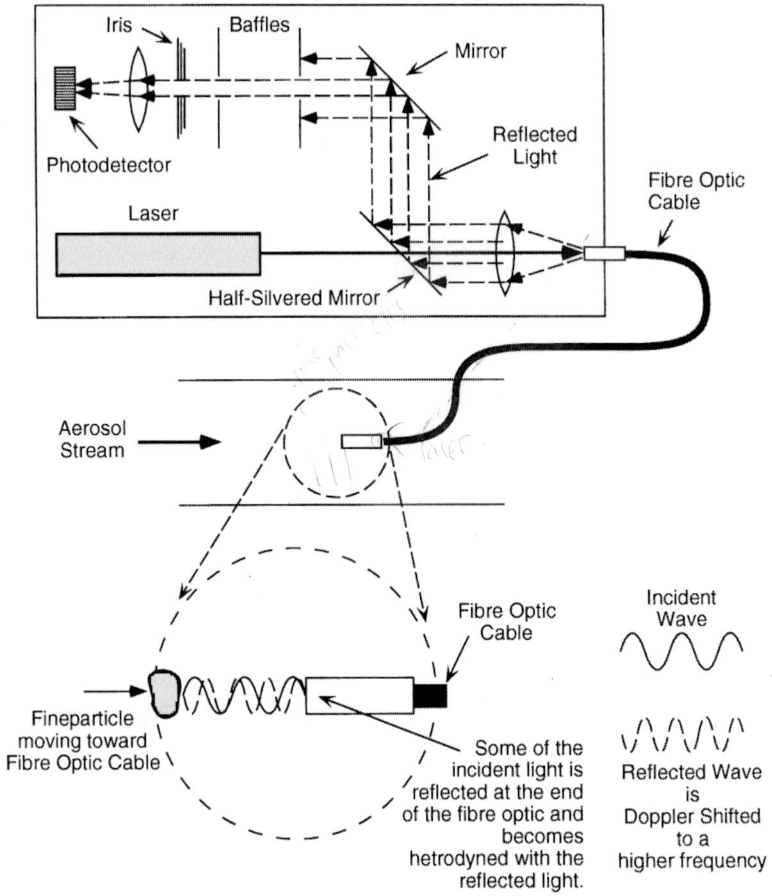

Figure 8.1. In the FODA system, light reflected from a fineparticle moving toward the end of a fibre optic cable is Doppler shifted. The reflected light is heterodyned with original light reflected from the end of the cable and used to measure the speed of the fineparticle.

velocity of the moving particle, also re-enters the fibre optic. In the fibre the Doppler shifted light is combined with the laser light reflected from the end of the fibre optic to generate a heterodyned signal.

The term heterodyning means unequal power. It is an electronic engineering term, the physical significance of which is perhaps more easily understood by looking at its acoustical equivalent. When tuning a guitar one plucks a string of the correct frequency and the adjacent string is fingered to theoretically produce the same note. If the two are not exactly the same, what is known as a beat frequency is heard. The guitarist tunes the guitar by tightening or loosening the tension on the string. When the string whose tension is being adjusted is at the same frequency as the reference string, the beat frequency disappears. The further away from the correct frequency that the second string is from the theoretical value the faster the beat frequency. The beat frequency of acoustics is the same as the heterodyning frequency of Doppler shifted light when it is mixed with the original frequency. Just as the beat frequency is much lower than the frequency of the vibrating string producing an audible sound. By heterodyning the signal, the optical signal to be measured is a much lower frequency than the frequency of the light. Measuring the Doppler shift in laser light directly, without heterodyning, is a virtually impossible task. On the other hand the heterodyned frequency of the Doppler shifted light is measured relatively easy with modern electronics.

If one looks at the optical system shown in Figure 8.1 the heterodyned signal moving back down the fibre optic cable is reflected by the semi reflective mirror and sent to a photodetector. In practice a whole population of particles can be moved past the end of the fibre by accelerating them into an inspection zone. For such a situation the large ones will be traveling with a slower velocity than the small ones and the range of Doppler shifted light results in a signal at the photodetector which measures the velocity and hence the size of all the moving fineparticles. Fourier analysis can be performed on this range of Doppler shifted light signals to deduce the size distribution of the moving fineparticles.

A different optical system was used by Yanta to achieve the same type of measurements. His system is shown in Figure 8.2 [3]. Two laser beams are generated by a beam splitter in the optical system. These beams intersect to create an inspection zone. With respect to the direction of flow, the light from beam A is going to give a lower Doppler shift than the beam B. As a consequence the light from A and B combine together in the photomultiplier tube generating a heterodyned signal from the two different Doppler shifts related to the velocity of the fineparticle. Although the basic measurement is actually a Doppler shift, in fact when one reads some of the literature on these methods it is very difficult to know where the Doppler effect comes into the system because the interpretation of the measurement is described in terms of Young's interference fringes. Beams A and B of Figure 8.2(a) actually produce a set of Young's interference fringes in the interrogation zone as shown in Figure 8.2(b). The fineparticle to be characterized moves through this set of fringes and produces an optical signal known in the literature as a Doppler burst. One of these Doppler bursts as recorded by Yanta is shown in Figure 8.2(c).

The frequency of the Doppler burst is used to calculate the velocity of the moving aerosol particle. Depending on the size of the particle, the Doppler burst can have the shape of either Figure 8.3(a) or (b). That of Figure 8.2(b) is a type of signal generated by a particle that is much bigger than the separation of the fringes. One can measure the overall size of the signal underneath the wave form of the Doppler burst to calculate the size of the particle as well as its velocity for an independent measure of size.

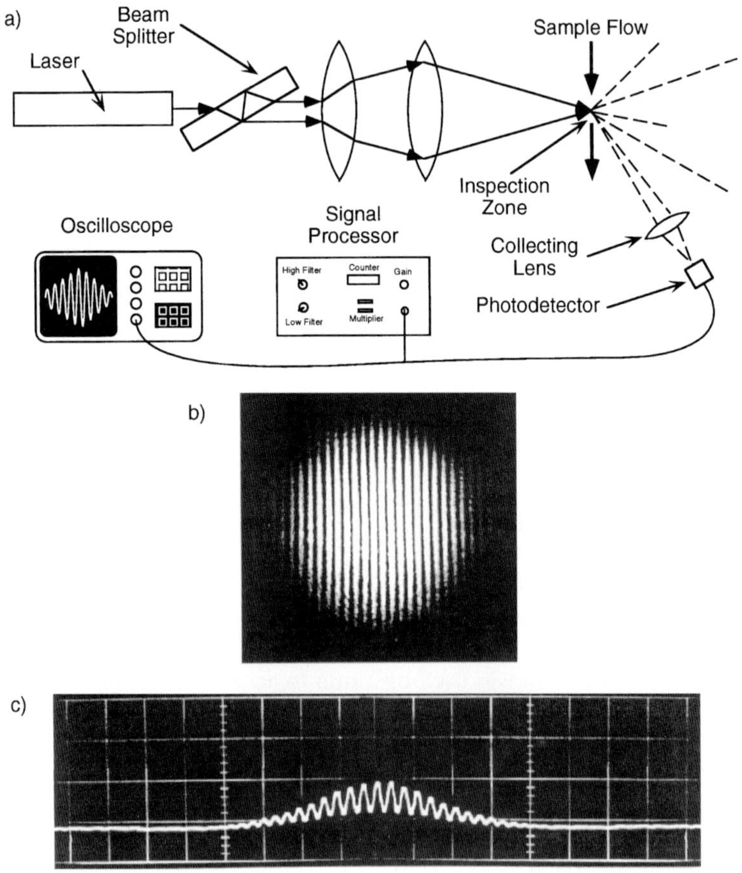

Figure 8.2. Yanta's equipment for size measurements using Doppler shifts. a) Arrangement of Yanta's equipment. b) Fringes created in the interrogation zone by interference of the two laser beams. c) Typical Doppler burst formed by a fineparticle passing through the interrogation zone.

However, there are difficulties for this use of the overall size of the Doppler burst. Note that the part of the signal which is modified by the oscillation is referred to as the pedestal of the Doppler burst [4].

Sometimes in spray research, or in a study of systems used to deliver therapeutic aerosols, it is useful to be able to measure both the velocity and size of a moving population of fineparticles. Several workers have pointed out that, to carry out this type of measurement, it is better to measure the Doppler shift to gain the velocity of the particles and then look at the phase shift of the Doppler shifted light to gain an estimate

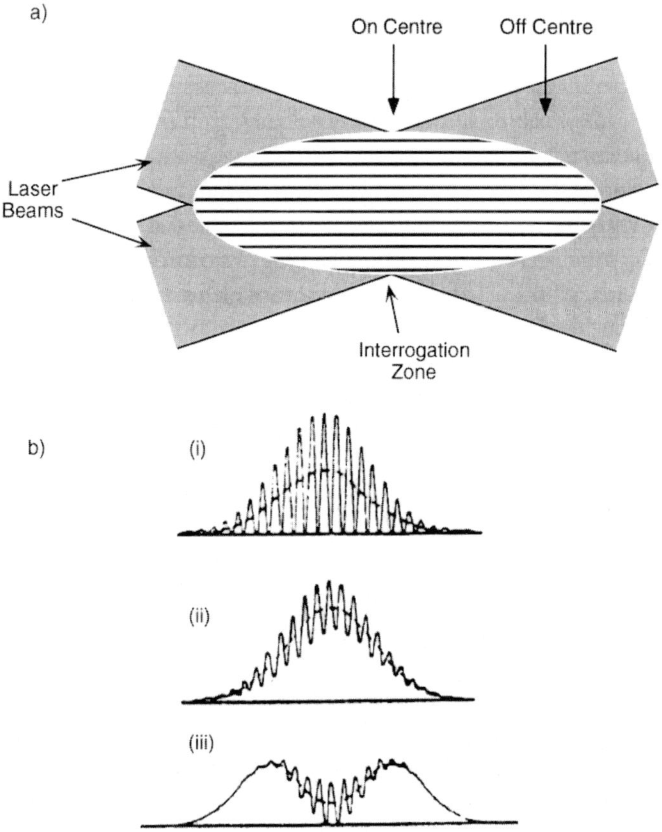

Figure 8.3. The Doppler burst produced by a fineparticle passing through the interrogation zone depends on both the size of the fineparticle with respect to the fringes, and the position with respect to the zone. a) Interrogation zone of a laser Doppler system. b) Bursts produced (i) by a fineparticle on centre (ii) by a fineparticle larger than the spacing of the fringes (iii) by a fineparticle passing the zone off-centre.

of the size of the particles. In Section 8.2 we will review methods which measure size directly by laser Doppler shifts. In Section 8.3 we will look at phase shifted Doppler signals to evaluate the size of fineparticles. In Section 8.4 we will look at the photon correlation methods used to measure the size of fineparticles smaller than 1 micron. It should be noted that the techniques for using Doppler shifted signals to evaluate size and velocity of fineparticles grew out of systems used to measure the velocity of air streams and from looking at the Doppler signals from dust in moving air streams. In Europe these instruments were known as laser Doppler anemometers (LDA) a technical term for a velocity measuring device based on the fact that in Europe, the anemometer is a device used to measure wind speeds. North American scientists prefer the term velocimeter. Therefore the techniques for measuring size based on Doppler shifted light are often referred to LDA's in Europe and LDV systems in North America. It should be noted that in general the data processing built into LDV systems involve complex mathematical systems but fortunately in commercially available machines all of the mathematics are encapsulated into operational algorithms. The operator of the instrumentation only needs to know the basis of the techniques and the potential limitations of the methodology. As in the case of diffractometers discussed in the previous chapter, vendors of the commercial machines have found it necessary to provide very comprehensive tutorial notes for the users of their machines. Useful information can be obtained by writing to the providers of such equipment listed in the various parts of this chapter. The basic optical system for various instruments discussed in this chapter may give a false impression of the simplicity of the implementation of the basic concepts but in practice expensive and sophisticated optical systems are required to reduce the noise in the experimental data. To provide the high speed data collection requires sophisticated electronic interfacing with data processing to generate the data display systems of the commercial machines.

8.2 Stream Counters Based on Doppler Shifted Laser Light

One of the earliest instruments of this type has already been described in Figure 8.2 devised by Yanta [3]. Some major manufacturers of LDV equipment utilize not only LDV to measure velocity but also incorporate equipment for measuring the phase of the Doppler shifted light to measure size of the particles when velocity and size are of interest to the investigation. Equipment manufactured by TSI and Dantec will be described in the next section. One of the most sophisticated pieces of equipment for measuring size by Doppler shift is a piece of equipment known as the E-SPART analyzer. This equipment was developed by Mazumder and co-workers [5, 6, 7]. The name E-SPART analyzer is an acronym created from the technical description – Electrical Single Particle Aerodynamic Relaxation Time Analyzer. The meaning of this technical description will become apparent as we describe its operation. The essential part of the equipment is shown in Figure 8.4(a). The equipment was designed to study aerosols and particles in the size range 0.3–70 microns aerodynamic diameter. The aerosolized fineparticles are fed into the inspection zone as shown in the figure. When the aerosol particles are in the inspection zone they are subjected to a sonic wave produced by the transducer shown. In response to this forced oscillation the large

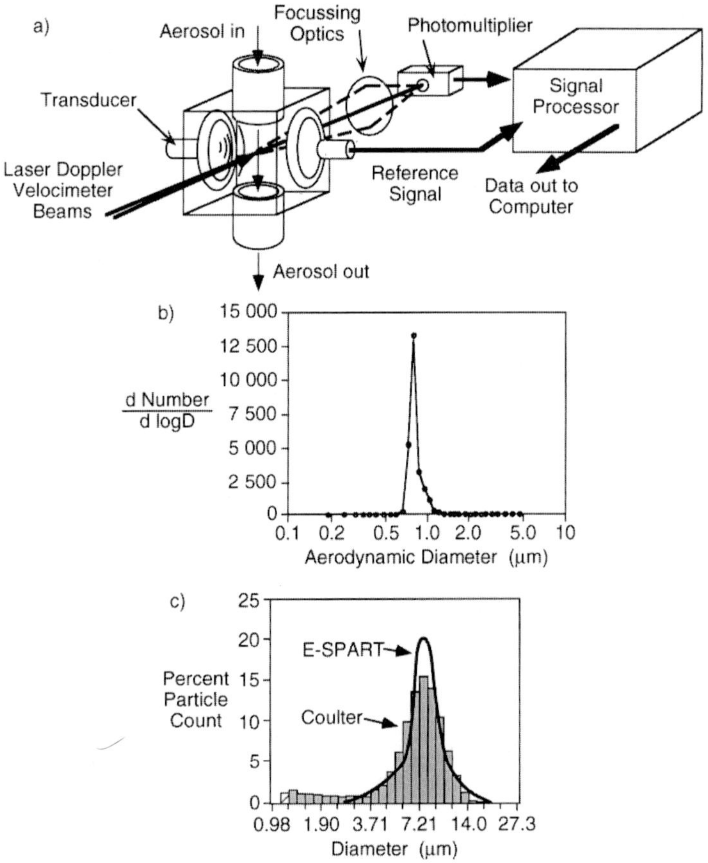

Figure 8.4. The E-SPART analyzer uses laser Doppler velocimetry to determine the size of fineparticles oscillated in the inspection zone by acoustic waves. a) Layout of the E-SPART analyzer. b) Size distribution for standard 0.8 micron polystyrene latex spheres. c) Comparison of the size distributions of dry ink determined by Coulter Multisizer and the E-SPART analyzer.

fineparticles will move more slowly than the smaller fineparticles. The movement is monitored with a dual laser beam which is essentially the same geometric configuration as that used by Yanta's and other similar instruments. When used in an appropriate format, the instrument can also measure the charge on the aerosolized fineparticles. The lag of a given size of fineparticle with respect to the acoustic driving frequencies is called the relaxation time and hence the name of the instrument. The instrument is calibrated using particles of known particle size such as polystyrene latex. For non-spherical particles the size data is described in terms of the equivalent spherical diameters. The typical data generated using this instrument is summarized in Figure 8.4(b) and (c). Because the toner fineparticles used in Xerographic printing are essentially spheres, one would anticipate that the data for the measurement for toner particles by the E-SPART analyzer and the Coulter Counter would agree as demonstrated by the data of 8.4(c). The equipment, which at the time of writing costs of the order of 120 000 U.S. dollars is available from the Hosokawa Micron Corporation [8]. The wealth of data that can be generated with this instrument in a short time is sure to lead to its widespread use in the study of therapeutic aerosols and efficiency of filtration systems used in the clean room industry and in studies of the efficiency of occupational health and hygiene equipment.

8.3 Phase Doppler Based Size Characterization Equipment

As was pointed out earlier, scientists working on the development of Doppler based systems for studying aerosols found that noise problems made it very difficult to accurately estimate size from the characteristics of the Doppler burst signal. However, the phase of the Doppler shifted light varies with size if the fineparticles are transparent and/or reflective. This has been exploited in measuring the size distribution of aerosol fineparticles. Discussions of the basic theory have been given by Bauckhage and by Naqwi [4, 9]. In North American commercial instruments exploiting phase Doppler measurements of size and velocity are available from TSI and Dantec [10, 11]. To measure the phase difference in the Doppler shifted light from the moving fineparticles it is necessary to use more than one detector. In Figure 8.5(a) the system described by Naqwi is shown [12, 13]. He studied the size distribution of an alumina powder. The size distribution of this powder by Coulter Counter is shown in Figure 8.5(b). Naqwi reports good agreement between measurements based on the phase Doppler shift. The distribution of the aerosol is deduced by converting the measured velocities into a size distribution.

In Figure 8.6(a) some interesting data reported by Bauckage on the simultaneous measurement of size and velocity of spray droplets, at a specific location within a spray nozzle system. using phase Doppler shift equipment, is shown [4]. To implement this type of measurement one needs specialized equipment but the power of the method is such that, particularly in the therapeutic aerosol field, we can anticipate further development of the technology. Thus in Figure 8.6(b) measurements on size distribution of a therapeutic aerosol of the drug albuterol, delivered from a metered dose inhaler used to treat asthmatic conditions, is shown. The area of size characterization of aerosols using phase Doppler techniques is a very active area of research and the scientific literature should be consulted to gain the latest information [14]. For example in a single issue of the journal Particle and Particle System Characterization, published in December 1997, there were two papers on Phase Doppler Anemometry and two on the light scattering properties of small fineparticles.

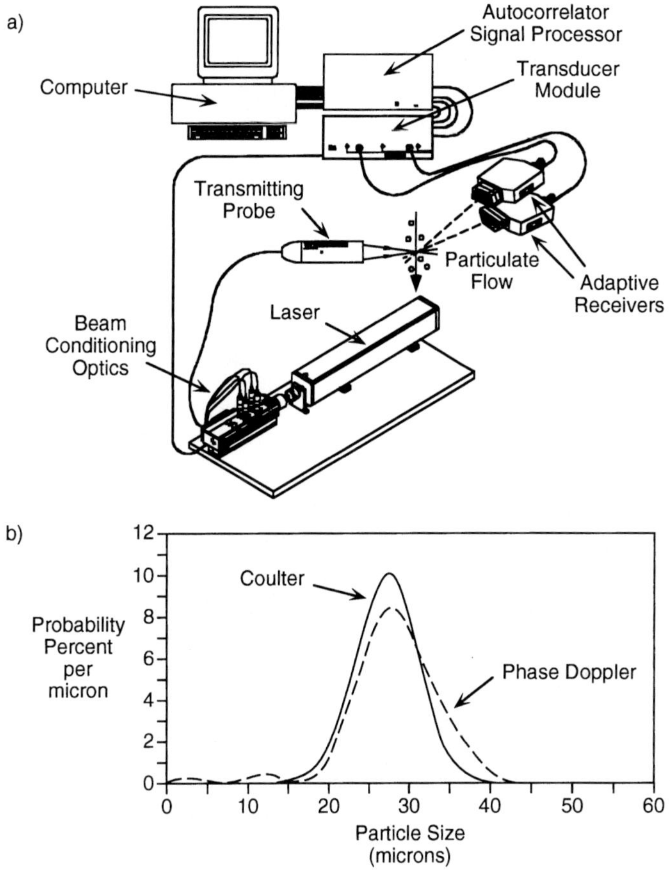

Figure 8.5. Naqwi has used Phase Doppler measurement to acquire the size distributions of various powders. a) The Phase Doppler system used by Naqwi. b) Comparison of the volume size distribution of an alumina powder obtained from a Coulter counter, to Phase Doppler measurements made by Naqwi. (Drawing and data provided by A. A. Naqwi and used by permission of TSI Inc.)

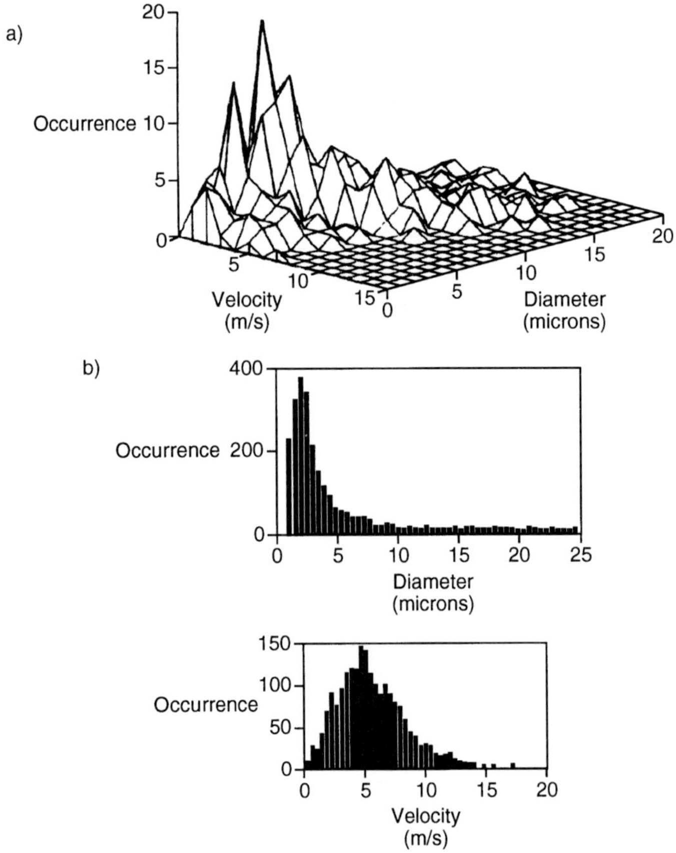

Figure 8.6. Laser Doppler systems have been used to determine the velocity and size distributions of fineparticles within a system concurrently. a) Data reported by Baukage and Schöne for spray droplets from a simple nozzle. b) Size and velocity distributions of Albuterol from a metered dose inhaler as reported by Rudoff et al. (Used by permission of Particle and Particle Systems Characterization.)

8.4 Photon Correlation Techniques for Characterizing Small Fineparticles

Forty years ago anyone interested in fineparticle characterization who had a science degree could cope with the theory of the techniques and was able to build relatively simple equipment to carry out the measurements. In the two most recent areas of fineparticle characterization – diffractometers (discussed earlier) and the method to be discussed in this section photon correlation spectroscopy (PCS), the relevant theories involve the use of concepts normally not studied until the post graduate level in an honours physics degree. The theory also involves electronic equipment and data processing concepts out of reach for people without a mathematical background. For these instruments the expensive electronic processing equipment and size evaluation equipment can truly be called a black box, from the point of view of the average operator.

As is the case for other sections in this book, whole books have been written on the methodology of photoncorrelation spectroscopy. All that we can attempt to do in this discussion is outline the basic concepts and direct the reader to sources of literature giving detailed information. Because the background required to understand the potential limitations of photon correlation spectroscopy would be outside the background of potential customers, many of the instrument vendors have prepared tutorial literature for the potential user. I have found material prepared by Lines, Nicoli and colleagues, and Weiner and colleagues, particularly useful [15, 16, 17]. Two excellent reviews of the technology and methodology, of photon correlation spectroscopy have been written by Finsey [18, 19]. The rate of growth of activity in this field can be appreciated from the fact that Finsey's review, written in 1994, had two hundred and ninety two references.

As already mentioned the technique of photon correlation spectroscopy is known by several names. In his first review of the technique Finsey calls the method dynamic light scattering (DLS) whereas in his second review he calls it Quasi-Elastic light scattering (QELS). In this book we use the term photon correlation spectroscopy (PCS) since it draws attention to the physics involved in the methodology. The terms dynamic light scattering and quasi-elastic light scattering both refer to the fact that, in the technique, the dynamic movement of the fineparticles because of Brownian Motion is studied as they scatter light. The basic system of the instrument is shown in Figure 8.7(a). Elings and Nicoli described the physical process going on in this procedure.

"A laser beam is focused into a cell containing a suspension of particles. A small fraction of the incident light is scattered by the particles and is collected at an

angle, usually 90 degrees, by a sensitive photomultiplier tube detector. Each fineparticle illuminated by the laser beam produces a scattered light wave whose phase at the detector depends on the position of the particle in the solution ..."

(note in their original discussion they use the term solution when they should mean suspension for an assembly of)

"... The total scattered intensity is the photomultiplier tube is the result of the superposition of all the individual scattered waves. The Brownian motion of the particles causes the relative phases of the light scattered from different particles to vary which in turn cause the intensity at the detector to fluctuate in time." [16]

Elings and Nicoli go on to state that

"although individual fluctuations occur randomly there is a well defined life time, τ, for their build up and decay, roughly equal to the average time required for a pair of fineparticles to change their separation by one-half the laser wavelength λ."

The actual output measured by the photomultiplier tube looks like the signal shown in Figure 8.7(b). The basic calculation that one carries out with this signal is to carry out a calculation known as autocorrelation of the data. This means that one takes the value at t and t + Δt and multiply them together. When this Dt is small the signal does not change very much and as one increases this value, the product of the calculation decays to produce a curve such as that shown in Figure 8.7(b). The decay rate of this curve – is the curve shown in Figure 8.7(c). It is known as the autocorrelation curve. It is a characteristic of this autocorrelation curve that it can be related to the Brownian motion. This in return can be related to the size of the fineparticles. As they get smaller the Brownian motion becomes more intense. The reader who is unfamiliar with the technique of autocorrelation processing of noisy signals will find a useful introduction to the subject in Heifti [20]. The term photocorrelation spectroscopy comes from the fact that one looks at a spectrum of Doppler shifted information using correlation techniques. Commercially available techniques from several vendors are available. [10, 11, 15, 16, 21, 22] In Figure 8.8(a) and (b) two typical results obtainable with this methodology as reported by Elings and Nicoli are shown [16]. On the development of uses of photo-correlation methods (PCS) Finsey in 1994 makes the following comments.

"Originating some 20 odd years ago from a research tool, in a form only suitable for experts, photon correlation spectroscopy has become a routine analytical instrument for the determination of particle sizes. Like other techniques it has strong and weak points. The major strong point is that it is difficult to imagine a faster technique for sizing micron particles. Average particle sizes and distri-

butions width can be determined in a few minutes without elaborate sample preparation procedures. The price to be paid for this advantage is the low resolution. Reasonably accurate resolution of the shape of the particle size distribution requires extremely accurate measurements over a period of 10 hours. PCS is very often an excellent choice for quality control where only an average size is sufficient. Some of the commercially available PCS equipment also allows the measurement of the time average scattered intensity as a function of the scattering angle and the inversion of such data for particle size distribution reveals more reliable particle size distributions."

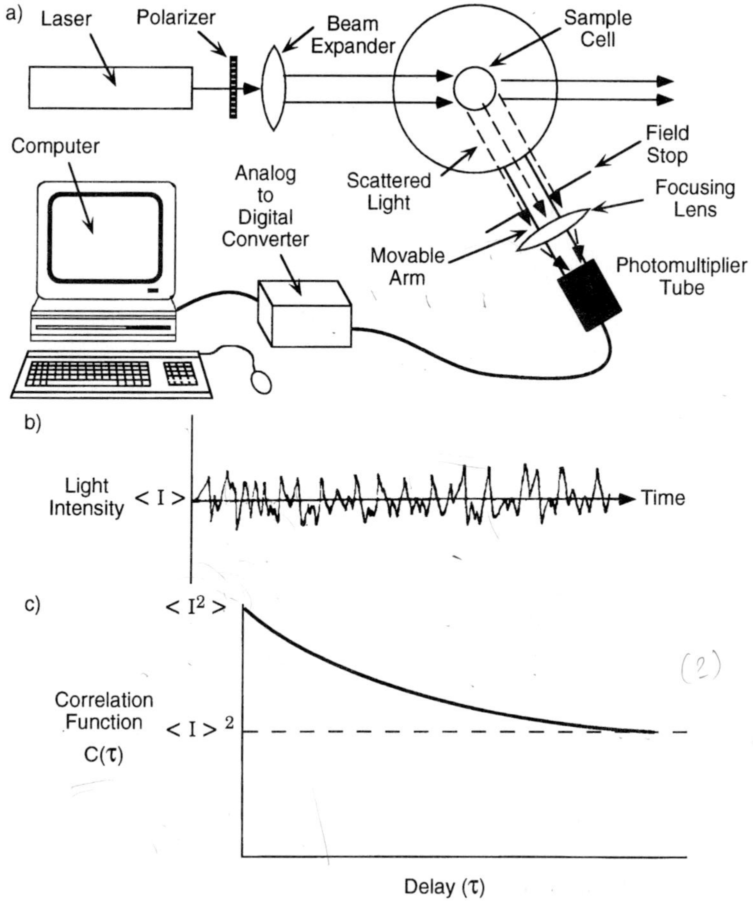

Figure 8.7. Nicoli used Dynamic Light Scattering to look at submicron fineparticles. a) Layout of the instrument used by Nicoli et al. b) Typical detector output from the instrument showing scattered light intensity about the average as a function of time. c) Autocorrelation of the scattered light intensity curve of (b).

As demonstrated by the graphs of Figure 8.8 one can in some circumstances deal with bimodal distributions which are well separated. A wide ranging continuous size distribution is difficult to characterize by PCS. In general, PCS is an applicable method from 0.003 to several microns. As in many other methods, data for non spherical particles use the size of an equivalent sphere diameter in the presentation of the data.

When using the technique Finsey cautions that preparation of a suitable dispersion is sometimes difficult. Thus he states,

> *"Much can be said about the dispersion procedure; it may involve the whole field of colloid chemistry. Practically, one has to choose the right liquid and/or dispersion agent and make sure that no coagulation or physical and chemical change of the dispersed phase occur. The suspension concentration should not be high enough to cause multiple light scattering"*

Finsey states that if higher concentration involving multiple scattered light is used,

> *"the particle size estimated by photon correlation spectroscopy becomes smaller due to the fact that the autocorrelation function decays faster for a multiple scattered signal than for a single scattered one. At the other end of the scale low concentrations can cause problems because there are additional intensity fluctuations in the recorded light caused by individual particles moving in and out of the measuring volume."*

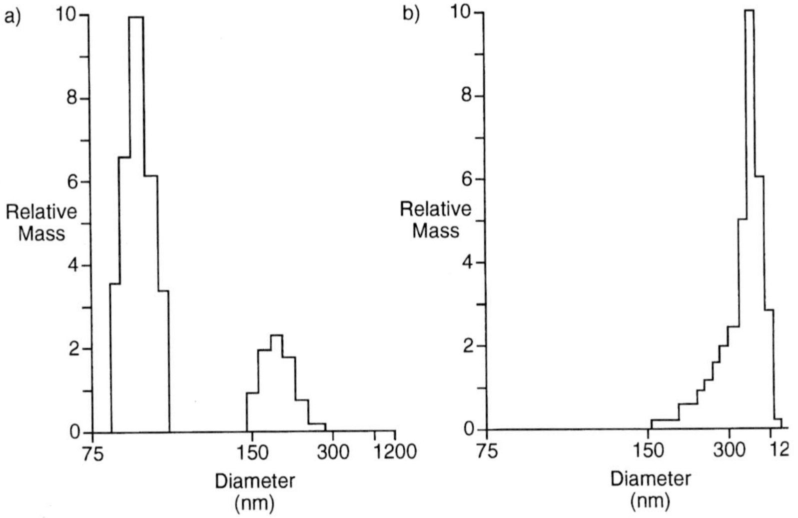

Figure 8.8. Typical data recorded by Elings and Nicoli using dynamic light scattering. a) Size distribution for a 3:1, by weight, mixture of 91 nm polystyrene latex spheres with 176 nm latex spheres. b) Size distribution of a broad, skewed population of polystyrene latex spheres.

One can also use fibre optics to measure Doppler shifted light in the back scattered mode to control polymerization processes in the production of latex spheres widely used in the coatings industry. Procedures are being developed to gain information by photon correlation spectroscopy using higher concentration suspensions.

References

[1] D. A. Ross et al, *Journal of Colloid and Interface Science*, 64:3, (1978) 533–542.

[2] See instruction manual for fibre optic Doppler anemometer, published by SIRE (Scientific Instrument Research Establishment) Ltd., Southhill Chislehurst, Kent, England.

[3] W. H. Yanta, "Measurement of Aerosol Size Distributions with a Laser Doppler Velocimeter", *National Bureau of Standards Special Publication 412, Aerosol Measurement*, proceedings of a seminar on aerosol measurement held at National Bureau of Standards, Gaithersburg, Maryland, May 7, 1974.

[4] K. Bauckhage, A. Schöne, "Measurement of Velocities and Diameters of Small Particles by using Fast Digital Circuits for the On-Line Evaluation of Laser Doppler Signals.", *Part. Charact.*, 2 (1985) 116–133.

[5] M. K. Mazumder, R. E. Ware, T. Yokayama, B. Rubin, D. Kamp, "Measurement of Particle Size and Electrostatic Charge of a Single Particle Basis in the Measurement of Suspended Particles by Quasi Elastic Light Scattering," John Wiley & Sons, New York (1982) 328–341.

[6] M. K. Mazumder, R. E. Ware, J. D. Wilson, R. G. Renninger, F. C. Hiller, P. C. McLeod, R. W. Raible, M. K. Testamen, "E-SPART Analyzer: Its Application to Aerodynamic Size Distribution," *J. Aerosol Sci.*, 10 (1979) 561–569.

[7] M. K. Mazumder, "E-SPART Analyzer: Its Performance and Application to Powder and Particle Technology Processes," *KONA*, 11 (1993) 105–118.

[8] The E-SPART analyzer is available commercially from the Hosokawa Micron Corporation, 10 Chatham Road, Summit, NJ 07901.

[9] A. A. Naqwi, "Sizing of Irregular Particles using a Phase Doppler System", *Part. Part. Syst. Charact.*, 13 (1996) 343–349.

[10] Dantec Measurement Technology A-S, Tomsbakken, 16–18, Dk-740, Skovlunde, Denmark. In the United States and Canada, Dantec Measurement Technology Incorporated, Mahwah, NJ 07430.

[11] For details of the phase Doppler equipment available from TSI contact the sales manager at TSI Incorporated, 500 Cardigan Road, P. O. Box 64394, St. Paul, MN 55164-9877, USA. It should be noted that some of the earlier work in this field was conducted by Bachalo and coworkers who operated a company called Aerometrics. In 1996 the Aerometric corporation merged with the TSI corporation.

[12] A. A. Naqwi C. W. Fandery, "Phase Doppler Measurements of Irregular Particles and Their Inversion to Velocity-Resolved Size Distributions", *Powder Handling and Processing*, 9:1 (Jan-March 1957) 45 and 51.

[13] R. J. Adrien, D. C. Bjorkquist, R. M. Fingerson, W. T. Lai, R. K. Mennon, A. A. Naqwi, "Optical diagnostic for single or multiple flows", TSI short course text 1997, Vol. 2, Lecture 32. Available from TSI Inc., P. O. Box 64204, St. Paul, Minnesota 55164, U. S. A.

[14] B. H. Kaye, *Soot Dust and Mist – An Introduction to Aerosol Science and Technology,* to be published by Chapman and Hall.

[15] Literature available from Coulter Electronic Northwell Drive Luton Beds LU3 3RH England.

[16] V. B. Elings, D. F. Nicoli, "Recent Advance in Submicron Particle Sizing by Dynamic Light Scattering," *American Laboratory* June (1984) literature available from Particle Sizing Systems, 75 Aero Camino, Santa Barbara, CA 93117.

[17] See Chapter 5 in B. Wiener, *Modern Methods of Particle Size Analysis,* Wiley, New York, 1984.

[18] R. Finsey, "Particle Sizing by Quasi Elastic Light Scattering," *Adv. in Colloid and Interface Sci.*, 52 (1994) 79–143.

[19] R. Finsey, "Particle Sizing in the Submicron Range by Dynamic Light Scattering," *KONA*, 11 (1993) 17–32.

[20] B. Hauli, G. M. Heifti, "Correlation Methods in Chemical Data Measurement," Chapter 4 in *Contemporary Topics in Bioethical and Chemical Chemistry,* Vol. 3, D. M. Hercules G. M. Hefti, L. R. Schnider and M. A. Evason editors, 1978.

[21] Malvern Instruments, 10 Southville Road, Southborough, MA 01772.

[22] Brookhaven Instruments Corp. 750 Blue Point Road, Holtsville New York 11742-1896 USA.

[??] R. C. Rodoff, S. V. Sankar, W. D. Bachalo, C. Sharky, "Aerosol and Nebulizer Characterization Performance of a Phase Doppler Based Particle Sizer", presented in the European Aerosol Conference Sept. 6–11 1992, Oxford United Kingdom. To contact Bachalo see reference 11.

[??] W. D. Bachalo and N. J. Houser, "Phase Doppler Spray Analyzer for Simultaneous Measurement of Drop Size and Velocity Distributions," *Part. Eng.*, 23 (1984) 390–583.

9 Characterizing the Properties of Powder Beds

9.1 Parameters Used to Describe and Characterize the Properties of Powder Beds

In many areas of powder technology, scientists need to characterize the properties of powder beds. Thus, in powder metallurgy and ceramics, the powder is poured into a container and then consolidated, either by heat or the application of pressure. The rate of consolidation, and the achievable consolidation are often governed by the structure of the spaces, known as pores, between the powder grains in the initial powder bed. When a powder is poured into a container, the volume of the freshly poured powder bed, divided into the weight of the powder is the apparent density of the powder bed or the aerated powder density.

A freshly poured powder bed can be initially consolidated by vibration so that the particles settle down into a more intimate contact. As the vessel containing the poured powder is tapped, the powder bed consolidates. Industrial standards have been set up to specify the amount of tapping, and the way in which the tapping is applied to the powder bed to achieve a consolidated bed and the result is known as the bulk density or the tap density of the powder. Commercial instrumentation is available for measuring the bulk density of a powder [1, 2, 3].

The ratio between the tap density and the apparent density is termed the Hausner ratio after Dr. Hausner, who first suggested the use of such a ratio, and to which he said is indicative of the magnitude of the friction between particles in the powder bed [4, 5].

An important property of a powder bed is the amount of space in the bed which is not occupied by powder grains. This parameter is known as the voidage, or the porosity of the powder bed, and is given by the relationship

$$\varepsilon = \frac{V_B - V_P}{V_B} = \frac{V_B - \frac{W_P}{\rho_P}}{V_B} = 1 - \frac{W_P}{\rho_P V_B} \qquad (9.1)$$

Where ε is the porosity, W_P the weight of powder, ρ_P the density of the powder grains, V_B the volume of the powder bed, and V_P the volume of the powder grains.

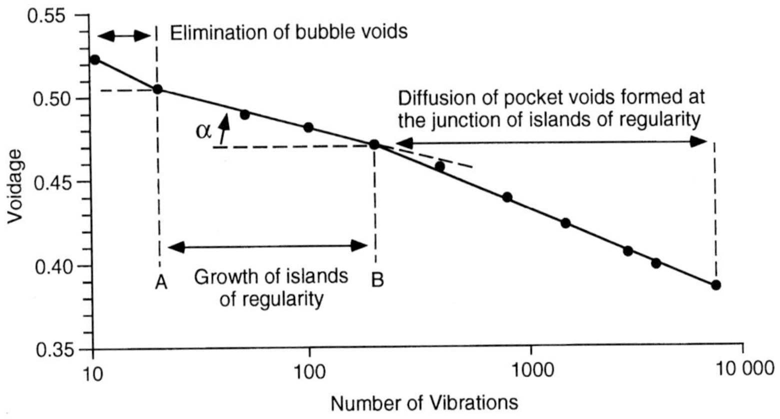

Figure 9.1. Work carried out by Kaye and co-workers demonstrates how the porosity of a freshly poured powder bed collapses under the influence of environmental vibration.

Sometimes the powder grains are themselves porous, and this internal voidage is not considered as being a contributor to the voidage of the powder bed. The way in which the porosity of a tapped powder bed changes is shown by the data summarized in Figure 9.1 from a study carried out by Kaye and co-workers [3].

The final density achievable by vibration usually depends upon the amplitude and frequency of the vibration applied to the powder bed. In this chapter, we will review the methods used for studying the porosity of a vibrated powder bed, and also a powder bed that has been consolidated by applied pressure, and other procedures such as sintering, or the cementing of the grains by recrystallization. In Chapter 10, we will look at further methods for determining the pore structure of a powder system based upon studies of the gas adsorption properties of the material.

An important property of a packed powder bed is the resistance it offers to a fluid moving through the system. This property is known as the permeability of the system. In the next section, the way in which the permeability of a powder bed can be characterized will be described and permeability measurements which have been used to characterize the fineness of a powder will be reviewed.

9.2 Permeability Methods for Characterizing the Fineness of a Powder System

One measures the permeability of a powder bed by measuring the resistance to flow offered by the bed when a given pressure is applied to drive a fluid through it. To interpret permeability data in terms of the surface area of the powder under test, it is common practice to use an equation originally developed by Kozeny and modified by Carman who considered the packed powder bed to consist of a set of tortuous capillary tubes [6, 7].

The Kozeny-Carman equation is usually written in the form

$$u = \frac{\varepsilon^3}{(1-\varepsilon)^2} \frac{\Delta p}{B\eta\ell S_W{}^2\rho^2} \tag{9.2}$$

Where

u = the average approach velocity of the fluid
ε = the porosity of powder bed
η = the viscosity of the fluid
Δp = pressure difference across the bed
ℓ = length of powder bed
B = a constant
ρ = density of powder
S_w = surface area per unit weight of the powder.

For the assumptions made in deriving this equation to be valid, the pressure difference across the powder plug is assumed small and the powder compact is presumed to be a uniform cylinder of length ℓ and cross-sectional area A.

Carman showed that for many materials the constant B, could be taken as 5. However, Carman stated clearly that the universal use of the value 5 will lead to error in certain cases [6, 7]. Because of the lack of alternative values, however, commercially available permeameters such as the Fisher Subsieve Sizer are calibrated for operation on the assumption that the value B = 5 is used in the Kozeny-Carman equation [8]. The unknown value of B for any specific powder is an immediate source of discrepancy between the surface area measured by permeability with that measured by other techniques. The only way to determine B exactly is to use an independent method of analysis to measure the surface area of the powder. B is sometimes called a tortuosity factor but it is my opinion that it is better to call the factor B a correlation factor and in this way firmly state the true nature of the adjustment being made to the equation.

Fowler and Hertel have investigated the values that the constant has for various powder systems [9]. It can be seen that in reality permeability methods of analysis are secondary methods in that the equation used to calculate surface areas has an empirical constant which has to be measured experimentally. It can now be seen that if measured permeability surface areas agree with those values as measured by independent methods this is not a vindication of the method but are an indication that the size distribution and/or the experimental conditions were similar to those used by Carman in his original work.

In some texts, permeability methods for characterizing fineness are referred to as surface area measuring devices. Relating the permeability of a powder bed to the surface area of the powder grains constituting the powder bed is a complex problem in fluid dynamics, and to achieve a formula linking the resistance to flow of a fluid through the powder bed to the surface area of the constituent powder grains requires the making of many simplifying assumptions. In fact it is necessary to accept the fact that in industrial practice one can only correlate the changing fineness of a powder to the measured permeability in an empirical manner. The position adopted in this book is that fineness characterization by permeability is an empirical procedure, one which has proven to be very useful, and could serve as a very useful procedure in the quality control of powder production. In some ways it is now the forgotten method of characterization. Note that the equipment used to measure permeability is described as a permeameter. In the development of the Kozeny-Carman equation it is assumed that the flow conditions through the packed powder bed are laminar and that the mean free path of molecules is small compared with the dimensions of the pores within the powder bed. For gases flowing through powders where all the grains are smaller than approximately 5 mm the effect of molecular movement in the gas should be taken into account. For flow under these conditions, Carman and Arnell derived a complex equation for use in calculating surface areas [8]. This equation, however, still contains the correlation factor which has to be determined empirically. If the purpose of the measurements is comparison between powders, then such sophistication of interpretative technique would seem to be unwarranted. However, for research investigations into physical behavior of very fine powder compacts, this factor should be taken into account [10, 11]. The Kozeney-Carman equation contains a porosity function so that theoretically the value of the deduced surface areas should be independent of the porosity of the powder plug. In practice it has been found that the measured value of the surface area deduced from permeability data is a function of the porosity of the plug. This fact is illustrated by data reported by Allen when studying a pharmaceutical powder [12].

ε	=	0.60	0.55	0.50	0.45	0.40	0.35	0.30	0.25
S_W	=	4500	4820	5150	5460	7300	9080	9080	7760

Edmundson and Toothill found that the measured surface area at 0.30 porosity was twice that measured at 0.60 porosity. In general, the measured surface area increases as the porosity decreases [13].

For many powders it has been found that for porosity below a certain limiting value, usually in the region of 0.45, there is a range of values where further reduction in porosity does not sensibly alter the measured surface area. If measurements are carried out at porosities below this so-called plateau, there may be a further increase in fineness. This last stage probably represents a real increase in fineness since the pressures which have to be applied to achieve the low porosity's could crush the individual grains of powder.

When carrying out surface area measurements by permeability methods it is convenient to standardize the porosity used at some value within the plateau region. It is some times claimed that a good reason for operating within the plateau region is because at these order of porosities, values of surface areas are in better agreement with those obtained by other methods. This is not a reason for operating within the plateau; it means using porosities of the order of magnitude found in the plateau when determining the magnitude of the Carman correlation factor. The real reason for operating within the plateau region is that, since in this region small changes in porosities do not significantly affect the magnitude of the measured surface area, effects due to small inevitable variabilities in the technique for preparing the plug of powder from run to run and operator to operator are minimized.

Carman suggested that the apparent increase in fineness of a powder with decreasing porosity of it's powder plug was due to the elimination of bridging of the powder grains as the powder is consolidated. Also that the plateau region corresponds to the consolidation of a structure from which gross pores have been eliminated. The existence of a curve of variation of measured surface area with porosity has usually been regarded as one more obstacle to be overcome in attaining precise surface area measurements by permeametry. However, the shape of the curve of variation, and the location of the plateau region, may be characteristic of the structure and behavior of the powder compact, as manifest in permeabilities, and may be as informative as the pressure volume compression data curves of gases. Latent in the powder compression/permeability data may be relationships as informative to the development of powder technology as Boyle's law and van der Waals' relationships have been for the understanding of gaseous behavior. Such information is currently being overlooked because of the desire to use permeability techniques as a method for surface area determination rather than as a useful technique for exploring the structure of a porous compact. See for example the data reported for powder mixtures by Kaye [14].

9.3 General Considerations

It is useful to classify permeability methods into two main groups. These are fixed pressure and variable pressure methods. In the former, a steady pressure is applied to a plug of powder and its pneumatic of fluid resistance properties are determined. In variable pressure methods the pressure drop across a powder plug is constantly changing and the rate of change of pressure is measured.

A difficulty common to all permeability methods is the preparation of the powder plug. It is usual to form the powder plug using a small hand-press consisting of a small cylindrical plunger which slides into the cylindrical permeameter cell. A basic system is shown in outline in Figure 9.2. The clearance between the plunger and the cylinder wall must be large enough to prevent the system seizing up if small particles of powder become trapped in between the cylinder wall and the plunger. In practice this means that the clearance has to be about 250 μm. As the powder bed is compressed the air leaving the bed must have a vent through which to escape. The plunger shown in Figure 9.2 has a central hole for this purpose; in some permeameters the air vent is a groove cut down one side of the plunger. When pressing the powder bed it is good practice to apply the pressure slowly and steadily so that air movement out of the bed is not sufficiently rapid to blow out the finer particles, which can cause size segregation within the powder bed. The plunger is often marked with an accurate linear scale so that the height of the powder bed can be measured directly. It is usually necessary to support the powder plug within the permeameter cell. A convenient support system is illustrated in Figure 9.2. When the metal cylinder is made, a small shoulder is left near the base of the cylinder. The powder bed is then supported by placing a perforated metal disc, covered by a matching filter paper disc, on this shoulder. In order to achieve an air-tight seal between the disc and the cylinder, the disc rests on a rubber ring placed on the shoulder so that when pressure is applied to the system a good seal is achieved. The dimensions of the perforated disc and the texture of the filter paper are selected so that the resistance of the support system to a fluid flowing through it is negligible compared to the resistance offered by the assembled powder bed. When assembling the powder system, a support ram is usually inserted at the opposite end of the cylinder so that the perforated disc is not distorted under the applied pressure. With some very fine cohesive powders the use of the two rams results in a self-supporting plug which does not require a support system at the base of the plug.

The technique used when placing the powder into the permeameter cell can effect the value or the measured surface area. Some workers place the total quantity of powder in the cell and apply pressure to the ram in one operation; others split the total quantity of powder into several increments and ram each portion separately as it is added to the permeability cell. The reason why these two experimental techniques will

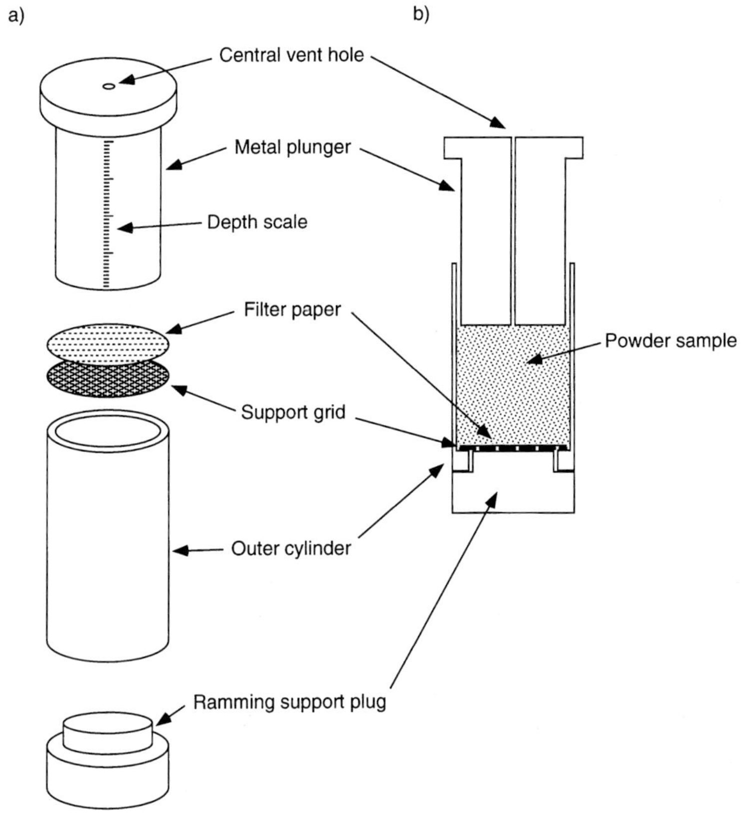

Figure 9.2. Apparatus used to assemble a powder plug for permeametry studies. a) Exploded view of the plug assembly apparatus. b) Sectional view of the assembled apparatus.

usually yield powder beds having different physical properties is that, when pressure is applied to the surface of a powder bed, the powder plug does not behave as if it were a rigid body. Instead of the applied pressure being transmitted directly downwards through the powder plug, the linked system of particles operate as arches and the pressure is shunted sideways onto the walls of the cylinder. The initial application of pressure will cause particles to move, but as a bridge forms the pressure is shunted sideways and voids located under the bridge, i. e. the arch, are protected from further consolidation. One of the reasons why the vibrational compaction of a powder using a floating ram is efficient is that the vibration causes the collapse of the bridges so that voids are more readily eliminated at the lower pressures. In the absence of vibration it is necessary to destroy the bridges with high pressures in order to eliminate the voids. It is not sufficiently realized how rapidly pressure is transmitted to the walls of the cell;

experiments conducted at IIT Research Institute show that within a depth of 1 cm from the surface of a powder system 3 cm across, there can be a very substantial drop in the applied pressure. From the above discussion it can be seen why the method of assembling the powder bed in increments will have a more uniform pore structure than that assembled in one operation. Isostatic compression of the powder plug discussed in Section 9.4 is a superior technique for assembling the plug but has not been widely used in classical permeability studies.

Another facet of the technique for assembling the powder bed on which opinions differ is the usefulness of rotating the plunger as pressure is applied to the powder bed. One practical advantage gained by rotating the cylinder is that usually it helps to keep the ram moving freely in the permeameter cell. Higher consolidation of the powder bed can be achieved by rotations coupled with applied pressure. The reason for this improvement is that the shear forces created in the bed during rotation help to eliminate the bridges which are preventing consolidation of the powder system. If the shear forces are too high they could cause gross faults to occur within the powder. These could act as leaks through the powder bed, resulting in a lower than anticipated fluid resistance for the whole system. Thus, in the extreme case in which the applied shear forces were sufficient to cause rotation of the powder plug in the permeameter cell, a leak along the length of the side of the powder plug could be created. Since powder plugs are usually assembled manually, and because application of shear forces could introduce another extra operator variable, it is probably a better experimental practice not to apply rotational shear forces. It is, however. essential to establish an adequate set of instructions with regard to the method of applying pressure to the powder compact since changes in measured surface areas with a change of operator can often be traced to the individual operators method of consolidating the powder bed. A case history will illustrate this point. In a control laboratory it was found that after a change of operator the measured surface area for a standard powder was consistently higher than before. It was found that the first operator assembled the bed in five increments and then achieved its final state of compaction by placing the permeameter cell-ram system into a laboratory vice and compacting to a fixed mark on the ram. The second operator was trained by the first operator without the personal supervision of the scientist in charge. The significance of the incremental packing was not pointed out to the second operator. The second operator found that he could ram the powder to the final volume in one operation using the laboratory vice. It was necessary to apply more force when achieving compaction using all the powder at once compared with that which had to be applied when assembling the powder incrementally but it was a lot less bother. The apparent increase in surface area came from the fact that, to achieve the given compaction in one operation, the top layer of the powder was much more highly compacted than in the former system. Thus, for a given powder, this highly compacted surface layer offered a high resistance to the flowing fluid and was interpreted as a higher surface area than that measured using the incrementally assembled system.

If relatively high pressures are applied to the powder bed, the bed should always be examined for cracks since under high stress a powder plug can behave as if it had elastic properties, and the release of the pressure can cause the powder plug to expand slightly causing cracks. Also, if the ram cylinder is a fairly close fit, rapid withdrawal of the ram can cause air movement, which again can create gross cracks within the powder bed.

When the fluid used in a permeameter is a gas, the humidity of the gas can affect the value of the measured surface area. It is usual to pass the gas through a drying tower before passing it through the powder bed. If this is done, however, it should be realized that, if the powder being tested is in equilibrium with a humid atmosphere, the dry gas flow through the plug will start to dry the powder bed and it may be necessary to wait some time for equilibrium rates to be established.

One universal rule which has to be followed in permeametry is that the flowing fluid must move downwards through the powder plug if the plug is not restrained mechanically at both ends. If the fluid is passed the opposite way, the powder plug will expand owing to the movement of the fluid. and false readings will be obtained.

9.4 Fixed-pressure permeametry

In a fixed-pressure permeameter the surface area of the powder is deduced from the measured resistance of a powder plug when fluid is flowing through it under a constant pressure head. One of the first permeameters was developed in the late 1930's by Lea and Nurse to study the fineness of cement [15]. It is useful to discuss their equipment in detail since its simplicity of design is helpful in understanding the physical significance of the various design elements used in the measurement of fineness using permeametry. The system they used is shown in Figure 9.3(a) For most analytical purposes, air is normally used but other gases such as helium, nitrogen, and argon have been used. When using air, a small pump is usually sufficient to supply the small pressure heads required. The placing of a surge tank before the drying tower and before the permeameter cells helps to smooth out pulses in the pressure arising from the action of the pump. The pressure changes across the permeameter cell and the flow meter are measured directly. The flow meter of the Lea and Nurse equipment was a uniform bore capillary tube 280 cm long. For the system used by Lea and Nurse, one can deduce the equation

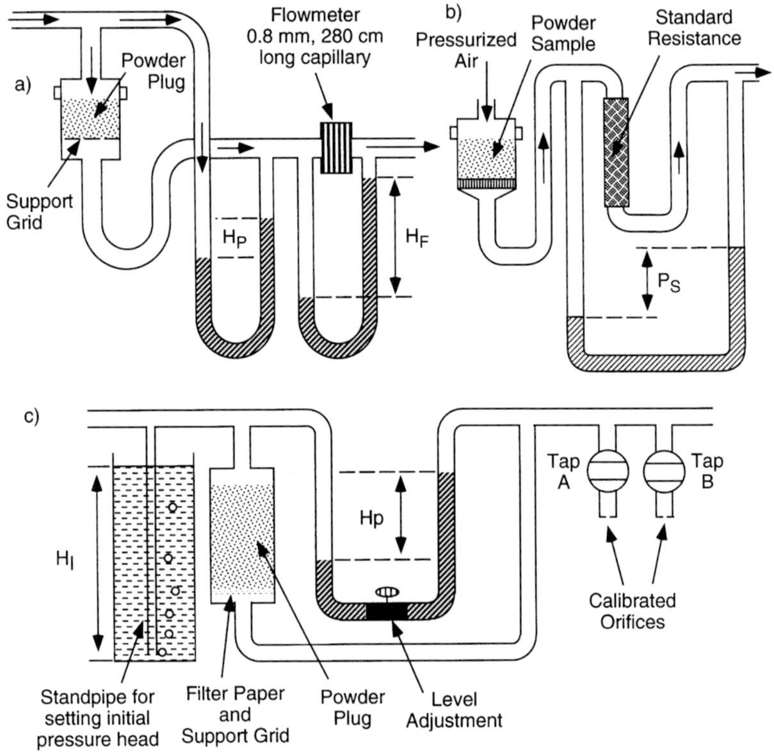

Figure 9.3. Pioneering designs of permeameters used to measure the fineness of powders led to a widely used commercial instrument known as the Fisher Subsieve Sizer. a) The Lea and Nurse permeameter. b) The Gooden and Smith permeameter used a column of fine sand as a standard resistance. c) The Fisher Subsieve Sizer uses a standpipe in a column of water to provide a constant pressure head.

$$S_W = \sqrt{\frac{H_P}{H_F} \frac{\varepsilon^3}{(1-\varepsilon)^2} \frac{8A}{\pi B \rho^2 r^4}}$$

Where H_p and H_F are the pressure measured in the manometers as shown. It should be noted that in the above equation there is no need to know the value of the viscosity of the fluid nor the exact value of the acceleration due to gravity in the laboratory where the experiments are being carried out. For a given flow meter and for a specific powder, the above equation reduces to the form

$$S_w = \sqrt{\frac{H_p}{H_F} \frac{\varepsilon^3}{(1-\varepsilon)^2} \frac{1}{BK}}$$

where A is a specific constant. In practice it is usual to standardize the porosity of the powder and thus further simplify the calculations. For most systems the value of B = 5 first proposed by Carman is used to calculate the surface area. An alternative procedure which can be very useful for control purposes is to calibrate the equipment using a standard powder, of the same type as the powders to be tested, of known surface area. If all measurements are made at the same porosities, then

$$S_W{}^2 = \frac{H_P}{H_F}\frac{\varepsilon^3}{(1-\varepsilon)^2}\frac{1}{BK}$$

and

$$S'_W{}^2 = \frac{H'_P}{H'_F}\frac{\varepsilon^3}{(1-\varepsilon)^2}\frac{1}{BK}$$

where the primed symbols denote the values of the variables for the standard powder. Therefore it follows that

$$S_W = \sqrt{\frac{H_P}{H_F}}\, S'_W \sqrt{\frac{H'_F}{H'_P}} \tag{9.3}$$

The last two factors of equation 9.3 are constants so that using this equation surface areas can be calculated very rapidly. It can be seen that one advantage of calibrating the permeameter, using a standard powder, when all the powders to be measured are of the same material, is that it is not necessary to assume a value for the correlation factor B.

By always carrying out measurements simultaneously on the powder bed and the flow meter, it is not necessary to know the absolute pressure in the permeameter system. However, the total pressure drop through the permeameter system should always be kept as small as possible because, for large pressure changes, it is necessary to allow for compression effects in the gas.

It is sometimes useful to think of permeameter pneumatic circuit systems in terms of the equivalent electrical circuits. For the various permeameters shown in the diagrams, the equivalent electrical circuits are shown.

The use of a capillary flow meter which can be calibrated theoretically is a useful feature of the Lea and Nurse permeameter but the presence of the 280 cm of capillary tube makes the equipment bulky for many laboratories. In later permeameters the capillary tube is replaced by a standard pneumatic resistance which has been calibrated using a flow meter for known characteristics. It should also be noted that if the use of the permeameter is restricted to samples of the same type of powder and equation 9.3

is used to calculate the surface area of the powders, then it is not necessary to know the magnitude of the standard pneumatic resistance.

In the permeameter designed by Gooden and Smith, the standard resistance was a column of fine sand held in a tube by two porous plugs [15]. Their equipment is shown in outline in Figure 9.3(b). Apparently only one manometer is used, but in fact it is necessary to know the pressure P at which air entered the powder bed. In this equipment the down stream end of the standard resistance was open to the atmosphere. The pressure drop across the powder bed is $P–P_S$ and that across the standard resistance is P_S. The appropriate formula for calculating the surface area is

$$S_W{}^2 = \frac{P - P_S}{P_S} \frac{\varepsilon^3}{(1-\varepsilon)^2} \frac{1}{BK} \tag{9.4}$$

where K is a specific constant involving the density of the powder and the magnitude of the standard resistance and P is the applied pressure.

For many years, from the 1940's through to the late 1980's, many industries used a piece of equipment based on the Gooden and Smith configuration in the quality control of powder production. The structure of the instrument known as the "Fisher Subsieve Sizer" is shown in Figure 9.3(c) [16]. The steady head of pressure is controlled and measured by using a standpipe pressure regulator. When operating this equipment it is essential to maintain the level of the fluid in the standpipe at the recommended level. After the air pump has been switched on, the rate of bubbling of the standpipe is adjusted so that the bubbles are formed in a slow steady stream. Under these conditions it can be assumed that the pressure at the entrance to the powder bed is due to the height of fluid H_1. The fluid in the manometer and the standpipe column are the same. In the Fisher Subsieve Sizer the standard resistance, fabricated from fine sand as originally used by Gooden and Smith, is replaced by two calibrated orifices. Either can be used in the pneumatic circuit of the permeameter by operating the taps A and B. The use of two resistances enables the sensitivity of the measurements to be varied, so as to make the pressure drops across the standard resistance and the powder bed of comparable magnitude. In the Fisher Subsieve Sizer a patented calculator chart was mounted behind the manometer. Before the calculator chart can be used, the lower arm of the manometer had to be adjusted to the level C and an adjustment screw is provided for this purpose. After the level is adjusted the calculator is so designed that the top of the fluid in the other arm of the manometer indicates an average particle size for the powder. This average value will be in error if the standpipe setting is not at the recommended level or the bubbling rate adjusted properly. The average particle size often quoted from investigations made with the Fisher Subsieve Sizer is calculated from the surface area of the powder assuming an equiva-

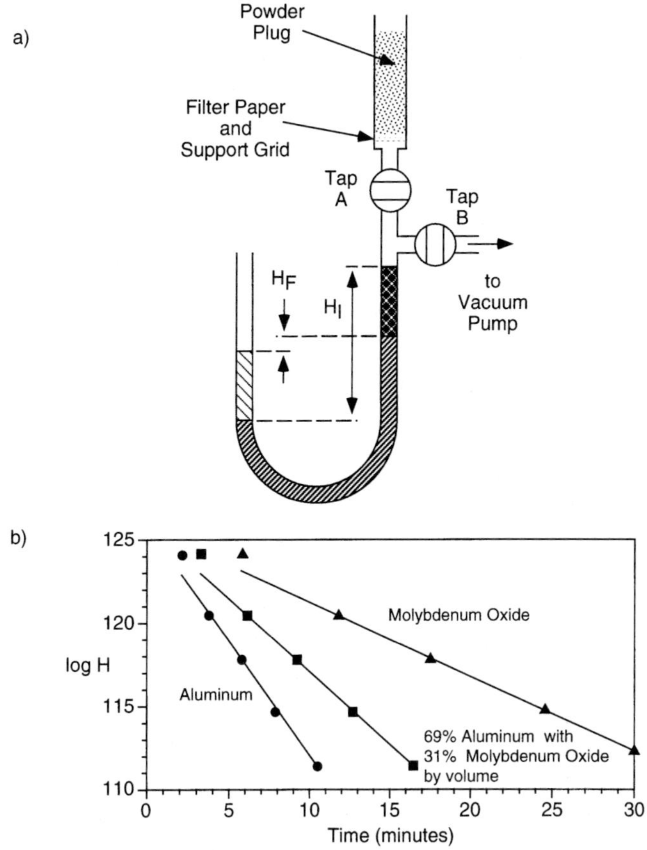

Figure 9.4. The Blaine Fineness Tester is used to measure the time required for a pressure head to dissipate by fluid flow through a plug of the powder under test. a) A simple Blaine Fineness Tester. b) Decay graphs for manometer differential height using three powder plugs of different material content.

lent powder composed of hard spherical grains of the same material as the powder and having the same surface area per gram. Other permeameters have been described by Dubrow and Niradka [17], Lötsch [18], Kaye and Jackson [19], and Rigden [20].

A variable pressure permeameter known as the Blaine Fineness Tester, was developed by Blaine [21]. It is shown in Figure 9.4. The basic measurement made when analyzing powders using a variable-pressure permeameter is the time required to achieve a given pressure drop in a manometer discharging through a powder plug. For specific manometer heights before and after discharge of air through the plug it can be shown that

$$S_W = \sqrt{\frac{1}{\log_e \frac{H_I}{H_F}} \frac{\varepsilon^3}{(1-\varepsilon)^2} \frac{A}{\alpha} \frac{\rho_F}{B\eta\ell\rho^2} T} \tag{9.5}$$

To carry out a measurement with equipment of the type shown, the experimental procedure is as follows. The powder plug is formed in the metal cylinder and mounted on the arm of the manometer, by means of a stopper with a central hole. When mounted, the cylinder taps A and B should be open to the atmosphere to prevent an air pulse moving up through the plug as the cylinder assembly is pressed together. If the pulse of air is not prevented it could cause coarse cracks in the plug and lead to erroneous results. With the plug in position, tap A is closed and the height of column of fluid in the left-hand side of the manometer raised the mark D by switching tap B so that the vacuum system is linked to the manometer. In variable pressure permeameters the setting of the fluid heights should always be arranged in this manner so that when the manometer is discharging towards equilibrium conditions the fluid movement is down through the powder plug to ensure plug stability. The level of the manometer fluid is raised slightly above the level at which readings are to be started, because when tap A is opened to permit fluid to move through the powder bed there is a short period as the manometer starts to move before steady conditions are attained. It is also necessary to make sure that the discharge rate of the manometer is not big enough for the manometer fluid to develop kinetic energy. To measure the time required for the manometer to fall from height H_I to H_F it is usually sufficient to time the movement directly, using a stopwatch. The difficulty inherent in the direct timing procedure is that, if the boundary is moving relatively swiftly, the personal error involved in stopping and starting a timing mechanism becomes a measurable fraction of the total time of fall. Alternatively, if the boundary is moving slowly it is difficult to decide exactly when the boundary coincides with the reference marks on the manometer. A more sophisticated experimental technique which can be used to improve the precision of the measurements is to time the arrival of the manometer boundary at a series of intermediate marks between the levels H_I and H_F. From equation. 9.5 it can be seen that for a given plug a plot of $\log_e H$ against t should be a straight line. Therefore, to improve the precision of the estimate of T for the time of fall from H to H_F a graph of $\log_e H$ against t is plotted and the best straight line drawn through the data. Since logarithms to the base e transform to those of base 10 on multiplying by the appropriate constant, ordinary logarithmic/linear graph paper can be used to plot the data. In Figure 9.4(b) typical sets of data for three different powder systems are shown for illustrative purposes. After the accurate value of the time of fall has been deduced, surface areas can be calculated using equation 9.5 by assuming the Carman value for the correlation parameter B. In normal use all the quantities under the square-root sign of equation 9.5 would be standardized so that for a powder of a given substance a direct simple relationship would exist between S_w and $T^{1/2}$. Spillane used this fact to design a permeameter for control analysis in which the time of fall was measured electrically and the dial of the timing mechanism was marked directly in surface area units of the powder under test [22].

As in the case of fixed pressure permeametry, if a series of measurements is going to be performed on one substance then the equipment can be calibrated using a powder of known surface area and then there is no need to assume a value for the factor B nor is it necessary to measure quantities H_F, H_1, α, η, r_2, A, g, ℓ, provided they are all fixed. When all these quantities are fixed and calibration is achieved using a powder of known surface area, the simple relationship for calculation the unknown surface area is

$$S_W = S_S\left(\frac{T}{T_S}\right)^{1/2}\left(\frac{\varepsilon^3/(1-\varepsilon)^2}{\varepsilon_S^3/(1-\varepsilon_S)^2}\right)^{1/2} \tag{9.6}$$

where the subscripts s denote values for the standard substance. It can be seen that, if standard porosity is used, the equation can be simplified still further to the form

$$S_W = S_S\left(\frac{T}{T_S}\right)^{1/2} \tag{9.7}$$

Ober and Frederick [23] have extensively studied the use of the variable-pressure permeameter in the cement industry and recommend that the porosity function $\varepsilon^3/(1-\varepsilon)^2$ of equation 9.6 should be replaced by the function $\varepsilon^3/(0.850-\varepsilon)^2$. As explained when discussing the physical significance of the porosity function in the Kozeny-Carman equation, the replacing of 1 by the constant 0.850 is probably specific to the substances studied and cannot be recommended as a general procedure for all permeametry studies. From equation 9.5 it can be seen that, if the times of fall are too short or too long, the range of the permeameter can be altered by altering the cross-sectional area a of the manometer tube. It is not always possible to work at constant porosity, and for specific laboratory studies it is not always possible to have a standard powder of known surface area. In those cases, when all experimental conditions are standardized, equation 9.5 reduces to the form

$$S_W = K'T^{1/2}\frac{\varepsilon^{1/2}}{(1-\varepsilon)} \tag{9.8}$$

where K' is a specific constant for the equipment and powder. For comparative purposes, permeability can then be expressed in arbitrary units of $\{\varepsilon^{1/2}/(1-\varepsilon)\}T^{1/2}$.

9.5 Cybernetic Permeameters for Quality Control of Powder Production

One of the industries where there is continuous research for automatic on-line control of the fineness of a product is the cement industry. For many years they have used the Blaine fineness number deduced from the data obtained using the simple permeameter discussed in the previous section. A problem with the classical Blaine method is the fact that it is labour intensive and, to some extent, operator dependent. The type of variation that is encountered in cement production is shown in the Figure 9.5. In this diagram the Blaine fineness tester data for a production run is compared to information generated by a diffractometer known as the Microtrac® [24, 25, see also Chapter 7].

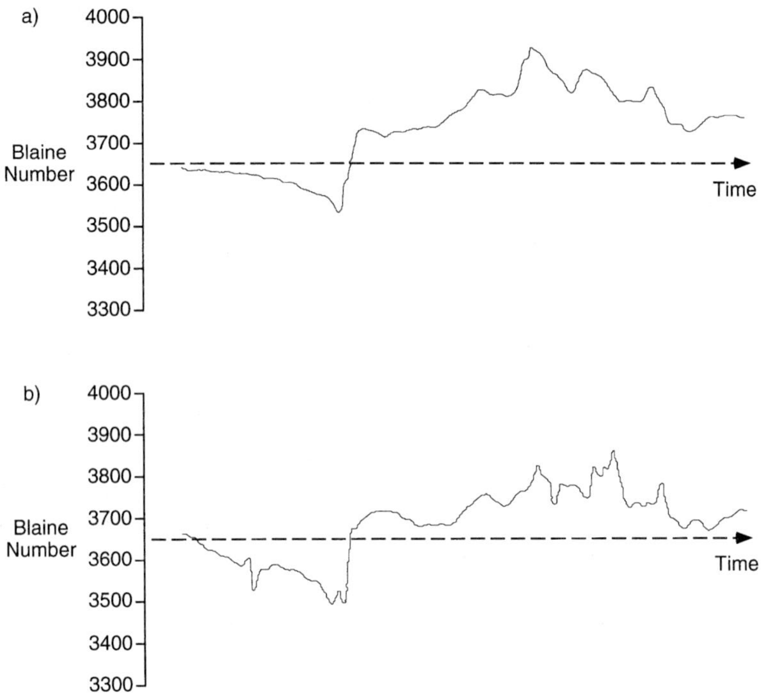

Figure 9.5. Air permeability data, expressed as a Blaine number, can be used to monitor the fineness of a product. The data above compares the fineness of a production run of cement powder, as measured by a Blaine Fineness tester, to data obtained from a diffractometer and converted to Blaine number. a) Data obtained from the Blaine Fineness Tester. b) Diffractometer data converted to Blaine number.

The information summarized in Figure 9.5 can be interpreted in two ways. First, the information generated by the Microtrac® could be used to achieve quality control of the cement powder although it appears to be slightly more resolved than the Blaine fineness tester. Vice-versa the other aspect of this data is that if one could automate the use of the Blaine fineness tester the equipment for the Blaine fineness tester is inherently an order of magnitude less, pricewise, than the diffractometers such as the Microtrac® (current prices are hard to evaluate but most of the diffractometers available commercially are of the order of 50 thousand dollars and upward, see discussion of diffractometers in chapter 7). In the fall of 1986 a cost benefit analysis carried out at Laurentian University for St. Mary's Cement Co. of Bowmanville, Ontario, concluded that, if development costs were not included, air permeability techniques such as the Blaine fineness tester, generate sufficient control information for many industries at costs which were lower by an order of magnitude than that of a comparable system based upon instruments such as the optical diffractometer [25].

As a result of this cost benefit analysis, a project was undertaken at Laurentian University with joint funding from St. Mary's Cement Co., and the research initiative fund of the Ontario Government. The guidelines for this project required the building of a prototype air permeability equipment in which the fineness of cement powder was measured in situ with a feedback time of less than 15 minutes and an accuracy of plus or minus 5 % on the fineness parameter of the Blaine number. The prototype for this equipment is described in the M.Sc. thesis of A. Hoffman, was delivered to the cement company at the end of the project [26].

At the core of the permeameter system for quality control developed in this project was a device developed by Kaye and co-workers called the softwall permeability cell [27]. As discussed in the previous section the major problems encountered in the undertaking of permeability measurements is the assembly of the powder plug. Kaye and Legault overcame this problem by developing a softwall cell in which the powder under test was compressed using isostatic fluid pressure. The cell that they used is shown in Figure 9.6 [27]. The advantages of isostatic compression of the plug is that, during the consolidation of the plug, the applied pressure is more uniform and is distributed along the whole length of the plug. A further advantage is that the height of the powder plug does not change during the compression. This simplifies the data processing. The second advantage of isostatic compression is that the application of hydrostatic pressure enables one to monitor the change in volume of the plug during the compression process as illustrated in the diagram another factor that facilitates data processing [28].

Another major problem with classical permeameter cells is that it is difficult to remove the powder plug from the cell after making the permeability measurements. When the plug is compressed isostatically it has a smaller cross-section at the completion of the assembly of the plug than prior to the experimental measurements so that subse-

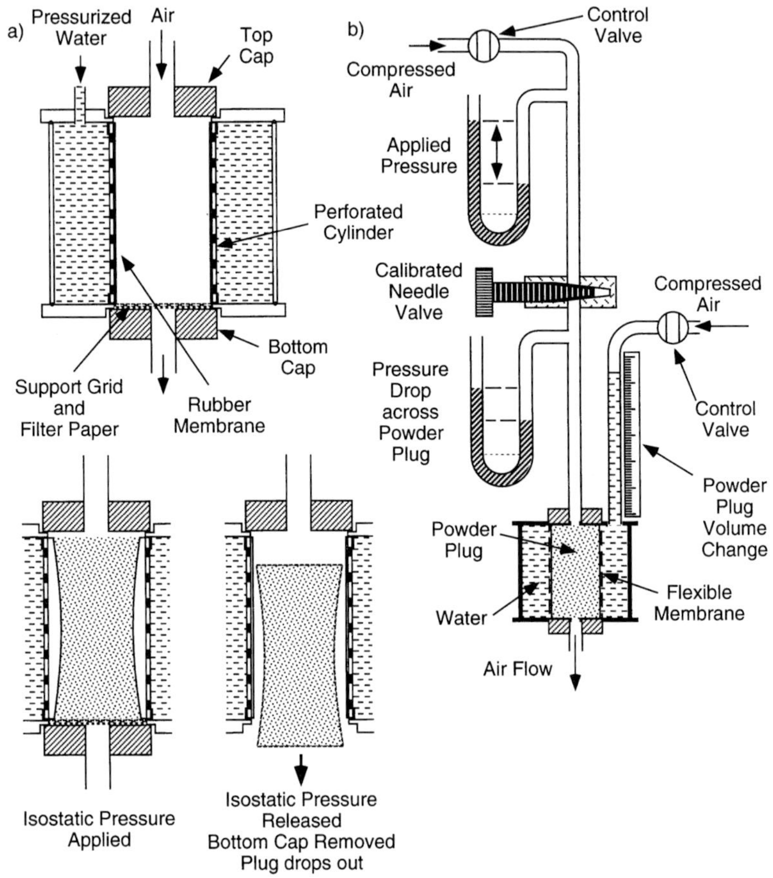

Figure 9.6. The soft-walled permeameter cell would revolutionize the use of permeametry measurements for monitoring the fineness of powders. a) The soft-walled permeameter cell allow for isostatic compression of a powder sample to a known porosity. When the pressure is released, and the bottom cap removed, the powder plug will drop out. b) A simple permeameter circuit which could be used on-line to provide feedback to a powder production line.

quent release of the hydrostatic pressure enables the smaller diameter plug to fall out of the base of the permeability cell [29]. Kaye and co-workers modified the softwall permeameter cell of Figure 9.6 so that the top and bottom of the cell could be opened and closed by movable elements and this cell was introduced into the circuit. (One of the major advantages of permeameter measurement is that the amount of powder used in a measurement procedure is relatively large and this simplifies automated sampling from the process line.)

To carry out a measurement, when the test sample was in place the variable pneumatic resistor was adjusted until the pressure drop over the powder plug equaled that over the variable resistor. In such a situation the two have the same pneumatic resistance (the electrical analog is the adjusting of a slide wire until the potential drop over the unknown resistor is the same of that of the known resistor.) In such a situation the dial or other display system of the variable pneumatic resistor can be calibrated directly in fineness units and the ultimate position of an indicator, or data from the system, could be used directly as a feedback control to adjust the operating parameters of the mill feed and the powder production unit.

One of the problems that can sometimes be encountered with powders is that cohesive powders, such as cement, do not always flow easily. Kaye and Legault showed that one can operate the system of Figure 9.6 with a cohesive powder by adding a silica flow agent to the system with subsequent adjustment of the calculation process to allow for the presence of the flow agent [27]. As technology moves towards design of on-line powder control systems the permeameter system of Figure 9.6 should not be overlooked.

It should be noted that the system of Figure 9.6 is actually potentially a cybernetic system for operating a powder production system and for this reason the system is referred to as a Cypermeter.

9.6 Determining the Pore Distribution of Packed Powder Beds and Porous Bodies

Many technologists are interested not only in the permeability of a powder system but also in the actual structure of the pores in the powder system. Sandstone or porous ceramics also have pore structures and permeabilities which are important in determining the physical properties of the system. In the pharmaceutical industry the porosity and permeability of tablets and in what are known as controlled release drug systems have physical properties determined by these pore structures. It should be noticed that the term pore can refer to either a connected or closed system. Thus the permeability of a porous structure is related to the interconnectiveness of pores whereas the activity of a catalyst may also be a function not only of the permeability of the system but also on the presence of pores in the individual particles which are said to be blind pores, i. e. not interconnected. The flow of oil through sandstone reservoirs, and of ground water pollution through a sedimentary rocks and soil, is also related to the pore structure of the system.

There are basically three techniques for studying the pore structure of a porous body. The first is what is known as mercury intrusion exploration of the pore structure. The second is the use of gas adsorption studies in which pore structure is derived from condensation isotherms. The third method is a study of sections made through the porous body or the consolidated powder system with subsequent image analysis. In this section we will first explore all mercury intrusion technology and then image analysis of sectioned material. A discussion of the method based upon gas adsorption studies will be deferred until Chapter 10.

The basic concept of mercury intrusion and/or the use of other liquids to study the pore structure of a consolidated mass of powder or a porous body, is based on the fact that for a non-wetting liquid, pressure is required to force the liquid into a capillary tube [29].

The physical basis of the method can be appreciated from the simple systems shown in Figure 9.7. If a closed, evacuated, capillary tube is inverted in a reservoir of mercury to which external pressure can be applied, then one can show that, if one increases the pressure above the surface of the mercury, the mercury is unable to enter the tube until

$$P = \frac{2\gamma \cos \theta}{R}$$

where P is the applied pressure, R is the radius of the tube, γ is the surface tension of the liquid, and θ is the contact angle of the mercury with the tube. This equation is known as the Washburn equation for the scientist who developed the equation from theory [29]. Mercury is widely used as an intrusion fluid and we will limit our discussion here to examples based on the use of mercury.

Figure 9.7 shows that when the pressure is sufficient to enter a tube of diameter R_1, the tube connected to R_1 will fill with mercury and the level of mercury in the reservoir will fall. The change in level can be measured and used to calculate the volume of the tube. As the pressure continues to increase, a pressure is reached at which the fluid can enter the narrower tube of diameter R_2 fitted with a bulbous reservoir. Once again changes in the mercury level can be used to measure the volume of each set of tubes.

To investigate the pore structure of a porous body, the internal spaces of the material are first evacuated of gas by placing it under a vacuum. It is then placed in a pool of mercury in a sealed, evacuated container. The pressure above the pool of mercury is then gradually increased, forcing the mercury into narrower and narrower channels in the porous body. The volume of mercury entering the pores with each increase in pressure is recorded. In the simplest model used to interpret the data, it is assumed that the pores are circular channels of uniform radius and that, as the pressure is increased, the change in volume of the mercury measures the volume of the holes of that

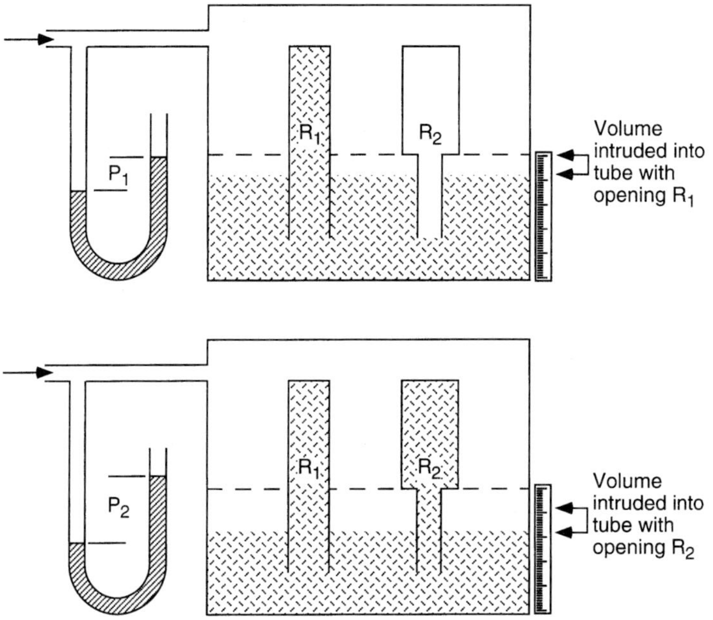

Figure 9.7. In mercury intrusion porosimetry, the volume of mercury entering the pores of a body at various applied pressures can be interpreted generally as data on the size distribution of the pores within a porous body. However the presence of large pores with small access throats can lead to misinterpretation of the true structure of the porous body.

size. However, as can be seen from the example of the tube of size R_2 in Figure 9.7, penetration through a narrow throat may lead to an extensive space behind the throat of much larger volume. From this simple illustration it can be seen that this throat-volume interpretation of the data is obviously a very simplistic model for characterizing the pore structure of a porous body. It would be better to say that the technique measures access throat diameter rather than pore size. However, in the absence of better interpreting models, extensive studies have been made of various types of porous bodies and the data interpreted using this simple model; are usually reported as "pore size distributions". In Figure 9.8(a) a mercury porosimetry data set for a relatively coarse powder with individual grains that were porous is shown. The data were taken from a study carried out by Orr [29].

It should be notice that in Figure 9.8(a) that the pressure applied on the mercury is shown on the top abscissa and that this has been interpreted as an access throat pore diameter in microns assuming appropriate values of θ and γ, on the bottom abscissa. If one replots that data of Figure 9.8(a) on log-log scales, the graph of Figure 9.8(b) is

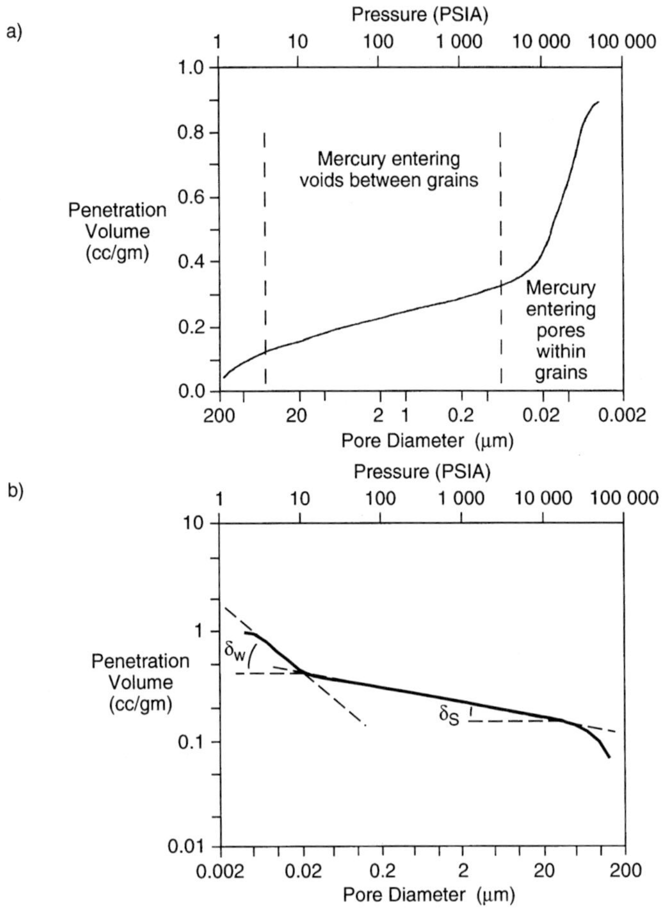

Figure 9.8. Data generated by mercury intrusion porosimetry can be reinterpreted using the concepts of fractal geometry, by replotting the data on log-log scales. The slopes of the linear regions of the resulting graph can be viewed as fractal dimensions in data space. a) Traditional presentation of mercury intrusion porosimetry data. b) Data of (a) plotted on log-log scales.

obtained. It has been suggested that this demonstrates that the invaded volume-access throat data for mercury porosimetry can be interpreted from the perspective of fractal geometry, since there are obviously two linear regions of the graph which are scaling functions. Thus when presenting these data in a scientific publication, I suggested that one could now use two fractal dimensions to describe the data [30, 31]. The value δ_1 to represent the invaded volume-access throat structure of the porous body from the perspective of the "between grain" voids, and δ_2 to represent the "within grain" voids. By using the term fractal dimension of "invaded volume-access to throat distribution" one avoids having to be specific about the physical significance of the data with respect to the pore size distribution of the material. The deduced fractal dimension is characteristic of the pore structure-ruggedness of the overall body but is not descriptive of the pore structure itself. Friesen and Mikula have suggested that one can interpret the fractal dimensions for the data such as those of Figure 9.8(b) by using an interpretive model based on Menger sponges [32, 33]. It has been suggested that the Appolonian gasket could be a better model for arriving at an interpretive hypothesis for explaining the physical significance of the fractal dimension for porous bodies derived from intrusion porosimetry data [31]. An important aspect of intrusion porosimetry is the measurement of the contact angle of the liquid being used and the solid particles being evaluated. Also it is necessary to have an accurate figure for the density for the powder grains. Several companies sell instruments to measure the contact angle and density of the grains. The equipment for measuring the density is called a pycnometer from a classical term for the density of a material. The companies that sell mercury intrusion equipment provide useful applications literature on both the intrusion equipment and equipment to measure contact angles and densities of the powder grains [34, 35, 36].

The next technique we shall look at which is used by technologists to study porous bodies, is to section them and look at the revealed pore distribution. For example if one looks at the system shown in Figure 9.9(a) it could be a section though a filter or a section though a piece of sandstone. It could also be the residual porosity in a sintered ceramic. The pore size distribution revealed in such a section can be described by a fractal dimension in data space. The view of Figure 9.9(a) is actually a simulated section through a paint system created by transforming a random number table into a system in which 2 of the digits in the random number table have been turned into black squares. Therefore the view of Figure 9.9(a) simulates a twenty percent porosity body. One can be interested in the pores in different ways, for example if one has added metal particles to a plastic body to create an electrically conductive material one would be interested in the clusters that are formed either by orthogonal or by diagonal connections. The two different types of clusters orthogonal and diagonal are shown in Figure 9.9(b). In Figure 9.9(c) the size distribution of the two different types of clusters created by the random dispersion simulated in Figure 9.9(a) are shown. It can be seen that both types clusters manifest a linear data relationship. It is interesting to note that the fractal dimension of these two data lines compare very well with theoretical value postulated by Schroeder in his study of clusters formed by the random juxtaposition in his deterministic chaos studies. The slope of these lines is a fractal dimension in data space. The fact that the data lines conform to the data space fractal dimension predicted by chaos theory can be used to determine if any given pore size distribution studied by sectioning a porous body is a random or structured. Deviations from the linear relationships of 2.05 would indicate a non-random genesis of the system being

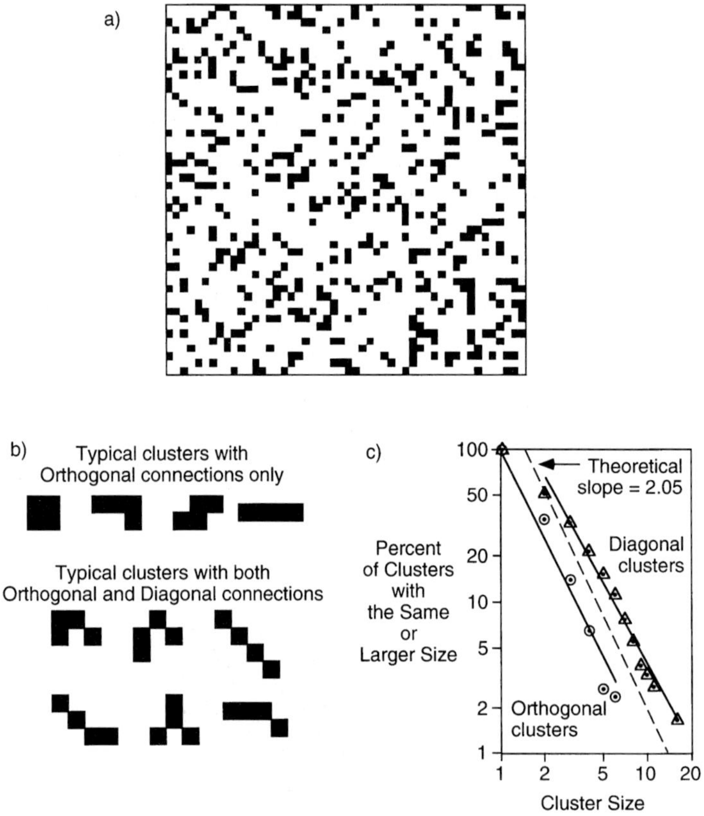

Figure 9.9. Clusters formed in a porous or pigmented body can be described by a hyperbolic function which constitutes a fractal dimension in data space. a) Simulated dispersion of monosized pores which occupy 20 % of the available space. b) Typical clusters which may be found in such a dispersion, displaying orthogonal and diagonal connections. c) Size distribution of the clusters present in (a) presented as a log-log plot.

studied. (See discussion in references 14, 30 & 31.) Technologists studying systems such as oil bearing sandstone and filter systems are interested in the possibility of complete tortuous paths through a body, a property which is known as percolation. In Figure 9.10, extension of the simulation study shown in Figure 9.9 to create the model of a three dimensional body is shown [14]. A more sophisticated model of the porous structure of an oil bearing sandstone to study the possibilities of connected pathways through the sandstone has been developed by Mathews and co-workers [37, 38]. The use of fractal geometry to characterize structure and properties of porous sandstone has been reviewed by A. H. Thompson and co-workers [39, 40]. Leuenberger and co-workers

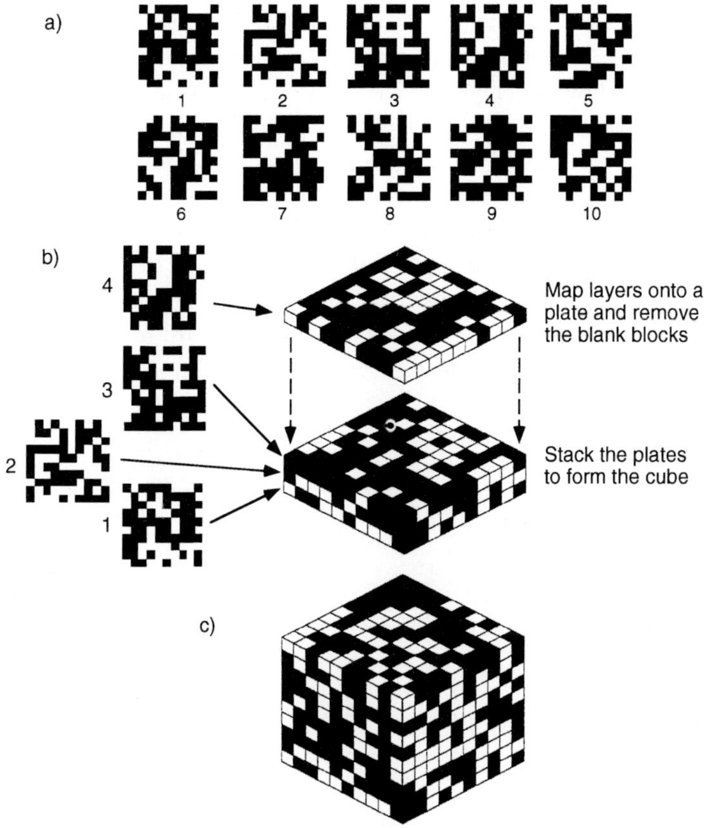

Figure 9.10. Extension of the simulation of Figure 9 into three dimensions results in a three dimensional porous body. The white cubes represent pores in the structure and connections between pores across several layers can result in a complete pathway through the cube. a) Layers generated from a random number table which will be used to create the block. b) Construction of the block from the layers of (a). c) Appearance of the completed cube.

have made extensive studies of the application of fractal geometry on the percolation and pore structure properties of tablets and controlled drug release systems [41, 42].

Another way in which the concepts of fractal geometry have been applied to the description of a porous body is to use what is known as a Sierpinski fractal. Mandelbrot, in his development of the concepts of fractal geometry, drew attention to some mathematical models originally developed by Sierpinski (accessible discussion of the original work of Sierpinski is to be found in Mandelbrot's book 41]. (See also reference 31.) Sierpinski's work dealt with the structure of porous bodies and the mathematical curve which is known as the Sierpinski's carpet. The basic concepts involved in drawing the Sierpinski carpet is shown in Figure 9.11(a). Sierpinski's carpet can have various structures depending upon the construction algorithm used to develop the curve, the particular one shown in Figure 9.11(a) is constructed in the following manner. The sides of the square are divided into three equal units. This defines the central square which is 1/9 of the total area. Imagine that this is then cut out to make the first stage of the Sierpinski's carpet. The same construction algorithm is then applied to the eight remaining sub-squares by the removal of their central squares as shown in the second stage diagram. The construction algorithm is then applied to each small remaining square area for the third stage. In theory if one carries out this construction ad infinitum or some mathematicians prefer to say ad nauseum (until you are sick of it), one is

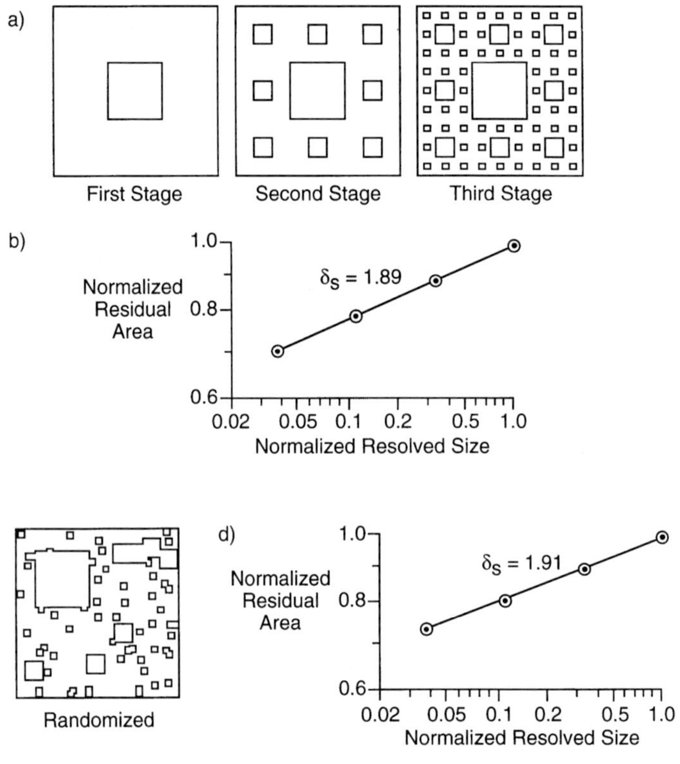

Figure 9.11. A mathematical curve known as a Sierpinski Carpet can be used as a model for important characteristics of a porous body. a) Repeated application of the algorithm for creating a Sierpinski carpet results in a figure with no area remaining. Shown are the first three steps in the construction. b) Determination of the Sierpinski fractal from the size distribution of the holes to stage three. c) Carpet of (a) with its holes randomized. d) The carpet of (c) disappears more slowly than (a) because some of the holes overlap.

left with a Sierpinski carpet which has no area defined by an infinite number of infinitely thin threads. A different Sierpinski's carpet could be constructed by dividing the sides of the primary square into fifths with the removal of one-twenty fifth from the centre of the square. Continuing this construction would lead to another Sierpinski carpet which also has no area but more infinite number of threads than the first Sierpinski carpet! A way of characterizing which structure one has used to develop a Sierpinski carpet is to study how fast the area disappears as one constructs the system. Thus in Figure 9.11(b) the residual area at each stage of the algorithmic construction is plotted against the size of the holes which are being used to create the carpet. (Note that the log scale on the ordinate is different log scale of that of the abscissa to show the decline of the area more clearly.) It can be shown that as the construction proceeds the disappearance of the carpet is a linear relationship on the log-log graph paper and that the slope of the line representing the disappearance of the carpet is a characteristic parameter which can be called the Sierpinski fractal. The Sierpinski fractal is calculated by subtracting the mathematical value of the slope of the data line from 2 (the basic dimension of a square). Beginners in the use of this type of this fractal dimension sometimes made the mistake of using the physical slope of the graph instead of the mathematical slope of the relationship which is distorted in a graphical presentation such of that of Figure 9.11(b).

At first glance the kind of structured model of a porous body represented by the third stage of the Sierpinski carpet of Figure 9.11(a), is unrealistic compared to the real system. However, if we take the areas of Figure 9.11(a) at stage three and randomize their location one obtains the statistical version of the Sierpinski carpet shown in Figure 9.11(c). This is described as a statistically self-similar version of an ideal Sierpinski carpet. Because some holes fall within other holes when one randomizes the location of the elements of the carpet, the carpet does not disappear as fast as the ideal curve to which it is self-similar as shown by Figure 9.11(d). One can appreciate from Figure 9.11(c) that the randomized Sierpinski carpet is indeed a realistic model for the appearance of the cross-section through a porous body.

To measure the Sierpinski fractal of the cross-section of a dispersion such as the picture shown in Figure 9.12(a) one commences ones measuring procedure by measuring only the largest holes present in the field of view and then one plots the fractional area of the carpet remaining when one subtracts the total area of these larger holes. This gives the second point on the graph of Figure 9.12(b). (In the diagram of Figure 9.12(b) the first point is the normalized area of the carpet without any holes present.) To generate the third point on the graph, one now considers the presence of the next smallest holes and when the area of these holes is measured and again subtracted from the data used to generate the previous point we have the third point of the graph and so on. The Sierpinski fractal calculated from the mathematical information (not the visual slope of the graph) is 1.89 since the mathematical slope was 0.11. It is interest-

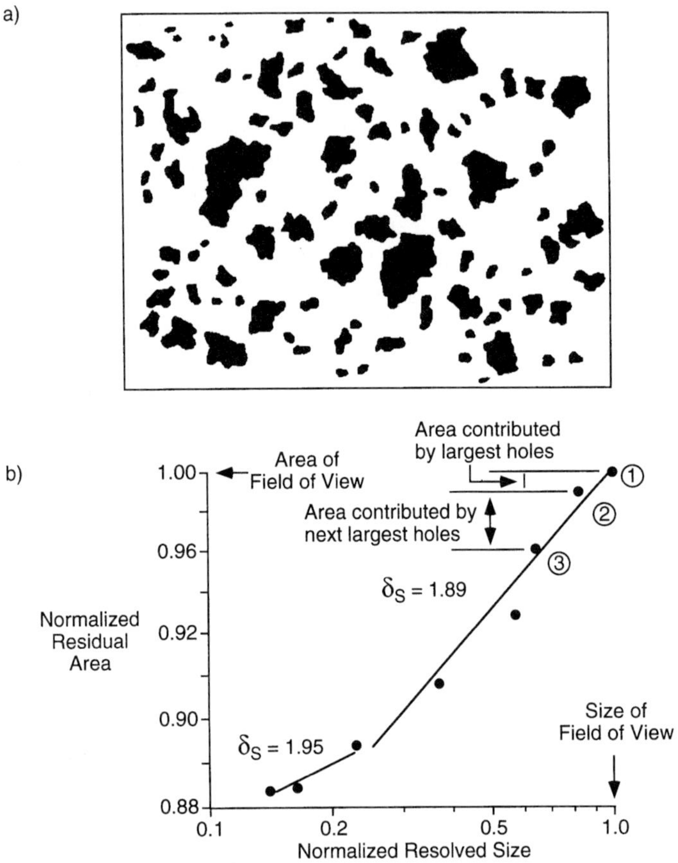

Figure 9.12. Sierpinski fractal for a section through concrete. The break in the data can be attributed to the resolution at which one starts to examine sand rather than gravel. a) Stylized appearance of a section through concrete. b) Graph of the data obtained by measuring the areas of the individual particles within the concrete matrix.

ing to note that there is a break in the data as this corresponds to the fact that after one has examined the gravel profiles which are larger now starts to examine the sand grains trapped in the matrix of the concrete. In Figure 9.13(b) the Sierpinski carpet plot of the data for the carbon black dispersion shown in Figure 9.13(a) is shown. It is interesting to note that again there is a break in the data line and that this corresponds to the fact that there are really two types of discrete particles present in the dispersion. These are the primary carbon blacks, which are dispersed in the matrix to generate the toner fineparticle, and in that process there is attrition of the extremities of the fractally structured carbon blacks generating what has been labeled as the dispersion

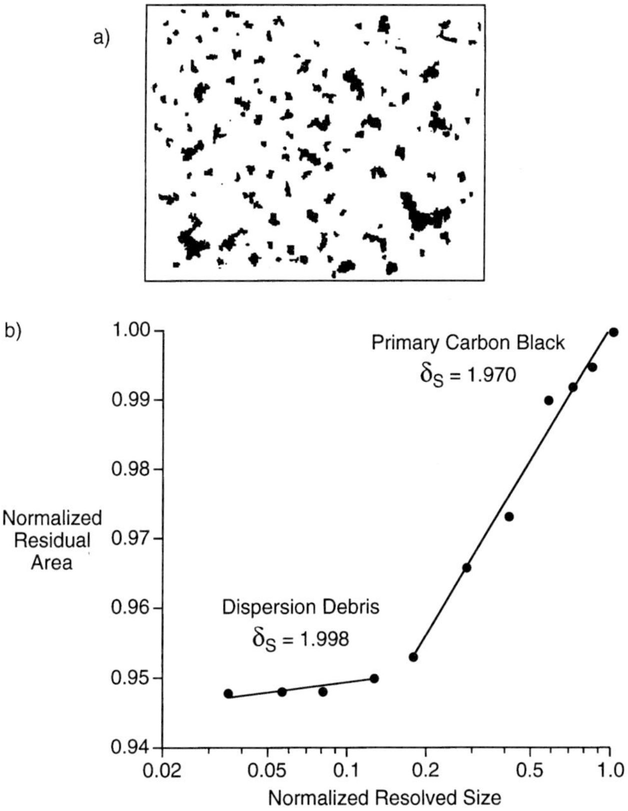

Figure 9.13. Sierpinski fractal calculation for a dispersion of carbon black in a plastic matrix shows that there are two distinct relationships. The first associated with the primary carbon blacks and the other associated with the small fragments broken away from the primary carbon blacks in the act of dispersing them. a) Stylized view of the carbon black dispersion. b) Log-log plot of the size distribution of the carbon blacks.

debris of Figure 9.13(b). For a full discussion of experimental procedures and useful examples of the utility of the Sierpinski fractal for describing porous and composite bodies see reference 31.

Basic ideas embodied in the construction of a Sierpinski carpet can be extended to three dimensions by taking a cube and dividing the cube into subportions. For example, the analog of the Sierpinski carpet shown in Figure 9.11 would be to divide the side of the cube into sets of three units generating a subcube at the centre of each side which is removed. One then continues the construction of the porous body by subdividing each residual cube and removing the centre subunit. The structure generated in this way is known as a Menger sponge [31, 43].

The existence of percolating pathways in a system such as a piece of sandstone can be studied experimentally by recording the movement of fluid invading the pore structures. Thus in Figure 9.14 direct measurement of fluid invasion in a piece of sandstone is shown. The movement of the fluid out from the injection point creates a fractal structure which is related to the pore structure of the material. In Figure 9.14(b) data processing generates the numerical estimates of the fractal dimension of the fluid front. In the same way it can be shown that if one drops a drop of liquid onto a porous pharmaceutical tablet the fluid will move out to create a fractal front related to the pore structure of the tablet [31].

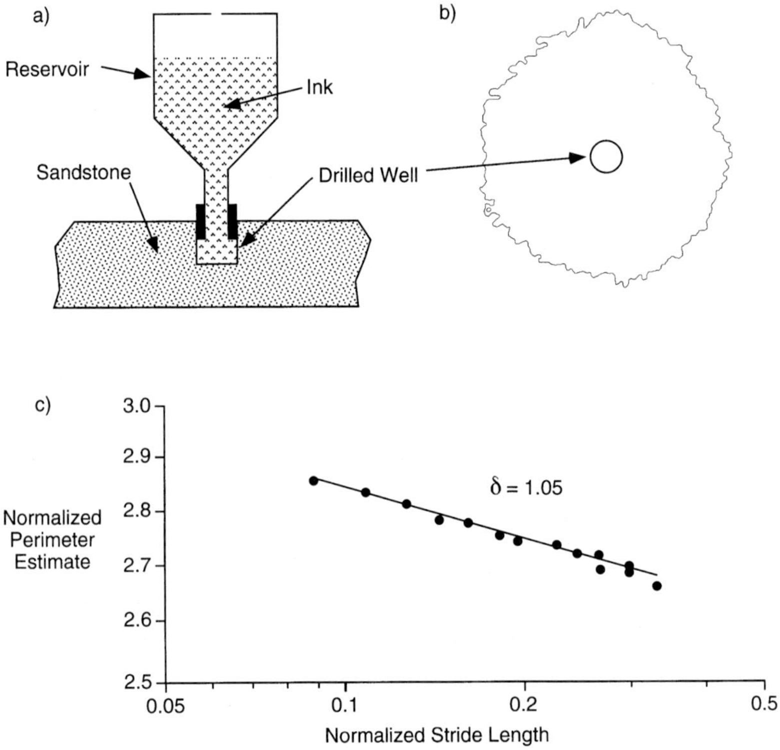

Figure 9.14. The fractal front created in a sandstone by an invading fluid, carries information on the pore structure of the rock. This information may be useful in methods used in secondary recovery of oil. a) Simple experimental setup for examining fractal fronts in sandstone. b) Trace of a front created by the method of (a). c) Richardson plot for the data generated for the front in (b).

References

[1] Hosokawa test equipment is available from Hosokawa Micron International Inc., 10 Chatham Road, Summit, New Jersey, USA 90071

[2] R. O. Gray and J. K. Beddow, "On the Hausner ratio and its relationship to some properties of metal powders", *Powder Technology*, 2 (1968–69) 323–326.

[3] see Chapter 3 on Powder Rheology in B. H. Kaye, *Powder Mixing,* Chapman and Hall, London, 1997.

[4] M. S. Mohammadi, N. Harnby, "Bulk density models as a means of typifying the microstructure and flow characteristics of cohesive powders," *Powder Technology,* 92 (1997) 1–8.

[5] H. H. Hausner, "Friction conditions in a massive metal powder", International Journal of Powder Metallurgy, 4, (1967), pp. 7.

[6] J. C. Carman, *Flow of Gasses Through Porous Media,* Butterworths, London, 1956.

[7] P. C. Arnell and J. C. Carman, "Surface area measurements of fine powders using modified permeability equation," *Can. J. Research*, 26A (1948) 128–136.

[8] B. H. Kaye, "Permeability techniques for characterizing Fineparticles", *Powder Technology* 1 (1967) 1.

[9] J. L Fowler and K. L. Hertel, "Flow of a gas through porous media," *J. Appl. Phys.*, 11 (1940) 496–502.

[10] D. T. Wasan, W. Wenek, R. Davies, H. Jackson and B. H. Kaye, "Analysis and Evaluation of Permeability Techniques for Characterizing Fine Particles; Part I Diffusion and Flow through Porous Media," *Powder Technology*. 14 (1976) 209–228

[11] D. T. Wasan, M. B. Bonado, S. K. Scood, R. Davies, M. Jackson, B. H. Kaye, and W. Wemek, "Analysis and Evaluation of Permeability Techniques for Characterizing Fine Particles; Part II Measurement of Specific Surface Area," *Powder Technology*, 14 (1976) 229–244

[12] T. Allen, *Particle Size Measurement,* Chapman & Hall, 4ed. 1993.

[13] I. C. Edmundson and J. P. R. Toothill. "Determination of specific surface area by airpermeation at low porosities," *Analyst*, 88 (1963) 807.

[14] B. H. Kaye, *Powder Mixing,* Chapman and Hall, London, 1997.

[15] E. L. Gooden and C. M. Smith, "Measuring the average particle diameter of powders" *Ind. Eng. Chem. (Anal. Ed.),* 12 (1940) 479–482.

[15] F. M. Leas and R. W. Nurse, "the specific surface of fine powders," *J. Soc. Chem. Ind.*, 58 (1939) 277.

[16] "Fisher Subsieve Sizer" is made by Fisher Scientific Co., 633 Greenwich St.,, New York 14, N. Y. and is no longer available commercially.

[17] B. Dubrow and M. Nieradka, "Determination of specific surface or sieve size powders," *Anal. Chem.,* 27 (1955) 302.

[18] P. Lötzsch, "Bestimmung der spezifischen Oberfläche von pulverörmigem Material durch eine Durchlässigkeitsmessung", *Chem. Tech. (Berlin),* 17 (1965) 414.

[19] B. H. Kaye and M. R. Jackson, "Design studies of permeameters for comparing fine powders," Proc. Conf. Particle Size Analysis, Loughborough, 1966. Brit. Soc. Anal. Chem., London, pp. 313–325.
[20] R. J. Rigden, *J. Soc. Chem Ind.* 66 (1947) 191
[21] R. L. Blaine, "A simplified air permeability fineness apparatus," *ASTM Bulletin.* 123 (1943) 51.
[22] F. J. Spillane, "An automatic direct-reading apparatus for determining the surface area of powders," *Analyst,* 82 (1957) 712
[23] S. Ober and K. S. Frederick, "A study of the Blaine Fineness Tester and a determination of surface area from air permeability data," Symposium on Particle Size Measurement, ASTM Spec. Tech. Publ No. 234, (1958) 279.
[24] "The particle size analyzer reduces grinding costs in Pit and Quarry," September 1977 – News article.
[25] The Microtrac system was developed and marketed by Leeds and Northrup. This branch of the General Signal corporation of North Wales, Pennsylvania was sold to one of the Honeywell companies and current information on this diffractometer is available from Honeywell.
[26] A. D. Hoffman, "A Soft-Wall Permeameter for Online Characterization of Grinding Circuits." M. Sc. Thesis, Laurentian University, Sudbury, Ontario, Canada (1989).
[27] B. H. Kaye and P. E. Legault, "Real-Time Permeametry for the Monitoring of Fineparticle Systems." *Powder Technology.* 23 (1979) 179–186.
[28] The commercial rights to the softwalled permeability cell are vested in Microscal ltd. 79 Southern Row, Kelsal, London England.
[29] C. Orr, "Application of Mercury Porosimetry Penetration in Material Analysis", *Powder Technology,* 3 (1969–70) 117–123.
[30] B. H. Kaye "Applied fractal Geometry and a fine particle specialist" Part 1-Boundaries in Rough Surfaces," *Part. Part. Syst. Charact.* 10 (1993) 99–110.
[31] See discussion in B. H. Kaye, *Chaos and Complexity – Discovering the surprising patterns of science and technology,* VCH, Weinheim Germany, 1993.
[32] B. H. Kaye, "Fractal Geometry and the Characterization of Rock Fragments", in Fragmentation, Form and Flow in Fractured Media, R. Engleman and Z. Jaeger (eds.), *Ann. Is. Phys. Soc.,* 8 (1986) 490–516.
[32] S. H. Ng, C. Fairbridge, B. H. Kaye, "Fractal Description of the Surface Structure of Coke Particles", *Langmuir,* 3:3 (May-June 1987) 340–345.
[33] W. Freisen, R. J. Mikula, Canmet Divisional Report ERP-CRL 86-128, available from Energy, Mines and Resources Canada, Canmet Technology Information Division, Technical Inquiries, Ottawa, Ontario, K1A 0G1, Canada.
[34] Micromeritics, 1 Micromeritics Drive, Norcross, Georgia, USA 30093-1877.
[35] Quantachrome Corp, 6 Aerialway, Syosset, New York, USA 11791.
[36] Porous Materials Inc. Cornell Business and Technology Park, 83 Brown Road Bldg, Iltiaca, New York, USA 14850

[37] G. P. Matthews, A. K. Moss, M. C. Spearing, F. Voland, "Network Calculation of Mercury Intrusion and Absolute Permeability in Sandstone and Other Porous Media," *Powder Technology*, 76 (1993) 95–107.

[38] G. P. Matthews, A. K. Moss, A. J. Ridgeway "The Effects of Correlated Networks on Mercury Intrusion Simulations and Permeabilities of Sandstone and Other Porous Media" *Powder Technology*, 83 (1995) 61–77.

[39] A. H. Thompson, A. J. Katz, C. E. Krohn, "The Micro Geometry and Transport Properties of Sedimentary Rock," *Advances in Physics* 36:5 (1987) 625–694.

[40] A. H. Thompson, "Fractals in rock Physics," *Annu. Rev. Earth Planet Science,* 9 (1991) 237–262.

[41] H. Leuenberger, A. D. Bonny, M. Kolb, "Percolation Effects on Matrix Type can Control Drug Release Systems," *International Journal of Pharmaceutics*, 115 (1995) 217–224.

[42] T. Leu and H. Leuenberger, "The Application of Percolation in Theory to the Compaction of Pharmaceutical Powders," *International Journal of Pharmaceutics,* 90 (1993) 213–219.

[43] B. B. Mandelbrot, *Fractal Geometry of Nature,* W. H. Freeman, San Fransisco, 1983.

10 Powder Structure Characterization by Gas Adsorption and Other Experimental Methods

10.1 Experimental Measurement of Powder Surface Areas by Gas Adsorption Techniques

The basic concept employed in gas adsorption studies of the structure of powders is to study the way in which gas molecules are adsorbed onto the surface of the powder under specified conditions so one can deduce the surface area of the powder. The technology of this method comes from the late 1930's. A very good review of the basic concepts has been written by Orr [1]. Although the basic principles of the method are simple, the interpretation of the data when dealing with complex powder systems can be very difficult. In this chapter we will outline the basic methodology and direct the reader to more comprehensive literature which can be consulted if the detailed structure of systems from a research perspective is of interest. Excellent books have been written describing the research methodology of gas adsorption for complex powder systems see for example references 2, 3, 4, and 5.

Routine gas adsorption studies can often be used when the aim of the investigation is quality control of powder systems. In the late 1950's I was involved in measuring the surface area of uranium dioxide powders using Krypton adsorption as a quality control method for the fabrication of nuclear reactor fuel rods.

Several countries have established standard procedures for gas adsorption studies, see for example British Standard 4359. The equipment specified in the British Standard for carrying out volumetric studies of gas adsorption is shown in Figure 10.1. Before we can discuss the operation of this equipment in detail it is necessary to establish some basic vocabulary and explore some of the important concepts involved in the interpretation of data.

In the experimental investigation of the adsorption properties of a powder the powder under study is placed in a sample jar which, as shown in Figure 10.1, can be connected by the appropriate position of valve V_1 to a vacuum pump. Before we can study adsorption properties of a powder we must make sure that the powder has been cleared of all previously adsorbed moisture and gases. Sometimes during this stage the sample is heated to accelerate the removal of the adsorbed gases. This process is known as out-gassing. A bottle neck in the surface area evaluation of a series of samples is the fact that each sample must be out-gassed. In commercial instruments there are often several sample vessels that are in parallel so they may all be out-gassed while any one par-

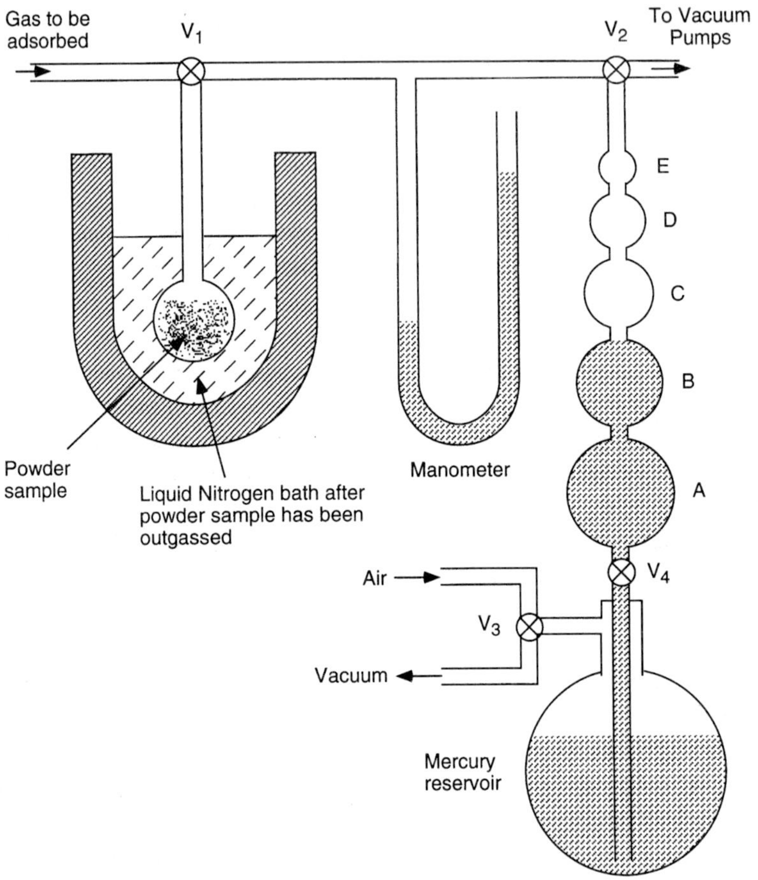

Figure 10.1. Basic configuration of the volumetric instrumentation for measuring the gas adsorbed onto the surface of a sample of powder.

ticular sample is being studied. In some instances the out-gassing period can take up to 24 hours.

After the out-gassing is completed, valve V_1 is switched so that a gas such as Helium is placed into the sample measuring system. The sample container is then placed in something such as liquid nitrogen so that the cooled powder sample can adsorb the gas from the container. The amount of gas adsorbed onto the powder can be measured by using either volumetric techniques or delicate electronic balances to weigh the change. By far the largest number or commercial instruments use volumetric techniques since the accurate weighing of the adsorbed gases requires very expensive equipment [1]. For this reason we will concentrate on describing volumetric methods for studying powder adsorption of the gas.

During this type of measurement, the volume of gas adsorbed by a unit mass of solid denoted by V_a/W depends on the equilibrium pressure, the absolute temperature, the solid, and the gas. The adsorbable gas will be below it's critical temperature and the pressure inside the equipment as measured by the manometer shown is expressed as a fraction of the saturation vapour pressure P_o. The resulting ratio P/P_o is called relative pressure. The amount of gas adsorbed can be calculated using the gas law relationships, the measured pressures, and known volumes of the equipment.

At the start of the experiment the volume of the powder sample must be deduced from the pressure inside the system during the initial gas conditions. To change the volume and the pressure inside the sample tube, the valves V_1 and V_2 are in such a position that by raising the mercury to the point A of the first bulb of the volumetric equipment the pressure in the system can be changed and measured. From the data the amount of gas adsorbed at the changing pressures is calculated. The second step is to raise the mercury to the point B and repeat the experiment and so on. From the series of adsorption measurements at one temperature, one can plot a graph of V_a/W vs. P/P_o; this graph is called the adsorption isotherm. Orr tells us that by 1970 tens of thousands of adsorptions had been determined for a variety of gases and solids. He states that the majority of the resultant isotherms are of one of the five types illustrated in Figure 10.2. It should be noted that sometimes adsorption of gases onto a surface of a powder can involve chemi-adsorption, in which there is a formation of bonds between the adsorbate and the adsorbent. In this situation it is difficult to measure surface area by a simple technique.

Within the group of isotherms which have been studied, by far the largest number are known to be Type 2 isotherms. In this discussion we will restrict ourselves to the interpretation of Type 2 isotherms using a theory known as the B.E.T. theory based on the work of S. Brunauer, B. Emmett, and E. Teller [5]. For a discussion of Type 1 isotherms, see reference 1. Type 1 is usually found in solids which have a porous structure.

There is an interesting story told about the development of this equation in the late 1930's. It is said that Brunauer and Emmett were having difficulty developing the theory and arranged a meeting over lunch with Teller, who later became famous because of his involvement in Atomic bombs and Hydrogen bomb theory. It is said that Teller worked out the basic equation for gas adsorption on a table cloth were the lunch was held and that Brunauer and Emmett had to buy the table cloth to take it back to the laboratory [5].

As Orr states, the great majority of solids are either Type 2 or 4 isotherms especially when adsorption is carried out at low temperature with inert gases. Analysis of these isotherms have been very successful in yielding the adsorbent surface area and their use constitutes a great deal of the activity in gaseous estimates of surface area. In Table 10.1 some of the saturation vapour pressures of gases used in gas adsorption studies

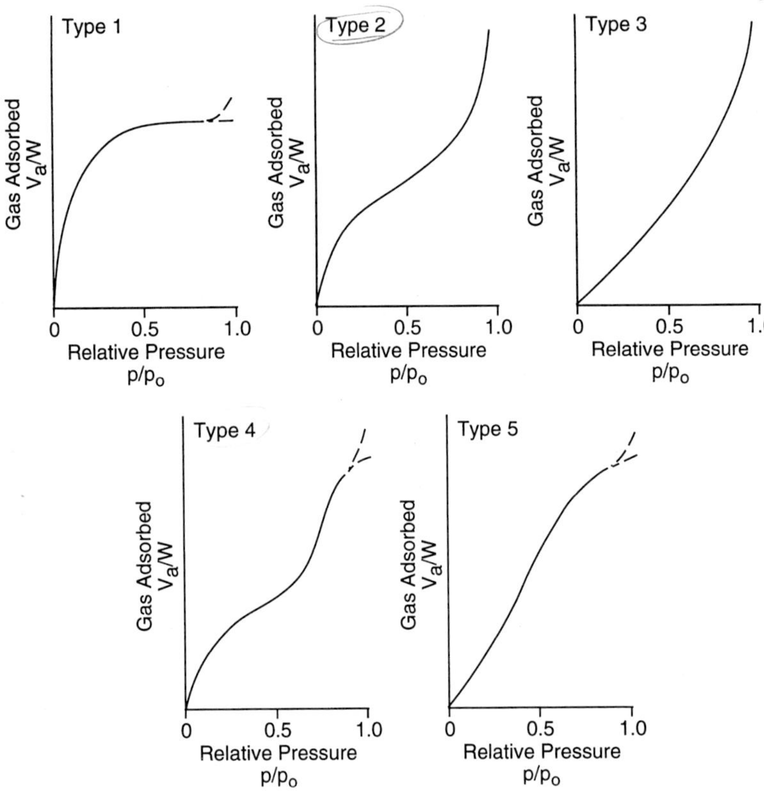

Figure 10.2. When studying surface area by gas adsorption, five basic types of adsorption isotherms may be manifest.

are shown. In all gas adsorption studies it is important to know the cross sectional area of the molecule of the adsorbent gas and in Table 10.2 cross sectional areas of many gases used in the studies are given. Using the B.E.T. equation one can show that the basic equation describing the Type 2 isotherm is

$$\frac{P}{V_a(P_o - P)} = \frac{1}{V_{mc}} + \frac{C-1}{V_{mc}} \cdot \frac{P}{P_0}$$

Where C is a constant. From the equation we see that a plot of $\left(\frac{P}{P_0 - P}\right)_{Vz}$ against P/P_0 will be a straight line of intercept $\frac{1}{V_{mc}}$ and slope $\frac{(C-1)}{V_{mc}}$. The calculation of

Table 10.1. Vapour (saturation) pressure as a function of absolute temperature.

	Vapor Pressure, Torr[1]		
Temperature, (K)	Krypton[2]	Argon	Nitrogen
76.0	1.825	160.29	646.468
76.5	2.049	–	–
77.0	2.297	188.74	729.398
77.5	2.872	–	774.270
78.0	3.203	221.31	818.847
78.5	3.567	–	866.464
79.0	3.966	258.45	916.204
79.5	4.402	–	968.136
80.0	4.879	300.64	1022.32
80.5	5.399	–	1078.83
81.0	5.966	348.42	1137.71
81.5	6.583	–	1199.05
82.0	7.253	402.34	1262.90
82.5	7.981	–	1329.33
83.0	8.770	463.02	1398.41
83.5	9.624	–	1470.21
84.0	10.548	528.88	1544.80
84.5	11.545	–	1622.25
85.0		592.37	1702.61

[1] Data from *Instruction Manual,* Micromeritics Instrument Corporation, Norcross, Georgia.

[2] As a subcooled liquid. Krypton is actually a solid in the lower temperature ranges, but liquid values are employed in surface area evaluation.

Table 10.2. Values for the area occupied by physically adsorbed molecules and the perfect gas law correction factor.
Data from *Instruction Manual,* Micromeritics Instrument Corporation, Norcross, Georgia.

Adsorbate	Temperature (K)	Area(s) per Molecule[1]	Perfect Gas Law Correction ($Torr^{-1}$)
Argon	78	13.8	11.4×10^{-5}
Argon	90	13.8	3.0×10^{-5}
Benzene	293	43.0	0
Carbon Dioxide	195	21.8	2.7×10^{-5}
Ethane	87	21.3	1.0×10^{-7}
Krypton	77	21.3	3×10^{-5}
n-Butane	273	44.4	14.2×10^{-5}
Nitrogen	78	16.2	6.6×10^{-5}
Oxygen	78	13.6	6.3×10^{-5}
Water Vapor	298	12.5	24×10^{-5}

[1] For values for other molecules, see McClellan, A. L. and Harnsberger, H. F., "Cross-Sectional Areas of Molecules adsorbed on Solid Surfaces", *J. Colloid and Interface Sci., 23,* 577–99 (1967).

the surface area perceived with the data points below point B on the basic isotherm shown in Figure 10.3(a) generates the graph shown in Figure 10.3(b).

Once the BET equation isotherm is obtained, the specific surface area of the adsorbent, B_w in square meters per gram, can be calculated using the appropriate adsorbate molecule coverage as given in Table 10.2 and by using the equation in the form

$$S_w = \frac{Ax10^{-20}\left(6.02x10^{23}\right)}{22.414x10^3(\text{slope} + \text{intercept})}$$

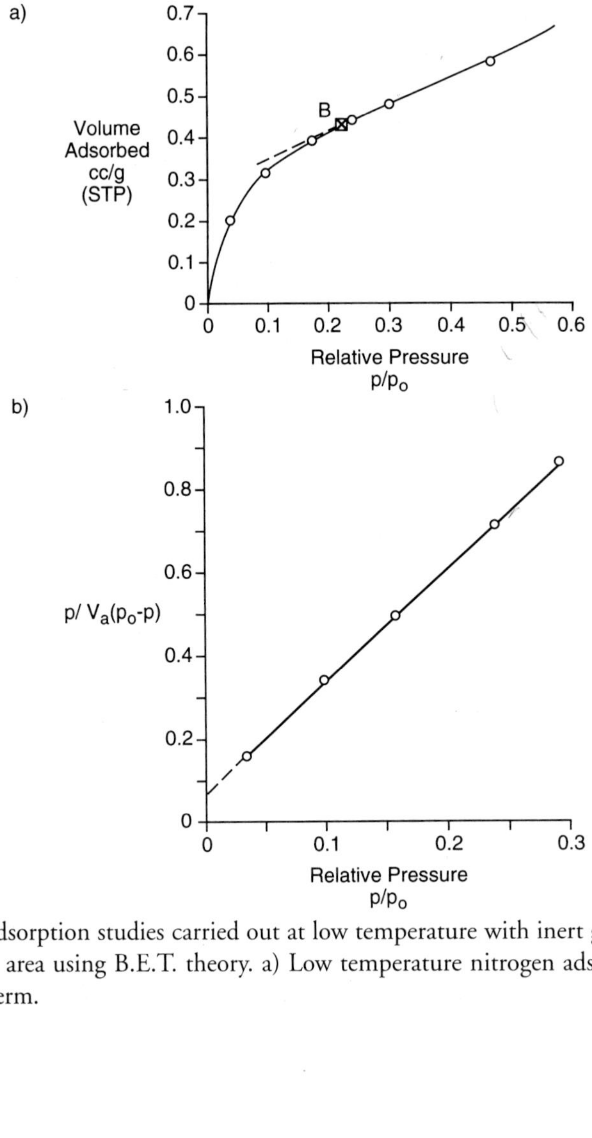

Figure 10.3. Adsorption studies carried out at low temperature with inert gases yield the adsorbent surface area using B.E.T. theory. a) Low temperature nitrogen adsorption isotherm. b) B.E.T. isotherm.

where the numerical values represent a unit conversion factor, Avagadro's number, and the molar gas volume. When nitrogen gas is adsorbed at the liquid nitrogen temperature, the equation reduces to

$$S_w = \frac{4.35}{\text{Slope} + \text{intercept}}$$

It should be noted (as Orr points out) adsorbed gas molecules do not occupy identical space on all solids; the occupancy depends somewhat on the atomic spacing of the solid. The values given in the tables are the best values for a variety of solids. For this reason primarily, results of any adsorption measurement cannot be considered absolutely accurate to better than +1 % even if measurements are carefully performed.

The calculations, such as the one given above for nitrogen adsorption, are all programmed into the data processing portions of the commercially available equipment marketed by several instrumentation companies [6–12].

To interpret adsorption isotherms using the BET theory it is usual to work with materials below 0.3 partial pressure for several reasons. On routine quality control of powders, it is often possible to work with just one point on the BET curve, assuming that the BET isotherm for those materials passed through the origin [1]. For some substances one can retrace the BET isotherm by reducing the pressures by pulling the mercury downwards in the system of Figure 10.1. The curve generated is called a desorption isotherm. Sometimes one finds that there is a lag in the desorption curve with respect to the adsorption isotherm as shown by the data of Figure 10.4(a).

It can be shown that the hysteresis of the type shown by the data of Figure 10.4(a) is due to the fact that the grains of the powder being studied have micropores. Thus, it is not so easy for the adsorbed material to leave the pores as it is to enter it, hence the hysteresis between the adsorption and desorption curves. In Figure 10.4(b) the pore size distribution generated by using the data of Figure 10.4(a) is shown [1].

Neilson and Eggertsen developed what is known as a flow technique for measuring the BET isotherm [13]. In this system a mixture of Helium and an adsorbate gas, usually Nitrogen, is passed through a small U-shaped cell containing the powder sample to be studied. At liquid Nitrogen temperature Helium will not adsorb on any surface while Nitrogen will physically adsorb on all substances. In an actual experiment the ratio of Helium to Nitrogen in the flowing stream being fed to the sample tube is measured using a thermal conductivity detector. The sample tube is then placed in liquid Nitrogen and Nitrogen is adsorbed onto the surface of the powder. The change in the thermal properties of the flowing stream of Helium and Nitrogen, as the Nitrogen is adsorbed, is measured generates an adsorption peak as shown in Figure 10.5(a). The liquid nitrogen is then removed from around the sample tube and the adsorbed gas then desorbs as the sample tube warms up to give the graphical output as shown in

Figure 10.5(a). It is usual in carrying out this experiment to use a sample of known surface area to give a calibration peak as shown in the diagram.

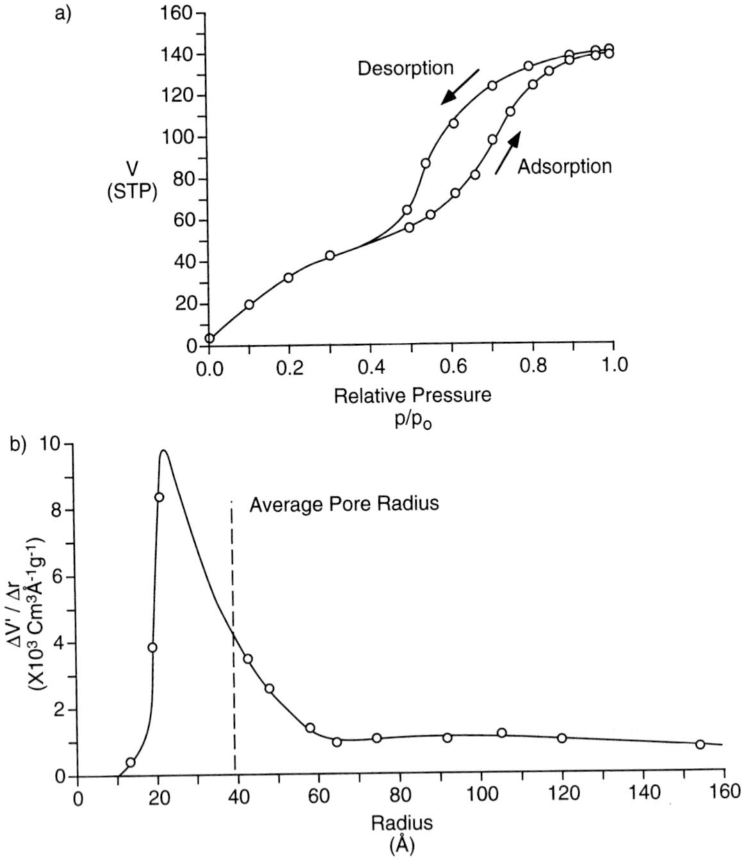

Figure 10.4. Typical data from gas adsorption analysis of a material. a) Adsorption-Desorption isotherm for an alumina powder. b) Pore volume distribution determined for a carbon black powder.

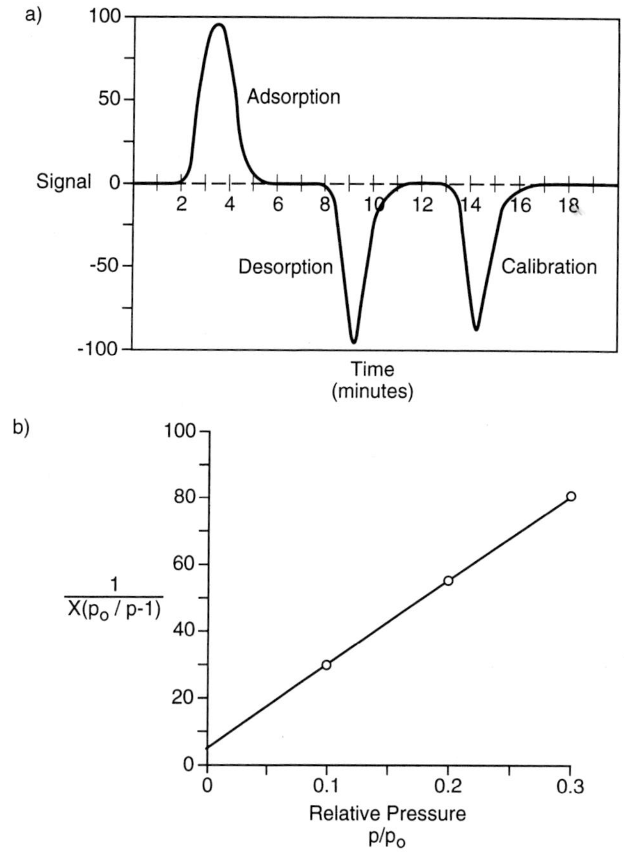

Figure 10.5. Neilson and Eggertsen have developed a flow technique for measuring the B.E.T. isotherm. a) Adsorption, desorption and calibration signals used to obtain one point on the B.E.T. plot. b) Typical B.E.T. plot for alumina with a surface area of 13.8 m^2g^{-1}.

10.2 Characterizing the Fractal Structure of Rough Surfaces via Gas Adsorption Studies

It was well known amongst workers studying the surface area of powder by gas adsorption that one can get different answers for the measured surface area by using different adsorbate gases. Meaning if one would measure the surface area with Krypton, one would not get the same answer as if one used Nitrogen. Various explanations were put forward as to why there was this difference between the measurements. Most of the arguments hinged on the fact that there was uncertainty in the actual cross-sectional area of the gas molecule when it was adsorbed onto the surface. The other main reason put forward was that the adsorptive bonding energy available at different sites on the surface of the solid has different abilities to adsorb the gaseous molecules. Soon after the development of fractal geometry, invented by Mandelbrot, investigators quickly recognized that the discrepancy between the various estimates of surface area based on different molecular explorations could possibly be explained by the fact that the surfaces being investigated were fractal [14]. Avnir was the first to point out that exploring rough structures using different sized gas molecules was equivalent to exploring coastlines of Great Britain using different stride lengths [15]. This basic concept is illustrated in Figure 10.6, the smaller the gas molecule, the more it can enter into the interstices of the roughness of the surface. Avnir showed that the fractal dimension of a rough surface can be studied by adsorbing different alcohols onto a rough surface as shown in Figure 10.6 [15]. There is some controversy over the fact that we can measure the fractal dimension of a surface by using different size adsorbent molecules, but the technique has been widely used [16–28].

Neimark has described a technique for assessing the fractal dimension of porous solids, by studying the adsorption of water onto the surface and using the concepts of capillary equilibrium to interpret the data [29]. Equipment for measuring the surface area of a solid by water adsorption is available from Bel Japan [30].

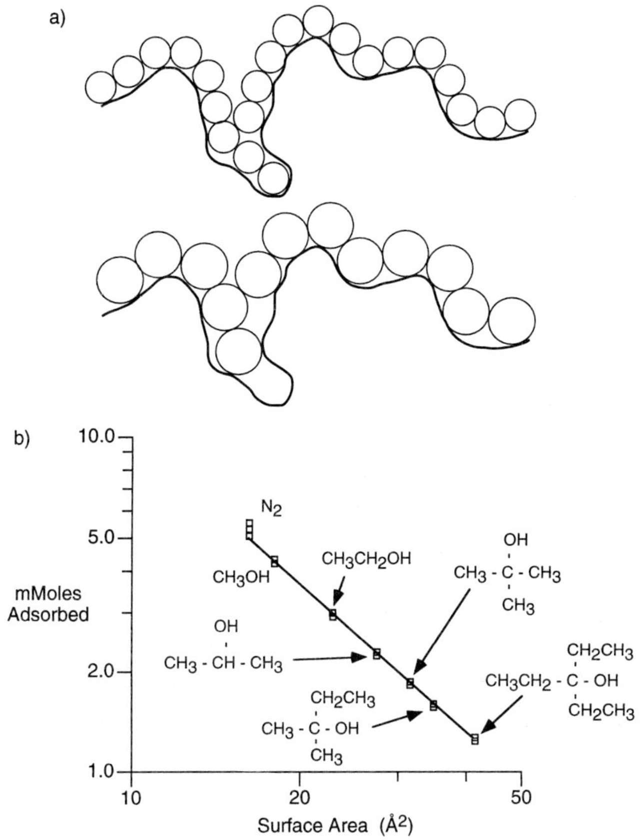

Figure 10.6. Appropriate interpretation of gas adsorption data can enable one to deduce the fractal dimension of a rough surface. a) In gas adsorption, the surface area of a powder is estimated from the number of molecules required to cover the surface. This estimate will vary depending on the size of the molecule used. b) The fractal dimension of a surface can be derived from surface area estimates obtained for a range of different sized molecules.

References

[1] Orr, Jr. "Surface Area Measurements," *Treatise on Analytical Chemistry,* I. M. Kolthoff, and P. J. Elving Eds., Interscience, John Wiley & Sons Inc.
[2] S. J. Gregg and K. S. W. Sing, *Adsorption Surface Area and Porosity,* 2nd Ed., Academic Press, New York, 1982.
[3] S. Brunauer, "Solid Surfaces and the Gas Solid Interface," *Advances in Chemistry*, No. 33, American Chemical Society, Washington D. C., 1961.
[4] S. Lowell, and J. E. Shields, *Powder Surface Area and Porosity,* Chapman and Hall, 2nd Ed., London England, 1984.
[5] S. Brunauer, B. H. Emmett, and E. Teller, "The Adsorption of Gases in Multimoleculars", *Journal of American Chemical Society*, 60 (1938) 309.
[6] Micromeretics, One Micromeretics Dr. Norcross, GA, USA 30093.
[7] Quantachrome Corp. 1900 Corporate Dr., Boynton Beach, FL, USA 33426.
[8] American Instrument Company Inc., 8030 Georgia Avenue, Silver Springs, MY, USA 20910.
[9] Perkin Elmer Coporation, Norwalk, CN, USA.
[10] Omicron Technology Corporation, 160 Sherman Avenue. Berkley Heights, NJ, USA 07922.
[11] Strohline, Dussledorf, Germany.
[12] Carlo Erba, Via Carlo Imbonati, 24-20159 Milan, Italy.
[13] M. Neilson and F. T. Eggertsen *Anal Chem,* 30 (1958) 1387.
[14] B. B. Mandelbrot "Fractals: Form, Chance and Dimension" Freeman, San Fransisco, 1977.
[15] D. Avnir, D. Farin, and P. Pfeifer, "Chemistry in Noninteger Dimensions Between Two and Three. Part 2. Fractal Surfaces of Adsorbance," *J. Chem. Phys.*, 79 (1983) 3566-3571.
[16] D. Avnir, and P. Pfeifer, "Fractal Dimension in Chemistry an Intensive Characteristic of Surface Irregularity," *Nouv. J. Chim.*, 7 (1983) 71–72.
[17] D. Avnir, D. Farin, and P. Pfeifer, "New Developments in the Application of Fractal Theory to Surface Geometric Irregularity," in *Symposium on Surface Science,* P. Braun et al. (Eds.) Technical University of Vienna, Vienna (1983) 233–236.
[18] P. Pfeifer, D. Avnir, and D. Farin, "Ideally Irregular Surfaces of Dimension Greater than Two, in Theory and Practice," *Surf. Sci.*, 126 (1983) 569–572.
[19] D. Avnir, D. Farin, and P. Pfeifer, "Surface Geometric irregularity of Particulate Materials. A Fractal Approach," *Journal Colloid Interface Science*, 103 (1985) 1112–1123.
[20] D. Avnir, D. Farin and P. Pfeifer, "Molecular Fractal Surfaces," *Nature* (London), (1984) 261–263.
[21] P. Pfeifer, D. Avnir, and D. Farin, "Scaling Behaviour of Surface Irregularity in the Molecular Domain: from Adsorption Studies to Fractal Catalysts," *J. Stat. Phys.*, 36 (1984) 699–716.

[[22] D. Farin, S. Peleg, D. Yavin, and D. Avnir, "Applications and Limitations of Boundary Line Fractal Analysis of Irregular Surfaces: Proteins, Aggregates, and Porous Materials," *Langmuir*, 1:4 (1985) 399–407.

[23] D. Avnir, D. Farin, and P. Pfeifer, "Fractal Dimensions of Surfaces. The use of Adsorption Data for the Quantitative Evaluation of Geometric Irregularity," Preprint provided by D. Avnir, Institute of Chemistry, Hebrew University of Jerusalem, Jerusalem, Israel.

[24] D. Avnir, "Fractal Aspects of Surface Science – An Interim Report," in C. J. Brinker (Ed.) *Better Ceramics Through Chemistry,* Materials Research Society, 1986.

[25] M. Silverberg, D. Farin, A. Ben-Shaul, and D. Avnir, "Chemically Active Fractals. Part 1 – The Dissolution of Fractal Objects," in R. Engleman and Z. Jaeger (Eds.) *Fragmentation, Form, and Flow in Fractured Media,* Proceedings of conference held at Neve Ilan, Israel, 6–9 January. 1986, *Ann.Is. Phys. Soc.*, 8 (1986) 451–457.

[26] C. Martine, V. Rives, and P. Malet, "Texture properties of Titanium Dioxide," *Powder Technology,* 46 (1986) 1–11. (In this paper the textural structure of Titanium Dioxide is measured using BET single point measurements.)

[27] M. K. Wu, "The roughness of aerosol particles – surface fractal dimension measured using nitrogen adsorption."

[28] See discussion in Chapter 7 of B. H. Kaye, *A Random Walk Through Fractal Dimensions,* Wiley-VCH, Germany, 2nd ed. 1993.

[29] A. Neimark, "A new approach to the determination of the surface fractal dimension of porous solids," *Physica A* 191 (1992) 258–262.

[30] Bel Japan Inc. 125 Ebie 6-Come Fukushima-Ku Osaka 553 Japan, Fax 06-454-9611.

Other Papers

A. Shields and S. Lowell, "The High Throughput Automated Instrument for Adsorption/ Desorption Isotherms," *American Laboratory*, (1984) 81–91.

R. W. Camp and H. D. Stanley, "Advances in Measurements of Surface Area by Gas Adsorption." *American Laboratory*, (Sept. 1991) 34–36.

B. A. Webb and J. P. Olivier and W. B. Conklin, "Model of the Physical Adsorption Isotherm," *American Laboratory,* (Nov. 1994) 38–44.

Olivier, Conklin, Szombathley, "Determination of Pore Size distribution from density functional theory – a comparison on Nitrogen and Argon results" *Studies in surface science and catalysts* (Characterization of Porous Solids 3) Elsevier, Amsterdam 1994 pp. 81–87.

Subject Index

Authors Index